T0135348

Foundations of Data Visualization

Min Chen · Helwig Hauser · Penny Rheingans ·
Gerik Scheuermann
Editors

Foundations of Data Visualization

 Springer

Editors
Min Chen
University of Oxford
Oxford, UK

Penny Rheingans
University of Maryland, Baltimore County
Baltimore, MD, USA

Helwig Hauser
Department of Informatics
University of Bergen
Bergen, Norway

Gerik Scheuermann
Institut für Informatik
Universität Leipzig
Leipzig, Germany

ISBN 978-3-030-34446-7   ISBN 978-3-030-34444-3   (eBook)
https://doi.org/10.1007/978-3-030-34444-3

This Springer imprint is published by the registered company Springer Nature Switzerland AG
The registered company address is: Gewerbestrasse 11, 6330 Cham, Switzerland

Foreword

Foundations of Data Visualization should become required reading for visualization researchers. In my Visualization Viewpoints article on Top Scientific Visualization Problems,[1] I encouraged visualization researchers to engage in what Bill Hibbard called "foundational problems" in visualization.[2] Researchers responded with exciting new papers describing theoretic frameworks, new conceptual models, ontologies, and taxonomies.[3] The editors of *Foundations of Visualization*, Min Chen, Helwig Hauser, Penny Rheingans, and Gerik Scheuermann, have assembled a who's who among international visualization researchers contributing to foundational visualization problems.

Foundations of Data Visualization begins by introducing important basic concepts related to visual abstractions, what measurement means in visual spaces, and knowledge-assisted models, including an information-theoretic perspective. The book then moves on to explore the fundamental mathematical and computer science underpinnings of visualization, illuminating many essential links between computer science and mathematical theory and visualization theory.

An observation variously attributed to Yogi Berra, Albert Einstein, Richard Feynman, and Jan L. A. van de Snepscheut goes, "In theory, there is no difference between theory and practice. But, in practice, there is." In a nod to this paradoxical truth, the book concludes with several chapters on empirical studies in visualization and multiple "real-life, feet on the ground" examples of visualization collaborations with university and industrial domain, scientific, engineering, and medical researchers, reminding us how visualization in its practice reaches into and influences virtually all disciplines and aspects of our lives.

---

[1] C. R. Johnson: Top Scientific Visualization Research Problems; IEEE Computer Graphics and Applications 24(4): Visualization Viewpoints, pp. 13–17, July/August 2004.

[2] B. Hibbard: Top Ten Visualization Problems; Proc. ACM SIGGRAPH 33(2), ACM Press, 1999, pp. 21–22; https://doi.org/10.1145/326460.326485.

[3] M. Chen, J. Kennedy: References for Theoretic Researches in Visualization; https://sites.google.com/site/drminchen/themes/theory-refs, 2017.

Likewise, as da Vinci understood the need for practitioners to study their own practices, whether the art of science or the science of art, so too did he comprehend the need to theorize those practices in order to understand them and hence to strengthen them. I cannot help but agree with him when he says, "He who loves practice without theory is like the sailor who boards ship without a rudder and compass and never knows where he may cast."

This book reminds us that, as practice evolves, theories of visualization emerge and themselves evolve. The book *Foundations of Data Visualization* proposes many "stakes in the ground" for future discussion and debate. I encourage all visualization researchers to read this volume and to engage and further the discussion—even to propose new theoretical ideas—on foundational visualization research.

August 2019   Chris R. Johnson
                                                                        University of Utah
                                                                      Salt Lake City, USA

Preface

Data visualization is the user- and task-oriented transformation of data from measurement or simulation, as well as from models (empirically crafted or machine-learned) into interactive images for exploration, analysis, and presentation. It has become an indispensable central part of the knowledge discovery process in many fields of contemporary endeavor. Since its inception about three decades ago, the techniques of data visualization have aided scientists, engineers, medical practitioners, analysts, and others in dealing with a wide variety of data. One of the powerful strengths of data visualization is the effective and efficient utilization of the human sensory and cognitive system to enable and support instructive exploration, complex analysis, and critical decision making, through the recognition of relevant patterns, the observation of unseen relations, and the identification of new connections with other data and complementing facts, concepts, theories, goals, and opinions, which are known to the users. Since vision dominates our sensory system, a significant amount of effort has been made to bring meaningful abstractions or other useful information to our eyes through interactive computer graphics. The foundations of data visualization should therefore address the fundamentals of visualization techniques, the intrinsics of visualization processes, the conceptualization of visualization users, their mind and tasks, and the principles of developing visualization applications. The interplay of these multidisciplinary foundations of data visualization and currently emerging, new research challenges in visualization constitute the broader basis of this book.

As the title indicates, this book focuses on the foundations of visualization as seen by about fifty experts from all areas of visualization, including scientific visualization, information visualization, and visual analytics, providing an in-depth discourse on a wide range of foundational topics, based on their broad expertise. The rapid advances in data visualization have resulted in a large collection of visual designs, algorithms, software tools, and development kits. We also commonly refer to a substantial body of work on mathematical methods in data visualization such as topological methods, feature extraction techniques, and information-theoretic solutions. However, we are still lacking a widely accepted and unified description of theoretical, perceptual, and cognitive aspects of visualization that would allow visualization practitioners to derive even better solutions—facilitated by a sound theoretical basis. With this book, we identify promising, related ideas and contribute to their further discussion, evaluation, validation, or falsification. Currently, many visualization researchers and developers employ empirical studies to decide if a visual design is more effective. They could benefit from a comprehensive theory that answers why one visual design is more effective than another and how a visual design can be optimized. Fortunately, the visualization community has accumulated a substantial amount of knowledge about the role of existing visualization techniques in specific analytic processes, but the generalization of such knowledge for explaining many phenomena in practice or guiding the development of new applications has been challenging, especially in terms of using mathematical theories and quantitative measures. Accordingly, progress in such principle research on data visualization would also contribute quantitative measures of visualization quality. In addition, the community seeks a better understanding of the merits and demerits of conducting different forms of empirical studies involving domain experts and may benefit from the development of theory-guided methodologies for evaluating visualization techniques and systems.

With the experience of delivering technical advances over the past three decades, it is timely for the visualization community to address these fundamental questions with a concerted effort. Such an effort will be critical to the long-term development of the subject, especially in building theoretical foundations for the subject. The community needs to develop suitable models for the whole visualization process from cleaning and filtering the data, analysis processing, mapping to graphical representations, to the perception and cognition by the human visual system and the interpretation by the human mind. While we have good empirical methods for evaluating visualization techniques and systems in applications, more effort will be necessary for using empirical studies to inform theory formation and to validate or falsify proposed theories. Such theories, once adequately validated, would in return provide the basis for more effective and efficient methods of evaluation. Modern visualization includes advanced numerical and algorithmic data processing, so the correctness of such processing requires a critical look at its assumptions, considering the application at hand. Only then, visualization can establish strong correlations between visualization algorithms and questions in the application domains. Further, uncertainty has received attention from the visualization community in recent years, but a full analysis of uncertainty at all stages of the established visualization pipeline is still not available. Theoretical foundations of uncertainty in visualization need to be related to uncertainty in the data, errors due to numerical processing, errors due to visual depiction, and, finally, uncertainty in human perception and cognition.

This book does not provide the absolute or final foundations of data visualization, and indeed no book could ever do. Nevertheless, to date, it is the most extensive collection of this discourse, by the visualization experts representing different areas of visualization, on four important foundational aspects of visualization:

Theoretical Underpinnings of Data Visualization: As the research field of data visualization evolves, and lots of individual contributions are made to a quickly growing corpus of the visualization literature, theoretic considerations become increasingly important. Explaining visualization and how it works become a pressing question as well as sorting out essential theoretic underpinnings of the visualization process. In Part I of this book, six chapters contribute a rich in-depth discussion of central theoretic questions in visualization research. A fabric of visualization (Chap. 1) is described, demonstrating the complex interaction of different aspects in data visualization, before the central topic of abstraction is addressed (Chap. 2). In Chap. 3, the question of what can we measure in visualization (and how) is discussed, before the focus is set on the role of prior knowledge in visualization (Chap. 4). Part I of this book then closes with two extensive chapters on important mathematical foundations of visualization (Chap. 5) and on essential concepts for mappings and transformations in data visualization (Chap. 6).

Empirical Studies in Visualization: While empirical studies provide useful means for evaluating visualization techniques and systems, it has become more common to use empirical studies to gain new insight about various fundamental questions about human perception and cognition in visualization. Meanwhile, there has been concern about whether empirical studies can serve as an effective and efficient means of evaluation in applications involving domain experts and aspiration for finding more cost-effective evaluation methods. The six chapters in this part address the topics on empirical studies from several perspectives that are rare in the visualization literature, including a survey of variables used in controlled and semi-controlled experiments (Chap. 7), a rational discourse on evaluation involving domain experts (Chap. 8), an in-depth discourse on evaluation in the form of long-term case studies (Chap. 9), an inspiring argument for using visualization as an analytical tool for handling data resulting from empirical studies (Chap. 10), a philosophical examination of different schools of thought that represent some most consequential hypotheses in visualization (Chap. 11), and a summary of the challenges and opportunities in empirical visualization research (Chap. 12).

Collaboration with Domain Experts: Many visualizations address questions and needs from researchers, engineers, or analysts. These users know the data, the underlying model, and the tasks well, usually even better than the visualization experts involved. This part discusses successful examples and draws conclusions from them in collaboration with such domain experts. The reader can also find some advice on how to find and start good collaborations. The three chapters in this part address the topic of collaboration with domain experts from practical and theoretical points of view. Seven successful case studies are described including learned lessons (Chap. 13), the view of industry in collaboration with universities is given special attention (Chap. 14), and more theoretical considerations about the collaboration between domain experts and visualization researchers are presented (Chap. 15).

Visualization for Broad Audiences: Besides domain experts, there is strong need for visualizations for broad audiences like the general public. A substantial part of science communication and public debate relies on effective visualizations allowing to understand and to draw conclusions for a lay audience on data. This part concerns the foundations of this specific challenge (Chap. 16), as well as remaining challenges (Chap. 21). It also includes descriptions of the goals, characteristics, and examples of visualization for broad audiences in four distinct settings: a research institute (Chap. 17), a large government agency (Chap. 18), a science center or museum (Chap. 19), and three different perspectives on educational settings (Chap. 20).

This book follows an inspiring, engaging, and energetic Dagstuhl Seminar in January 2018 on the topic. The editors and all the authors are very grateful to Schloss Dagstuhl, its staff, and its funding organizations for the unique opportunity to hold the seminar there. Without this support and great atmosphere, this book would not have been possible. The section of Acknowledgment details our gratitude to all seminar participants, authors, reviewers, the individuals, and organizations that helped to produce this book.

Oxford, UK   Min Chen
Bergen, Norway   Helwig Hauser
Orono, USA   Penny Rheingans
Leipzig, Germany   Gerik Scheuermann
February 2020

Acknowledgements

Fig. 1 Attendees of the Dagstuhl Seminar 18041 on *Foundations of Data Visualization*, which was held on January 21–26, 2018. *Source* Schloss Dagstuhl, Leibniz-Zentrum für Informatik GmbH, Oktavie-Allee, 66687 Wadern, Deutschland; Postprocessing: H. Hauser

The preparation of this book started mainly during the Dagstuhl Seminar 18041 on *Foundations of Data Visualization*, which was held on January 21–26, 2018. We are grateful to all attendees of this event (Fig. 1), showing an unprecedented amount of enthusiasm and contributing large collaborative efforts for collecting, analyzing, and consolidating the foundations of the subject which is so dear to every attendee— *visualization*. We acknowledge especially the contributions made by those attendees who took part in discussions, shared their views—some even started their writing—but eventually are not listed as the contributors of the chapters in this book. They include: Hamish Carr, Eduard Gröller, Hans-Christian Hege, Nathalie Henry-Riche, Gordon Kindlmann, Laura A. McNamara, and Jarke van Wijk.

We are very, very grateful to all the authors of the twenty chapters in this book and appreciate their collaborative spirit and their enormous endurance in completing their ambitious writing plans. We value tremendously the contributions made by all co-authors recruited by the attendees of the Dagstuhl Seminar to provide their crucial knowledge. We are certain that the observation, analysis, critique, insight, and vision offered by the authors will collectively have a profound impact on the development of visualization as a scientific and technological subject.

We are also in debt to the hardworking reviewers of all chapters in the book. They include: Alfie Abdul-Rahman, Jim Ahrens, Fabian Bolte, Michael Böttinger, Roxana Bujack, Soumya Dutta, Christian Heine, Ingrid Hotz, Tobias Isenberg, Barbora Kozlikova, Helene-Nicole Kostis, Alark Joshi, Heidi Lam, Liz Marai, Krešimir Matković, Laura A. McNamara, Silvia Miksch, Margit Pohl, Ivan Viola, Thomas Wischgoll, and Ross Whitaker. Their comments, critiques, and suggestions have been indispensable in improving the quality of the book, and their time, effort, knowledge, and wisdom are deeply appreciated. In addition, the editors of this book have also acted as anonymous reviewers for some chapters.

The book was prepared primarily on a ShareLaTeX host at Leipzig University, Germany. Christian Blecha has provided vital technical support to the editors and to the authors throughout the writing period of this book. We cannot thank him enough for his great effort and marvelous skill. We also thank Michael Böttinger for his advice and assistance in dealing with some difficult LaTeX issues.

We also record here our enormous gratitude to Helen Desmond (Springer) for her advice and patience throughout this book project. We appreciate the support offered by the Springer teams for managing the Springer LaTeX style and for typesetting, cover design, Web access, and so on.

Last but not least, on behalf of all attendees of the Dagstuhl Seminar 18041 on *Foundations of Data Visualization*, we would like to take our hats off and thank once again the brilliant Dagstuhl staff, who (as always!) made the activities surrounding the highly challenging research topic very enjoyable and productive. Especially, we thank Director Prof. Dr. Raimund Seidel, Susanne Bach-Bernhard, Annette Beyer, Jutka Gasiorowski, Michael Gerke, Thomas Schillo, and Dr. Michael Wagner.

Oxford, UK   Min Chen
Bergen, Norway   Helwig Hauser
Orono, USA   Penny Rheingans
Leipzig, Germany   Gerik Scheuermann
February 2020

Contents

Part I Theoretical Underpinnings of Data Visualization

1 **The Fabric of Visualization** 5
G. Elisabeta Marai and Torsten Möller

2 **Visual Abstraction** 15
Ivan Viola, Min Chen, and Tobias Isenberg

3 **Measures in Visualization Space** 39
Fabian Bolte and Stefan Bruckner

4 **Knowledge-Assisted Visualization and Guidance** 61
Silvia Miksch, Heike Leitte, and Min Chen

5 **Mathematical Foundations in Visualization** 87
Ingrid Hotz, Roxana Bujack, Christoph Garth, and Bei Wang

6 **Transformations, Mappings, and Data Summaries** 121
Ross Whitaker and Ingrid Hotz

Part II Empirical Studies in Visualization

7 **A Survey of Variables Used in Empirical Studies
for Visualization** 161
Alfie Abdul-Rahman, Min Chen, and David H. Laidlaw

8 **Empirical Evaluations with Domain Experts** 181
Krešimir Matković, Thomas Wischgoll, and David H. Laidlaw

9 **Evaluation of Visualization Systems with Long-Term Case
Studies** 195
Bernhard Preim and Alark Joshi

10 **Vis4Vis: Visualization for (Empirical) Visualization Research** 209
Daniel Weiskopf

11 **"Isms" in Visualization** 225
Min Chen and Darren J. Edwards

12 **Open Challenges in Empirical Visualization Research** 243
Caroline Ziemkiewicz, Min Chen, David H. Laidlaw,
Bernhard Preim, and Daniel Weiskopf

Part III Collaboration with Domain Experts

13 **Case Studies for Working with Domain Experts** 255
Johanna Beyer, Charles Hansen, Mario Hlawitschka, Ingrid Hotz,
Barbora Kozlikova, Gerik Scheuermann, Markus Stommel,
Marc Streit, Johannes Waschke, Thomas Wischgoll, and Yong Wan

14 **Collaborations Between Industry and University** 279
Daniela Oelke and Ariane Sutor

15 **Collaborating Successfully with Domain Experts** 285
Mario Hlawitschka, Gerik Scheuermann, Christian Blecha,
Marc Streit, and Amitabh Varshney

Part IV Developing Visualizations for Broad Audiences

16 **Reflections on Visualization for Broad Audiences** 297
Michael Böttinger, Helen-Nicole Kostis, Maria Velez-Rojas,
Penny Rheingans, and Anders Ynnerman

17 **Reaching Broad Audiences from a Research Institute Setting** 307
Michael Böttinger

18 **Reaching Broad Audiences from a Large Agency Setting** 319
Helen-Nicole Kostis, Miguel O. Román, Virginia Kalb,
Eleanor C. Stokes, Ranjay M. Shrestha, Zhuosen Wang, Lori Schultz,
Qingsong Sun, Jordan Bell, Andrew Molthan, Ryan Boller,
and Assaf Anyamba

19 **Reaching Broad Audiences from a Science Center
or Museum Setting** 341
Anders Ynnerman, Patric Ljung, and Alexander Bock

20 **Reaching Broad Audiences in an Educational Setting** 365
Penny Rheingans, Helen-Nicole Kostis, Paulo A. Oemig,
Geraldine B. Robbins, and Anders Ynnerman

21 **Challenges and Open Issues in Visualization
for Broad Audiences** 381
Michael Böttinger, Helen-Nicole Kostis, and Anders Ynnermann

Contributors

Alfie Abdul-Rahman — King's College London, UK

Assaf Anyamba — Universities Space Research Association and NASA Goddard Space Flight Center, Greenbelt, MD, USA

Jordan Bell — University of Alabama in Huntsville, AL, USA

Johanna Beyer — Harvard University, Cambridge, MA, USA

Christian Blecha — Leipzig University, Germany

Alexander Bock — Linköping University, Norrköping, Sweden

Ryan Boller — NASA Goddard Space Flight Center, Greenbelt, MD, USA

Fabian Bolte — University of Bergen, Norway

Michael Böttinger — DKRZ, Hamburg, Germany

Stefan Bruckner — University of Bergen, Norway

Roxana Bujack — Los Alamos National Laboratory, Santa Fe, NM, USA

Min Chen — University of Oxford, UK

Darren J. Edwards — Swansea University, UK

Christoph Garth — Technische Universität Kaiserslautern, Germany

Hans Hagen — Technische Universität Kaiserslautern, Germany

Charles Hansen — University of Utah, Salt Lake City, UT, USA

Helwig Hauser — University of Bergen, Norway

Mario Hlawitschka — Leipzig University of Applied Sciences, Germany

Ingrid Hotz — Linköping University, Norrköping, Sweden

Tobias Isenberg — Inria Saclay, Palaiseau, France

Alark Joshi — University of San Francisco, CA, USA

Virginia Kalb — NASA Goddard Space Flight Center, Greenbelt, MD, USA

Helen-Nicole Kostis — Universities Space Research Association and NASA Goddard Space Flight Center, Greenbelt, MD, USA

Barbora Kozlikova — Masaryk University, Brno, Czech Republic

David H. Laidlaw — Brown University, Providence, RI, USA

Heike Leitte — TU Kaiserslautern, Germany

Patric Ljung — Linköping University, Norrköping, Sweden

G. Elisabeta Marai — University of Illinois at Chicago, IL, USA

Krešimir Matković — VRVis Zentrum für Virtual Reality und Visualisierung Forschungs-GmbH, Vienna, Austria

Silvia Miksch — TU Wien, Vienna, Austria

Torsten Möller — University of Vienna, Austria

Andrew Molthan — NASA Marshall Space Flight Center, Huntsville, AL, USA

Daniela Oelke — SIEMENS AG, Munich, Germany

Paulo A. Oemig — New Mexico State University, Las Cruces, NM, USA

Bernhard Preim — Otto-von-Guericke-Universität Magdeburg, Germany

Penny Rheingans — University of Maine, Orono, ME, USA

Geraldine B. Robbins — ASRC/AFSS NASA Goddard Space Flight Center, Greenbelt, MD, USA

Miguel O. Román — Universities Space Research Association (USRA), Columbia, MD, USA

Gerik Scheuermann — Leipzig University, Germany

Lori Schultz — University of Alabama in Huntsville, AL, USA

Ranjay M. Shrestha — Science Systems and Applications, Inc., Greenbelt, MD, USA

Eleanor C. Stokes — University of Maryland, College Park, MD, USA

Markus Stommel — TU Dortmund, Germany

Marc Streit — Johannes Kepler University Linz, Austria

Qingsong Sun — Science Systems and Applications, Inc., Greenbelt, MD, USA

Ariane Sutor — SIEMENS AG, Munich, Germany

Amitabh Varshney — University of Maryland, College Park, MD, USA

Maria Velez-Rojas — CA Technologies, Santa Clara, CA, USA

Ivan Viola — King Abdullah University of Science and Technology, Thuwal, Saudi Arabia

Yong Wan — University of Utah, Salt Lake City, UT, USA

Bei Wang — University of Utah, Salt Lake City, UT, USA

Zhuosen Wang — University of Maryland, College Park, MD, USA

Johannes Waschke — University of Applied Sciences Leipzig, Germany

Daniel Weiskopf — University of Stuttgart, Germany

Ross Whitaker — University of Utah, Salt Lake City, UT, USA

Thomas Wischgoll — Wright State University, Dayton, OH, USA

Anders Ynnerman — Linköping University, Norrköping, Sweden

Caroline Ziemkiewicz — Forrester Research, Inc, Cambridge, MA, USA

Part I
Theoretical Underpinnings of Data Visualization

Thinking theoretically about data visualization includes a variety of perspectives, of which several are addressed in the following chapters. Holistic attempts to formulating a theory of data visualization, for example, may result in help with explaining visualization and how it works. Seeing visualization as a rich compound of aspects leads to theoretic considerations of particular questions, for example, relating to the user (and her/his tasks) in visualization, or to mathematical and technological concepts that enable the successful communication of the data to the user. As an increasing amount of research on visualization unfolds–data visualization is still a relatively young discipline, compared to many others such as physics, chemistry, etc.– the field is seeking principal explanations as well as theoretical foundations. Clearly, one may not expect to arrive at the all-explaining, ever-standing theory without any intermediate steps–scientific theories develop over time as every falsification of an older theory invites the formulation of a new and better theory. In that sense, the following chapters amount to a serious contribution to this still young discussion about theoretic considerations in visualization and it is expected that this contribution stimulates further thoughts and new work on next-level theoretic foundations of data visualization.

In Chap. 1, "The Fabric of Visualization", Marai and Möller describe the landscape of visualization foundations in terms of a Human aspect, a Systems aspect, and a Formal aspect–together with the domains on which these visualization foundations are based on. They emphasize that visualization is rooted in a rich variety of different fields and that their contribution shapes a multi-aspect fabric of visualization theory– still young and sparse. Together with thoughts about the essential evaluation of theoretic considerations–in visualization as well as in general–they explain that this book should be seen much more as a starting point of theoretic thoughts about visualization than as a final answer.

In Chap. 2, "Visual Abstraction", Viola, Chen, and Isenberg address the central concept of abstraction in visualization and provide a formal footing for the notion of abstraction. They discuss the different roles of abstraction in visualization and how they relate to the different domains that visualization is rootet in. After reviewing the notion of abstraction as known from related fields, they then put special emphasis on visual abstraction, before working out an information-theoretic analysis of abstraction as a process.

In Chap. 3, "Measures in Visualization Space", Bolte and Bruckner discuss the important notion of measurements–here with a special focus on visualization. Clearly, measurements, may they be taken from parts of a visualization solution, or from entire visualization approaches, quantitatively or qualitatively, provide an essential basis for a useful reasoning about visualization, for evaluation, comparison, and for prediction. For four types of measurements–quality metrics, metaperceptual processes, perceptual characteristics, and structure-oriented measures–they discuss also their empirical measurability and their descriptive power, concluding that more work is needed–not only to formulate useful measures, but also to work out their measurability.

In Chap. 4, "Knowledge-Assisted Visualization and Guidance", Miksch, Leitte, and Chen direct the reader's focus towards one of the most critical reasons for why visualization is often such a successful mechanism for helping users with their tasks, i.e., the user's tacit knowledge from a priori processes. They describe how knowledge-assisted visualization provides solutions for incorporating implicit and explicit knowledge into visualization solutions, including also information-theoretic considerations about the visualization process, targeted at supporting users during decision making. Moreover, they also address the non-trivial topic of guidance in visualization and how it can help users to bridge knowledge gaps during an interactive visualization session.

In Chap. 5, "Mathematical Foundations in Visualization", Hotz, Bujack, Garth, and Wang provide a large-scale review of mathematical concepts in visualization–demonstrating impressively, how wide-spread and important the role of mathematics and its many different subfields, including calculus, linear algebra, differential geometry, topology, statistics, and others, is in visualization. This overview addresses many mathematical topics that form an essential basis of many visualization solutions and understanding them sufficiently well amounts to a reprequisite for designing and for understanding visualization.

In Chap. 6, "Transformations, Mappings, and Data Summaries", Whitaker and Hotz take a detailed look at visualization as a mapping from data to interactive graphics, discussing the different types of data and providing a rich overview of crucial aspects of transformations, mappings, and summaries that constitute essential parts of successful visualization processes. They review a large variety of different methods and techniques, providing a useful overview of typical approaches to central steps in the visualization pipeline.

On the Role of Mathematics in Visualization

Hans Hagen and Roxana Bujack

Visualization, as a part of computer science as well as a part of mathematics, relies on both fields, not only as a foundation, but also as a source of crucial concepts and tools. Mathematics is an important foundation of science, because it provides a logical and quantitative approach to the scientific process. Mathematics in science can be thought of an analogy to words and their substance in a language.

> Mathemathics is the queen of the science.
>
> *Carl Friedrich Gauss*

There is no consensus on the definition of mathematics, not even on the question whether it is an art or a science. Also, we know today from Gödel's incompleteness theorem that it cannot be completely embedded into a firm axiomatic framework, because every sufficiently powerful axiomatic system has undecidable statements. But this does not make it any less crucial for science. It provides concepts, like quantity, structure, space, and change to study tools, methods, and approaches—not at the least in visualization—that facilitate practice.

> Without mathematics, there's nothing you can do. Everything around you is mathematics. Everything around you is numbers.
>
> *Shakuntala Devi*

Mathematical concepts are a fundamental part of nearly all branches of science. What would be

- Computational Fluid Dynamics without *vector fields,*
- Magnetic Resonance Imaging without *tensor fields,*
- Special Relativity without *hyperbolic geometry,* or
- General Relativity without *differential geometry?*

Through the tools of mathematics, we are able to precisely describe and formalize observed phenomena. This allows us to make reproducible statements that have a clear meaning across boundaries of fields. Since visualization relies on a multitude of scientific fields and on top of that serves a large number of different application sciences, it is crucial that it uses an underlying shared language. Mathematics is the only reasonable candidate, because it has an operational vocabulary that all sciences already know and that is general enough as well es also precise enough to work across fields, cultures, and spoken languages.

> The book of nature is written in the language of mathematics.
>
> *Galileo Galilei*

Often mathematics, inspired by one area, proves useful in many other areas, and joins the general stock of mathematical concepts. In visualization, we use it to describe sources, kinds, and transformations of data.We quantify the validity and limits of the output. A mathematical theory may even predict phenomena that have not been observed experimentally yet and thereby guide intuition and future scientific exploration. Even the purest mathematical concepts often turn out to have practical applications.

> How can it be that mathematics, a product of human thought independent of experience, is so admirably adapted to the objects of reality?
>
> *Albert Einstein*

A rigorous mathematical statement of the goal can directly steer us to a specific established method or implementation from other fields, e.g., casting a graph layout as an optimization problem. Further, it opens up the complete toolset that mathematics, physics, perceptual sciences, color theory, and computer science have to offer, which helps us to do our work and research far more efficiently.

> In mathematics the art of proposing a question must be held of higher value than solving it.
>
> *Georg Cantor*

All in all, mathematics is an essential tool for all sciences; and vice versa, all sciences provide inspiration and insight into mathematics.

> Number rules the universe.
>
> *Pytagoras*

To wrap up, one additional quote is added from our discussion about the meaning of mathematics for visualization at Schloss Dagstuhl in January 2018:

> The more math, the less blah blah.
>
> *Hans Hagen*

Chapter 1
The Fabric of Visualization

G. Elisabeta Marai and Torsten Möller

Abstract The visualization theory foundations draw on several domains, from signal processing to software design and perception. This chapter describes the landscape of visualization foundations along three aspects: a Humans aspect, a Systems aspect, and a Formal aspect, along with the domains the visualization foundations are rooted in. This chapter further provides definitions for the visualization, theory foundation, theory, model, and concept terms, and a discussion of theory granularity, from grand theories to middle-range theories and to practice theories. The chapter further discusses several challenges related to the theory fabric of the visualization that result from the diversity of our roots. The chapter ends with a discussion of possible evaluation criteria for theory components, with respect to the range of theories and models, from mathematical frameworks to guidelines and best practice advice presented in this book.

1.1 Visualization: Definition and Essential Aspects

Since the year 2000, the term "visualization" appears to be ten times more frequently used in books than "computational biology" and about five times more frequently than "compilers" [6]. No doubt, this use is compounded by the dual dictionary definition of the term, where visualization is defined [3] as either: (1) [*mass noun*] The representation of an object, situation, or set of information as a chart or other image (e.g., "video systems allow visualization of the entire gastrointestinal tract"), or [*count noun*] a chart or other image that is created as a visual representation of

G. E. Marai (✉)
University of Illinois at Chicago, Chicago, USA
e-mail: gmarai@uic.edu

T. Möller
University of Vienna, Vienna, Austria
e-mail: Torsten.Moeller@UniVie.ac.at

© Springer Nature Switzerland AG 2020
M. Chen et al. (eds.), *Foundations of Data Visualization*,
https://doi.org/10.1007/978-3-030-34444-3_1

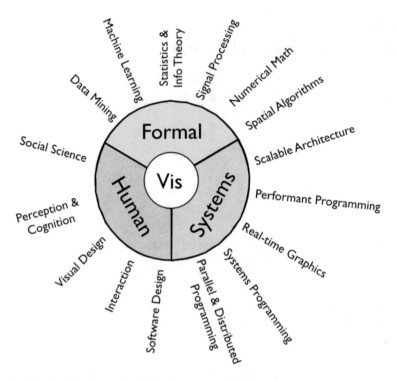

Fig. 1.1 Visualization theory foundations draw on multiple domains

an object, situation, or set of information (e.g., "3D visualizations for architectural design") or (2) the formation of a mental image of something (e.g., "visualization is a helpful technique for relieving stress").

In fact, the emphasis on visualization as a computing discipline started relatively recently in 1987, with the publication of "Visualization in Scientific Computing," a special issue of Computer Graphics [14]. Since then, the term has been continuously revisited and redefined, to clarify, for example, that interactive visualization was distinct from digital weather animations.

In this book, we refer to visualization along its first dictionary interpretation above, and as more precisely defined by Munzner: "Computer-based visualization systems provide visual representations of datasets designed to help people carry out tasks more effectively" [16]. In this textbook definition of visualization, we note the explicit reference to computer-based systems, to the analysis tasks involving datasets, and to humans who perform the tasks and their visual system. In fact, these three axes capture the essential aspects on which the theory foundations of visualization stand: the Systems, Formal/Analysis, and Human aspects.

In turn, these aspects draw on several domains, from signal processing to perception. Figure 1.1 illustrates the landscape of visualization foundations along the three aspects (Systems, Formal, and Humans), as well as the domains the visualization foundations are rooted in, with their respective connections to specific disciplines: the System aspect connects to Computer Science and Engineering; the Formal aspect connects to Formal Sciences; and the Humans aspect connects to Social Sciences. Clearly, some of these roots are further connected to each other—for example, software design in visualization involves both Systems and Humans, and data mining is related to data base programming (Systems), as well as to machine learning and statistics (Formal).

1.1.1 Terminology

At the center of Fig. 1.1 lie the existing visualization theory components (including principles, frameworks, models, guidelines), whether specific to the visualization field or borrowed from related fields. Before we discuss the challenges related to these theory components, as well as possible criteria for their evaluation, we need terminology. In the following sections, **theory foundation** denotes the set of concepts, theories, and models on which the practice of visualization is based, as well as a system of ideas intended to explain phenomena or observations related to visualization. A **concept** is an abstract idea or a general notion. A **theory** provides a description of concepts and their relationships, in order to help us understand a phenomenon or observation. A **model** is a system or prototype used as an example to follow or imitate; unlike a theory, a model usually involves some meaningful arrangement or sequence of concepts. A **theory component** is any aspect of a theory foundation above, be it a concept, a theory, or a model.

Furthermore, following similar definitions in other fields [20], theories can have different granularity, spanning the formal to practice space. **Grand theories** have the broadest scope and present general concepts and propositions or principles. Grand visualization theories consist of conceptual frameworks intended to be pertinent to all instances of visualization. Theories at this level may both reflect and provide insights useful for practice but are not designed for empirical testing. **Middle-range theories** are narrower in scope than grand theories—they are simple, straightforward, general, and consider a limited number of variables and limited aspect of reality, they present concepts at a lower level of abstraction, and guide theory-based research and visualization practice strategies. The functions of middle-range theories are to describe, explain, or predict phenomena and observations. Middle-range theories offer an effective bridge between grand visualization theories and visualization practice. One of the hallmarks of middle-range theory compared to grand theories is that middle-range theories are more tangible and verifiable through testing. **Practice theories** have the most limited scope and level of abstraction and are developed for use within a specific range of visualization situations. Visualization practice theories may provide guidelines for visualization design and implementation and predict outcomes and the impact of visualization practice. The capacity of these theories is limited, and they analyze a narrow aspect of a visualization phenomenon or observation. Visualization practice theories are usually defined to an exact community (e.g., broad audiences, domain expert audiences, etc.).

1.2   Theory Foundation Challenges

Because the visualization theory foundations draw on several domains (Fig. 1.1), it is important to be aware of the effect these roots have on the theory fabric of visualization. First, while several books, including this collection, seek to formulate theory components in the form of concepts, theories, and models, it is important to be aware that, with such complex roots, the landscape of visualization theory is sparsely populated. For example, it is unlikely that exactly three types of visual analysis workflows exist, although only three are currently documented ("Overview-first", "Search-first", and "Details-first") [8, 12, 24]. Second, given the root diversity of visualization, theoreticians should be open to multiple points of view. In fact, as a result of their adaptation to the visualization research context, some theories and models may disagree with each other, some may complement each other, some may be incorrectly framed into, transferred to, or applied to the visualization field, and some concepts may be duplicated. Third, given that the visualization field itself is still evolving, we should be prepared to revisit our theory foundation periodically. For example, from a practical perspective, some of the visualization theories are based on observations and may be later contradicted by other observations. Fourth, we note that while some of the visualization roots are mature (e.g., numerical math), others are still, themselves, evolving at a fast pace (e.g., social science). In consequence, there is always a chance that visualization theory components will ignore important new developments in our root disciplines. For these reasons, the visualization field should never overlook or dismiss a challenge to our theory foundation. Furthermore, given the currently high entry bar for new theory contributions, we could also carefully consider how to evaluate and challenge an existing theory component.

In this section, we illustrate these issues by examining first the source of one example foundational component, showing how that component entered the visualization field, what evidence supported it at the time, and how it was later challenged.

1.2.1   A Trajectory: User- and Activity-Centered Design

To illustrate some of the challenges associated with having multiple roots as a field, let us consider the historical background and trajectory of a core theory concept in visualization, that of domain characterization in the visualization design process.

Visualization relies significantly on data from other domains. As a consequence, as we train in visualization research, we also train into interdisciplinary collaboration. Whether we seek to design a novel technique, or adapt or extend an existing one, the users' needs, goals, and constraints need to be first discussed. In the visualization literature, this first step is known as "characterizing the domain", and it is particularly important, since subsequent layers in the design process depend on it [16]. This step is also notoriously difficult: From one end, designers may lack the domain knowledge to extract or even understand the domain experts' needs; from the other end, the domain expert may not be able to articulate those needs, or have the time to apprentice a visualization researcher.

While the concept of domain characterization exists in both the software engineering and in the interaction design literature (as "requirements engineering"), specific models of this process have been constructed in the visualization field, perhaps because visualization design relies on the human visual system and depends heavily on data. These models tend to rely on the user-centered design or human-centered design (HCD) paradigm introduced by human–computer interaction research. In this paradigm, as described by Don Norman in his "The Design of Everyday Things" 2002 book [18], we (the designer) start the design process by observing the user, we generate ideas, then prototype, after which we test the prototype with the user, and reiterate through this process. The core of this paradigm is that a deep, detailed knowledge of the *user* is necessary in order to design.

Yet several aspects of this HCD paradigm, as adapted into the visualization field, come at odds with either other roots of visualization, or with empirical observations in the field. For example, if, as proposed in the HCD-derived literature, the value of a visualization is measured in its number of users [27], then the relative value of a visual computing project commissioned by the two researchers who will find a cure for Alzheimer's disease would be really questionable, despite its transformative impact. A step further, the software design literature emphasizes writing functional specifications (a layman description of the function of a software system, without any implementation or design details) before prototyping [25]; that essential stage does not appear in HCD models, despite the fact that most visualization systems are a form of software. Furthermore, HCD models of domain characterization also lead to starkly lower rates of project success than those observed in software engineering [13]. Last but not least, visualization domain characterization models based on the HCD paradigm emphasize a so-called visualization triad, Humans-Data-Tasks, and in doing so tend to lose sight of the user workflows and context of the user activity. Workflows are not just sequences of tasks; they are sets of interrelated processes [19].

Interestingly, by 2005 Norman was already cautioning the interaction community against HCD: "HCD has become such a dominant theme in design that it is now accepted by interface and application designers automatically, without thought, let alone criticism. That's a dangerous state—when things are treated as accepted wisdom" [19]. Around that time, Norman started advocating for an alternative model, called activity-centered design (ACD). Rooted in Russian and Scandinavian Activity Theory, ACD focuses on activities, not on the individual person: "...because people are quite willing to learn things that appear to be essential to the activity, activity should be allowed to define the product and its structure" (Norman, The Design of Everyday Things Revised and Expanded, 3rd ed) [17]. Because activities are performed by humans, ACD can be regarded as an enhancement of human-centered design. However, note that ACD specifically ranks activities before users, and by extension, before data and users. Furthermore, activities are a higher-level concept than tasks: With increasing granularity, users have activities (problems) and tasks. An activity is a high-level structure such as "go shopping" or "understand the relationship between E. coli genomes", while a task is a lower-level component of an activity such as "drive to market", "find a shopping basket", "use a shopping list to guide the purchases", respectively, "load the complete E. coli dataset (673 genomes)", "locate an ortholog cluster in the 673 genomes", "examine the gene neighborhood of the ortholog cluster" [1, 13], etc. An activity is a collected set of tasks, but all performed together, potentially as part of a workflow, toward a common high-level goal. In contrast, a task is an organized, cohesive set of operations directed toward a single, low-level goal [13].

Most notably, the concepts underlying ACD resemble the software engineering emphasis on the functionality of a software system. In particular, because designers and domain experts can agree, during the initial requirement engineering stage of the design, on the activities to be supported by a system, these functionality-related requirements can be verified and formally approved by the "user" before the ideation stage. After all, a functional specification describes the functionality and features of a product, and it does not concern itself with how the product is implemented, the underlying algorithms, exact interactions, or visual encodings used. Therefore, functional specifications can effectively ensure that the designers are not solving the wrong problem. They can also help the designers avoid situations where the way the data are shown does not fit correctly the user workflow—**before** the prototyping stage.

While the interaction design domain was coming to terms with the ACD paradigm, the visualization field continued to develop theories and guidelines using the HCD paradigm and the Humans-Data-Tasks triad for at least another decade. A first model unifying via ACD the interaction design roots and the software design roots of domain characterization was proposed only in 2017 [13]. In this activity-centered model, functional specifications explicitly capture the user activities and workflows, as determined during the requirements session, in the form of designer-written scenarios. Asking the user to review these scenarios is a unique opportunity to verify that the visualization designers are not solving the wrong problem, before the prototyping stage. When evaluated on a set of 75 visualization projects, this ACD visualization model correlated with a 63% success rate, compared to a 25% success rate using HCD models [13], marking a wealth of missed opportunities in the field.

The example described above illustrates how theory components gleaned from across the fabric of visualization may disagree with each other, and how they, and the entire field, may benefit from being reconciled under a visualization-specific framework. Given the heavy evidence accompanying the 2017 illustrative model (the experience of many young designers, as opposed to the experience of one to a few authors, in the case of earlier guidelines; or in contrast to considering the explicit incorporation for the first time of user workflows into a model as sufficient merit, as argued below), the example also illustrates the high entry barrier to new theory components. In general, once theory components are established in the visualization field (e.g., the "Overview-first" mantra [24]), alternatives (e.g., the "Search-first" mantra [8], and later the "Details-first" mantra [12]) are introduced with difficulty.

1.3 Evaluation of a Theory Component

Once a theory component takes root within a field, the entry barrier for new theory components goes high. In particular, in our experience, appropriately or not, new models and theories are often required during the review process to provide, along with the theory component itself, some form of evaluation of that element. This request comes in contrast to earlier published works; for example, the "Overview-first" mantra was not accompanied by evidence [24].

One way to address this conundrum is by considering several possible evaluation criteria for theory components. We note that theory components can be supported by empirical evidence, or be mathematically provable. Neither form of evidence is infallible: Empirical evidence may be contradicted by later observations, and mathematical proofs often make assumptions (which by definition are statements taken to be true, without proof). In the assessment of the science philosopher Karl Popper, a theory in the empirical sciences can never be proven, although it can be falsified [7]. For example, the statement "All swans are white" can hold true in certain parts of the world and be falsified once a black swan is observed.

With this observation in mind, in the visualization literature, a model or theory has been acceptably supported by as little as one to a few concrete examples coming from the experience of one to a few authors [11, 15, 22, 23] and sometimes by no evidence [24]. A theory component has also been acceptably supported by evidence from other reports in the literature, by direct comparison against an accepted alternative theory component [13], or by mathematical proof [2]. In general, the visualization literature captures some of the different types of evaluation and concepts used in the field [9, 10], without addressing the theoretical underpinnings of these types and concepts.

In principle, there are many possible evaluation criteria for theory components. Because this topic has not been explicitly discussed in the visualization community, it is worth looking at how other disciplines handle the same issue of theory evaluation. For example, one of the disciplines with keen interest in this topic is nursing. As in visualization, nursing theory includes both theories unique to the field and theories that have been borrowed from related sciences by practitioners to explain and explore phenomena specific to nursing. Furthermore, nursing theories also span a wide range, from grand theories to practitioner guidelines, and they are also largely based on observations and phenomena, and on a mix of quantitative and qualitative data. The seminal nursing theory evaluation work of Fawcett and Rizzo-Parse [5, 20, 21] proposes as evaluation criteria the following elements: significance, internal consistency, parsimony, testability, empirical adequacy, and pragmatic adequacy [4], which we briefly discuss below.

Significance: The criterion of significance focuses on the context of the theory component. This criterion requires justification of the importance of the theory component to the discipline and is met when the origins of the theory component are explicit, when antecedent knowledge is cited, and when the special contributions made by the theory component are identified.

Internal consistency: This criterion focuses on both the context and the content of the theory component. The criterion requires all philosophical claims, conceptual model, and concepts and propositions, to be consistent with each other, the linkages between concepts to be specified and that no contradictions in propositions are evident. The concepts also need to reflect semantic clarity (e.g., explicit definitions are given) and semantic consistency (the same term and the same definition are used consistently for each concept in the entirety of the author's discussion).

Parsimony: This criterion assesses whether the content of a theory component is stated clearly and concisely. The fewer the concepts and propositions needed to fully explicate the theory component, the better.

Testability: Theory components may be amenable to direct empirical testing. Such an approach would require the concepts to be observable and the propositions to be measurable. The criterion of testability for middle-range theories may be met, for example, when specific instruments or experimental protocols have been developed to observe the concepts, and statistical techniques are available to measure the assertions made by the propositions. Descriptions of personal experiences may be used, although they are not mandatory, to evaluate the testability of grand theories.

Empirical adequacy: This criterion requires the assertions made by the theory component to be congruent with empirical evidence, determined by means of a systematic review. If the empirical data conform to the theoretical assertions, it may be appropriate to tentatively accept the assertions as reasonable or adequate.

Pragmatic adequacy: This criterion focuses on the utility of the theory component for visualization practice. The criterion requires that practice theories be used in the real world of visualization practice, while the extent to which a grand theory or a middle-range theory meets this criterion could be determined, for example, by reviewing all descriptions of the use of the theory in practice.

This list of potential evaluation criteria is not exhaustive; other criteria could be heurism (the amount of research and new thinking stimulated by the theory; whether other theorists quote the theory and use it as a springboard to create their own theories), tests-of-time (a theory's durability over time), and so on. Furthermore, some of the criteria above might be more appropriate for specific theory components; for example, pragmatic adequacy (or actionability) seems a good fit with the guidelines and best practices described in the latter half of this book, while parsimony and internal consistency may be more appropriate for the mathematical frameworks and formal models described in the first half. Overall, we note the need for a wider range of evaluation criteria of theory components, and for a better understanding of where specific evaluation criteria may apply.

1.4   Conclusion

The visualization field draws on a multitude of domains, connected to different branches of science and engineering. Accordingly, many of the visualization theory foundations draw from principles in these domains and sciences. In this chapter, we organized these aspects along three main axes: Systems, Formal and Humans, and we used these axes to describe the fabric of the visualization field and its ties to multiple science and engineering domains. Building theory foundations for visualization is the collective responsibility of the visualization community. Our different roots mean that our different subcommunities have different contributions to this space, and at the same time, that our resulting space coverage is in consequence sparse.

As a result of their adaptation to the visualization research context, we noted that some of the resulting theories and models may disagree with each other, some may complement each other, and some may be incorrectly framed into, transferred to, or applied to the visualization field. These complications affect the fabric of visualization and lead in some cases to terminology overloading and duplication. The field is still very young and fragmented, and there is a need to reconcile conflicting views in the existing theory landscape. To maintain growth and intellectual diversity, we furthermore need to keep an open mind with respect to existing guidelines, accept challenges to our existing theory components, and lower the entry barrier for alternative theories.

There are multiple resources available that discuss theory components of visualization, including software design and user-centric design frameworks, mathematical and systems engineering frameworks, and perceptual and cognitive frameworks [16, 26, 28]. This book itself contributes additional theory components, and is not an exhaustive resource either. All these existing and proposed theory components can be evaluated along a multitude of criteria. While mathematical frameworks may score highly along parsimony and consistency criteria, guidelines and best practice advice also have complementary value along pragmatic adequacy criteria. In general, the field stands to benefit from a wide range of theory contributions and from a wider range of evaluation criteria.

References

1. Aurisano, J., Reda, K., Johnson, A., Marai, E.G., Leigh, J.: BactoGeNIE: a large-scale comparative genome visualization for big displays. BMC Bioinform. **16**(11), S6 (2015)
2. Chen, M., Jaenicke, H.: An information-theoretic framework for visualization. IEEE Trans. Vis. Comput. Graph. **16**(6), 1206–1215 (2010)
3. Dictionary, O.: Oxford dictionaries. Language Matters (2014)
4. Fawcett, J.: Criteria for evaluation of theory. Nurs. Sci. Q. **18**(2), 131–135 (2005)
5. Fawcett, J., Desanto-Madeya, S.: Contemporary Nursing Knowledge: Analysis and Evaluation of Nursing Models and Theories. FA Davis, Philadelphia (2012)
6. Google: Google Books N-Gram Search. http://books.google.com/ngrams/. Accessed 05 Apr 2019

7. Gower, B.: Scientific Method: An Historical and Philosophical Introduction. Psychology Press, London (1997)
8. van Ham, F., Perer, A.: "Search, show context, expand on demand": supporting large graph exploration with degree-of-interest. IEEE Trans. Vis. Comput. Graph. **15**(6), 953–960 (2009)
9. Isenberg, T., Isenberg, P., Chen, J., Sedlmair, M., Möller, T.: A systematic review on the practice of evaluating visualization. IEEE Trans. Vis. Comput. Graph. **19**(12), 2818–2827 (2013)
10. Lam, H., Bertini, E., Isenberg, P., Plaisant, C., Carpendale, S.: Empirical studies in information visualization: seven scenarios. IEEE Trans. Vis. Comput. Graph. **18**(9), 1520–1536 (2011)
11. Lloyd, D., Dykes, J.: Human-centered approaches in geovisualization design: investigating multiple methods through a long-term case study. IEEE Trans. Vis. Comput. Graph. **17**(12), 2498–2507 (2011)
12. Luciani, T., Burks, A., Sugiyama, C., Komperda, J., Marai, G.: Details-first, show context, overview last: supporting exploration of viscous fingers in large-scale ensemble simulations. IEEE Trans. Vis. Comput. Graph. **25**(1), 1–11 (2019)
13. Marai, G.E.: Activity-centered domain characterization for problem-driven scientific visualization. IEEE Trans. Vis. Comput. Graph. **24**(1), 913–922 (2018)
14. McCormick, B.H., DeFanti, T.A., Brown, M.D.: Visualization in scientific computing: a synopsis. IEEE Comput. Graph. Appl. **7**(7), 61–70 (1987)
15. Munzner, T.: A nested model for visualization design and validation. IEEE Trans. Vis. Comput. Graph. **15**(6), 921–928 (2009)
16. Munzner, T.: Visualization analysis and design. AK Peters/CRC Press, Natick (2014)
17. Norman, D.: The Design of Everyday Things: Revised and Expanded Edition. Basic books, New York (2013)
18. Norman, D.A.: The Design of Everyday Things. Basic Books Inc, New York (2002)
19. Norman, D.A.: Human-centered design considered harmful. Interactions **12**(4), 14–19 (2005)
20. Parse, R.R.: Nursing Science Major Paradigms, Theories, and Critiques. W B Saunders Co, Philadelphia (1987)
21. Parse, R.R.: Parse's criteria for evaluation of theory with a comparison of Fawcett's and Parse's approaches. Nurs. Sci. Q. **18**(2), 135–137 (2005)
22. Pretorius, A.J., Van Wijk, J.J.: What does the user want to see?: what do the data want to be? Inf. Vis. **8**(3), 153–166 (2009)
23. Sedlmair, M., Meyer, M., Munzner, T.: Design study methodology: reflections from the trenches and the stacks. IEEE Trans. Vis. Comput. Graph. **18**(12), 2431–2440 (2012)
24. Shneiderman, B.: The eyes have it: a task by data type taxonomy for information visualizations. In: Proceedings of IEEE Symposium on Visual Languages, p. 336 (1996)
25. Spolsky, J.: Painless functional specifications part 1: why bother? In: Joel on Software, pp. 45–51. Springer, Berlin (2004)
26. Tufte, E.R.: The Visual Display of Quantitative Information, vol. 2. Graphics Press, Cheshire (2001)
27. Van Wijk, J.J.: The value of visualization. In: Visualization, 2005 (VIS 05). IEEE, pp. 79–86 (2005)
28. Ware, C.: Information Visualization: Perception for Design, 3rd edn. Morgan Kaufmann, Burlington (2012)

Chapter 2
Visual Abstraction

Ivan Viola, Min Chen and Tobias Isenberg

Abstract In this chapter, we revisit the concept of abstraction as it is used in visualization and put it on a solid formal footing. While the term "abstraction" is utilized in many scientific disciplines, arts, as well as everyday life, visualization inherits the notion of data abstraction or class abstraction from computer science, topological abstraction from mathematics, and visual abstraction from arts. All these notions have a lot in common, yet there is a major discrepancy in the terminology and basic understanding about visual abstraction in the context of visualization. We thus root the notion of abstraction in the philosophy of science, clarify the basic terminology, and provide crisp definitions of visual abstraction as a process. Furthermore, we clarify how it relates to similar terms often used interchangeably in the field of visualization. Visual abstraction is characterized by a conceptual space where this process exists, by the purpose it should serve, and by the perceptual and cognitive qualities of the beholder. These characteristics can be used to control the process of visual abstraction to produce effective and informative visual representations.

2.1 Definitions

The term *abstraction* often lacks a precise definition in many fields. While several fields have defined the term for their own purposes, there is only a vague under-

I. Viola (✉)
King Abdullah University of Science and Technology, Thuwal, Saudi Arabia
e-mail: Ivan.Viola@KAUST.edu.sa

M. Chen
University of Oxford, Oxford, UK
e-mail: Min.Chen@OERC.ox.ac.uk

T. Isenberg
Inria Saclay, Palaiseau, France
e-mail: tobias.isenberg@inria.fr

© Springer Nature Switzerland AG 2020
M. Chen et al. (eds.), *Foundations of Data Visualization*,
https://doi.org/10.1007/978-3-030-34444-3_2

standing of its meaning that is shared by all fields. Some scientific disciplines and scholarly fields have adjusted the vaguely understood meaning to fit the needs of the respective discipline or field. In this chapter, we first present our key definitions related to visual abstraction, and we then provide the justification for the definitions. In giving these definitions, we revise our previous set of definitions relating to the concept of abstraction [19], based on new discussions related to, and insights from, our further literature study. Terminology related to abstraction has been adopted from Leppänen [9] and is discussed in Sect. 2.2.7.

Definition 2.1 An **abstraction** is a process that transforms a *source thing* into a less concrete *sign thing* of the *source thing*. Abstraction uses a concept of *point-of-view*, which determines which aspects of source thing should be preserved in its sign thing and which should be suppressed.

Definition 2.2 A **data representation** is a sign thing that stands in digital form for a *referent thing* from reality or another *sign thing*, using data structures or *concept things*. Similarly, a **visual representation** is a *sign thing* that stands for a *referent* from reality or another *sign thing* so that it can be visually perceived and cognitively processed by a human observer.

Definition 2.3 Visual abstraction is a particular type of abstraction where the sign thing is visual, while the source thing is either non-visual or visual. A visual representation results from a process of **visual abstraction** if such transformation intentionally disregards certain aspects of data representations.

Definition 2.4 The abstraction process also involves a **point-of-view** component defined through the task, which the visualization process aids to accomplish. This task is represented as a combination of *targets* on which particular *actions* are performed.

Proposition 2.1 *The amount or **significance of abstraction** of a thing can be, in computer or signal representations, quantified by means of information theory.*

Definition 2.5 A **meaningful visual abstraction** is a visual abstraction such that, for a given point-of-view and for a given *purpose* or *goal*, key aspects of the underlying *referent thing* are preserved in the visual representation so that the cognitive load when perceiving it as a stimulus is significantly reduced.

Definition 2.6 A **visualization** is a process that transforms data representations of a thing from reality into visual representations. Visualization is a process that is intended to be a meaningful visual abstraction process. The designers of visualization processes must understand the point-of-view component and tasks. Otherwise, they would not reach the full meaningfulness intended.

Definition 2.7 An **abstraction axis** is the *perceived* sequence of visual representations that are assembled by the designer of a visualization system to illustrate a given point or series of points about reality. Each of the building blocks of an abstraction axis is the result of an individual abstraction process to a visual representation. Each transition between two successive abstraction axis building blocks can but does not have to remove information, some can also both remove and add information based on chosen blocks specific abstraction. If two or more abstraction axes are constructed such that they affect independent aspects of the visual representations, they can be combined into an **abstraction space** that observers can explore.

2.2   Flavors of Abstraction

The notion of what is abstract and what is concrete is a fundamental discussion in philosophy, without a clear consensus. In its simplest terms, an abstract object has no physical referent, while concrete objects have physical referents. Reiterating Frege's writings, "The Thought" [7] is even stronger in restricting what an abstract thought is: "An object is abstract if and only if it is both non-physical and non-mental." An object is acknowledged as mental when "it exists at a time if and only if it is the object or content of some mental state or process at that time." This statement implies that an abstract object is an object if and only if it cannot be found in nature, cannot be constructed, and one cannot even form a mental image of it.

Another definition of abstract objects is that they lack causal powers [17]. This means that abstract objects cannot affect other objects in any way. An empty set is such a case of an abstract entity as it does not have any causal powers. The definition of abstract entity is often so strict that some philosophers deny the existence of an abstract entity as such. However, there seems to be better agreement on what an abstraction is: "It is a distinctive mental process in which new ideas or conceptions are formed by considering several objects or ideas and omitting the features that distinguish them" [17]. Lewis [10] proposed that "abstract entities are abstractions from concrete entities. They result from somehow subtracting specificity, so that an incomplete description of the original concrete entity would be a complete description of the abstraction." In the rest of the chapter, we use the term abstraction aligned with these definitions to only describe a process, as we have also done in our own definitions at the beginning. The entity after abstraction is, in our case, denoted as a representation. We do not enter the dispute of whether it is an abstract entity or not. In such a way, we build on the part that philosophers agreed upon, while we avoid the terminological controversy. Before we look at the use of abstraction in visualization, let us first consider its occurrence in related arts and sciences.

2.2.1   Abstraction in the Arts

In the arts, the term *abstract art* refers to non-figurative artwork, where the intent is to develop art beyond depiction of natural or man-made objects. The composition may exist with a degree of independence from visual references in the world [1]. This art movement started during the early twentieth century and emerged from figurative art.

Artists such as Picasso, Mondrian, Kandinski, and many others originally depicted natural objects. The beginning of non-figurative art started with a deep analysis and observation of the creative process, where the graphical elements that composed the rendering became themselves the subject of study. The natural objects were gradually represented through collection of simpler geometric primitives. The artists searched for an expression of minimal set of visual elements that are still able to carry the figurative meaning. But they did not stop there. Artists further experimented with the graphical elements beyond recognizability of any corresponding figure from the rendering itself. Interestingly, one can sometimes discover a correspondence to their earlier works where a particular figure is still recognizable, thus transitively the figure can be imagined in the fully abstract art with such aids as well. It indicates that the artists still had a particular figure in mind, when rendering a particular art, while, without the prior work context, this figure would not be discovered by another human observer. This gradual process, which transformed figurative art into what is now called abstract art, is abstraction.

2.2.2 Abstraction and Generalization in Cartography

In cartography, depending on a chosen scale for a map and its type/target audience, a subset of information is selected, the elements to be depicted are simplified, and their depiction is adjusted. For example, streets can be shown with a much larger width than in reality, yet fine details of their path are removed. When zooming out, important elements and landmarks in the map are depicted, while generally less relevant elements are suppressed. At a particular level of scale, for example, the post office, a religious place, a building of historical significance, a bridge over the river, or the main streets are clearly depicted in the map, while similar objects in terms of spatial dimensions are abstracted into very simplified representations, if they are shown at all. The field has created a solid vocabulary and guidelines on how certain elements should be depicted and when should they be visible. In cartographic visual language, the umbrella term for guidelines of how different scales should depict certain information is *map generalization* [2]. We discuss the specific meaning of the term *generalization* below, but other principles such as *grouping* or *classification* are applied here as well. In prior work, these concepts are considered as distinct abstraction principles and we discuss their specifics below.

2.2.3 Abstraction in Shape Analysis

In shape analysis, the term *abstraction* typically refers to a skeletonization or extraction of topological features that represent essential characteristics of the underlying shape [6, 8]. Here, abstraction preserves the key properties of the geometric components such as their connectivity. The levels of detail of these abstracted representations are controlled through measures like persistence: this measure determines which structures are too small for particular scale to justify their validity and which are grouped into other larger-scale structures. Such abstracted representations facilitate the extraction of hierarchies in shapes to facilitate geometric linkage, multi-scale representations, and—importantly—the topological representation is much sparser and facilitates an unobstructed clear view on the key geometric properties. The same holds for the topology of flow data, where a flow field is *classified* into points and regions of certain uniform properties such as sinks, sources, and separatrices (curves or surfaces) that partition the flow according to its long-term behavior.

2.2.4 Mathematical Abstraction

The term mathematical abstraction refers to a process of transforming a specific real-world situation into generalized form using mathematical formalism.[1] The specifics which do not affect the solution to a given problem are removed so that, in the end, only a set of key elements with properties and relations to each other remains, which can be expressed formally. Problems to solve in mathematics class are frequently expressed as real-world situations. The tasks are to abstract from the real-world specifics and apply a mathematical formalism that provides the answer to the given problem. The development of mathematics and physical sciences has advanced through mathematical abstraction into Euclidean geometry, algebra, and analysis. These developments have been possible due to humans being capable of thinking in an abstract way.

2.2.5 Abstract Thinking

School students are trained in abstract thinking by being challenged to solve a specific real-world problem. To be able to do so, they are trained to abstract from the case specifics by extracting only the essential components so that a formal solution can be calculated and, finally, interpreted back for the specific real-world scenario. Abstract thinking is, according to cognitive psychology [14], the most complex stage in the development of cognitive thinking, where generalizations and concepts are used in the thought process. From a set of observations, hypotheses can be formed and logical reasoning can lead to conclusive statements [14].

---

[1] https://en.wikipedia.org/wiki/Abstraction_(mathematics).

2.2.6   Abstraction in Object-Oriented Design

In computer science, the term abstraction achieves yet another flavor of its meaning. In object-oriented design, the most frequently used programming methodology, it primarily relates to the definition of classes and methods that cannot be instantiated. Typically, classes and methods are hierarchically grouped into increasingly abstract constructs such that implementations of particular functionality can be shared among many different elements. While for most of these classes it is possible to create instances, an abstract class is a construct that itself cannot be instantiated but which organizes the functionality into a comprehensive representation. The class hierarchy as the outcome of such abstraction gives a clear understanding of differences in functionality among various classes as well as what they have in common. It also facilitates further extensibility of existing code to support new cases that were not considered in the initial software design.

2.2.7   Abstraction Ontology

In the area of information and knowledge modeling, a particularly interesting past work closely relates to our own investigation. Leppänen [9] distinguishes between first-order and second-order abstraction. First-order abstraction is associated with primary things, while second-order abstraction acts upon a predicate that defines the primary things. An example of a primary thing is *sedan* with several predicates, among others a *color*. The result of the abstraction of a sedan would be a *car* or a *vehicle*, which corresponds to first-order abstraction. Let us assume that an instance of a sedan is painted with a particular blue, for example, *Maya Blue*. This predicate can also be abstracted to *light blue* or *blue*, a process which is of the second-order abstraction type and is also termed as predicate abstraction.

Importantly, Leppänen defines four elementary abstraction principles: classification, generalization, composition, and grouping. First, classification is defined through the term *isInstanceOf* or that instances are *typeOf*. The opposite to classification is instantiation. Second, generalization is a principle of abstraction where the differences of subtypes are suppressed to fit a supertype. This refers to an *isA* relationship and the antonym to generalization is specialization. Third, composition is a principle of abstraction in which a whole concept is composed of part concepts. These parts are abstracted to form a whole object. This refers to a *partOf* relationship and its opposite is the decomposition. Finally, the last principle of abstraction is grouping which relies on a *isMemberOf* relationship and whose opposite is individualization. For example, a particular person can be a member of a political party. This abstraction includes aggregation, set membership, and association. Both first-order and second-order abstractions can benefit all four elementary abstraction principles. In all cases, an important property to highlight is that abstraction is associated with an intentional and controlled loss of information.

Leppänen's work stresses the importance of the concept *point-of-view* that plays crucial role during the abstraction process. When using classification on a *thing* termed, for example, *Margaret Thatcher*, the abstraction along classification would lead to entity *female* or *UK Prime Minister*. If we would be using grouping, the abstraction would lead to *Conservative Party*. In case the composition principle is used for abstraction of *UK Prime Minister*, the outcome would be *UK Government*. Therefore, things might generally have many different kinds of abstractions as things from reality are typically embedded in complex and intertwined abstraction hierarchies.

In his work, Leppänen combines philosophical and semiotic standpoints. In the context of semiotic frameworks [16], they refer to three kinds of *things*: a *concept thing*, a *referent thing*, and a *sign thing*. Concepts are mental constructs, words of mind, and form basic components of human knowledge. A referent is an element of reality that relates to the concept. A reality describes a set of anything that exists or can possibly exist, physically or virtually. A sign is anything that can stand for something else, including symbols, text, or images. As such it is a representation of a concept. These concepts are used below in the discussion of abstraction in visualization. We applied the same terminology in our definitions from the start of the chapter, but we added the concept of a *source thing* (Definition 2.1).

2.2.8   Summary of Abstraction in the World Outside Visualization

The intuitive understanding of abstraction has been reinforced by this brief excursion into various fields and that stand and argue for abstraction. We can observe that the term is not used uniformly and that it is frequently exchanged with other terms. The recurrent pattern is that abstraction relates to formation of some higher-order constructs or representations that are result of a transformation of lower-level entities. The lowest entities are more tangible, while the higher levels of the abstraction hierarchy are further removed from tangibility and become more mental constructs and concepts (defined as *the constituents of thoughts* [17]) that, in one way or another, allow humans to recognize certain characteristics clearer than the lower-level representations. The ability to abstract seems to be one of the core properties of humans, present while shaping the entire body of analytical knowledge humankind has formed throughout our history.

2.3   Abstraction for Visualization

Let us now investigate how abstraction manifests itself in visualization. We propose that abstraction is equally central to visualization as it is to other areas in which analytic reasoning is the core part of a processing workflow. Visualization is the process of transforming the digital representation of data into visual representations that are exposed to a human viewer (Definition 2.6). It takes advantage of the fact that most humans are extremely efficient in comprehending information presented as a visual stimulus. Naturally, this stimulus has to be well designed to convey the intended information (Definition 2.5). This aspect is the main concern of the visualization mapping stage of the visualization pipeline. Visualization is omnipresent in studying various real-world phenomena, conveying structures, methods, or concepts. In visualization, the abstraction process guides the transformation into visual representations (Definitions 2.2 and 2.3), similar to the process of abstract thinking. In some sense, it serves as an extension of the working memory, where needed information can be instantaneously accessed. We thus first clarify the meaning of abstraction in visualization and then discuss its core properties.

To bring visualization into the context of semiotic frameworks, the *sign* is termed as *representation*, both digital and visual, and the *referent* is the studied phenomenon from *reality* (Definitions 2.1 and 2.2). The *concept* is what relates to the *referent* and can be conveyed through the representation. In visualization, abstraction is performed at least in three stages: first, the abstraction of the reality into data representations; second, the data representation is, through abstraction, transformed into visual representations; third, a visual representation is transformed into a mental model or a memory representation through the perceptual and cognitive processes of the human observer.

Abstraction has occurred if the quantum of information before the abstraction is higher than in the representation after abstraction, while some aspects of the original representation are preserved and become more prominent (Definition 2.1). In case there is no intended information loss, we refer to a more general term *transformation* or *mapping*. For example, several simultaneous abstraction processes that individually work on different aspects of the *things* could be combined, some work in a positive direction (removal of information) and others in a negative direction. This could lead to composite *transformation* or *mapping* that transfer one representation into another, with information loss and information gain at the same time.

2.3.1   Task Abstraction

Visualization is driven by a particular *intent*. There is a reason behind a visualization, even in the casual scenarios. This intent defines the *point-of-view* (Definition 2.4), which, as a controlling mechanism, can steer how abstraction changes the representations. In the visualization literature, Munzner [15] describes a hierarchical framework into which specific individual visualization usage scenarios can be abstracted. On the highest level, Munzner classifies the tasks as a combination of an *action* upon a *target*. The *action* class is instantiated into *analyze*, *search*, and *query*, which can be further instantiated into lower-level classes of *actions*. The *target* is instantiated into *data* in general, *attributes*, *networks*, and *spatial data* which are further instantiated into more detailed targets. It is the combination of the *action* and *target* that would define the *point-of-view* to guide the abstraction process.

2.3.2   Data Abstraction

Munzner [15] also defines various types of *data* and *data sets* for visualization. Data types are *items*, *attributes*, *links*, *positions*, and *grids*. Data set types are *tables*, *networks*, *trees*, *fields*, *geometry*, *clusters*, *sets*, and *lists*. All these types are *concept things* (Definition 2.2). The data abstraction here refers to the transformation from the real-world phenomenon, the *referent thing*, into data structures (*concept thing*) and digital representations (*sign thing*), to facilitate an efficient and automatized computational processing. This task of data abstraction is somewhat similar to the mathematical abstraction process. In both cases, we end up with a formal representation on which standardized mathematical or computational machinery can be applied.

The initial data abstraction is typically performed during the acquisition process. Either real-world observations are made and digitally stored in a particular data representation or even a mathematical model is formulated based upon these observations. Both forms are data representations abstracted from the thing that exists in reality, and these representations have been achieved through a classification process.

The result of the initial data abstraction is frequently further abstracted into another data representation to promote a particular *point-of-view*, neglecting unimportant aspects of the original data representation. As such, the filtering operation is typically applied, which might be considered to relate to map generalization and as such corresponds to the generalization abstraction principle. Once the data representation contains the relevant data prominently, a conversion into data representation is performed that can efficiently be visually represented.

2.4   Visual Abstraction

After series of data abstractions and transformations (the latter when no information loss happens), in visualization, the data is transformed/abstracted into visual representations. A visual representation is then shown on a display, perceived, and further cognitively processed by a human observer. The visual abstraction process that generates this visual representation can be performed in many ways: In the case of kernel density estimation plots or clustering techniques, for instance, data can be visually abstracted using a composition principle such that smaller elements become a part of higher-order representations. In the case of volumetric scalar fields, the voxel values can be classified into color and opacity ranges. By this, some voxels become abstracted into types such as air, soft tissue, or hard tissue. Level of detail techniques would typically relate to composition or grouping; in atomistic visualization, individ-

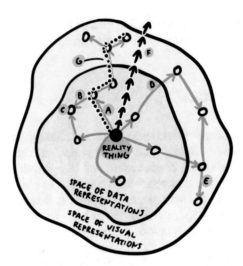

Fig. 2.1 Abstraction space in which a *thing* from reality is gradually transformed into visual representations: **a** initial abstraction into a digital form, **b** data abstraction into new data representation, **c** different data abstractions can lead to identical data representation, **d** visual abstraction transforms the data representation into a visual representation, **e** visual abstraction transforms one visual representation into another visual representation, **f** the abstraction space encodes less and less information from the original *thing* from reality. The further from center the more sparse the representation is. **g** the dotted line conveys a visualization pipeline that can be seen as a composite visual abstraction

ual atoms become member of particular molecules, which in turn become members of certain compartments, up to cells. In many cases of particular visual abstraction, it can be simultaneously argued for different abstraction principles, and there might be more principles than those proposed by Leppänen [9].

Munzner [15] provides a conceptual framework according to which visual representations or encoding can be categorized. This framework is rather extensive; however, on the low level, the visual encoding can abstract data representations through two key aspects. The first aspect is the graphical *mark* that positions each data element: *points*, *lines*, and *areas*. These marks further encode quantitative information or their mutual relationship is conveyed through various perceptual channels: *position*, *shape*, *color*, *size*, and *angle*. These basic low-level perception-driven visual elements can be combined to create rich spectrum of possible visual representations. These visual representations can be used to *encode*, *manipulate*, *compare*, or *reduce* the data in the visual representation space.

Data representations and visual representations can be ordered according to how much they abstract a particular phenomenon. The abstraction process is depicted in Fig. 2.1. We can see the abstraction space related to one *thing*, one entity from reality. There are several ways how the *thing* can be abstracted into a digital form. After this first stage of the process, the data representation has been abstracted from the *thing*. There could be several data abstractions applied, under which the data becomes sparser and sparser so that the information sought by the user or intended by the visualization creator becomes gradually clearer. Sometimes, even the series of abstractions can take different paths yet still result in the same data representation. In practice, however, such data abstractions would only apply to a given path to a particular visual representation, as most visualization systems will maintain their original datasets to allow users to also observe different visual representations, which would be the result of a different sequence of data abstractions.

After the sequence of data abstractions, the data representation is still non-visual. If we apply a visual mapping to such data (whether with intentional loss of information or not), we achieve a visual representation that can be viewed on a display. But even visual representations can be further transformed into sparser visual representations by means of visual abstraction. The more far away in the abstraction space, the less information from the reality is preserved. If we concatenate a path from the reality to the final visual representation, we can see a visualization pipeline. In case we perform a transformation so that the distance between the original representation and the reality and the target representation and the reality is the same, we do not perform an abstraction. If the target representation is closer to the reality than the original representation, we perform an inverse operation to abstraction. Yet this inverse abstraction only happens in the eyes of the beholder, as we always remove information along the path from reality to visual representation.

2.4.1 Meaningful Abstraction

It is not clear whether an abstraction has to be meaningful or whether its only condition is a loss of information. What if, for example, a high-dimensional data set is projected onto fewer dimensions? Projection is, in principle, a valid abstraction. But what if we project only every second data element and create a confusing data representation in which only half of the data set is projected onto lower-dimensional space. Is such a meaningless projection also an abstraction? From the information-theoretic point-of-view, we have lost a certain amount of information, so it can be considered as an abstraction. To differentiate us from this view, we should define the term *meaningful abstraction* for those abstractions that are useful in some application contexts (Definition 2.5).

Visual mapping may result in a representation with an equal amount of information, however, more visually confusing than the previous representation. For example, it is known that humans have difficulties with identifying portraits of known faces if they are rotated by 180° from the natural portrait orientation [18]. From the information-theoretic point-of-view, the rotation does not remove information from the image, but there is a significant difference in cognitive load between these two representations. Such a rotation is consequently not a meaningful visual mapping. The same holds for two visual representations of a graph, a node-link diagram and an adjacency matrix. When one visual representation is transformed into another, no information is lost. Yet the cognitive load for viewers differs between these two representations. Building on the term of *meaningfulness*, a visual representation can be more meaningful (or effective) for a particular intent than another visual representation. Visual abstractions that lead to these representations might be ordered or perhaps even quantified in how meaningful they are.

The concept of *meaningfulness* in terms of visual abstraction processes is tightly related to visual perception processes. In principle, a meaningful visual abstraction makes the job of visual processing simpler so that less of a cognitive processing needs to be invested, for a given purpose or goal, in comprehending the abstracted visual representation to understand the intended aspects of the reality. Therefore, visual abstractions relevant to visualization will need to result in lower cognitive load when comprehending the abstracted representation, for the chosen intention. Therefore, the *meaningful visual abstraction* has to pass two conditions: the target visual representation has to formally contain less information and the cognitive load has to be lower. The perceived information, if not increase, should decrease at most linearly with the cognitive load.

At this point, we solidify the previous discussion and define some key terms in visualization. Abstraction is a process; it is a transformation along which some information is intentionally lost to give prominence to the **higher-level** information within. The abstraction process results in a representation. For pure data abstraction, it results in a data representation, while, when visual abstraction is involved, it results in a visual representation. These abstractions can be considered as meaningful as long as they are benefiting particular application example or purpose. The meaningfulness property is scoped by the set of meaningful applications. A visual representation is the result of a visual transformation. When information is intentionally lost and the cognitive load is lower, while the perceived information loss is, at most, linear with the cognitive load difference, then we consider the visual abstraction as meaningful. Visual mapping and visual encoding, while both having their distinct meaning, can be used interchangeably with visual transformation. Visual metaphor operates on the concept of *analogy*. It presents a sign thing of a different referent thing from reality than the one originally regarded. This way visual mapping associates properties of one referent thing to another referent thing. An example of a visual metaphor is Chernoff faces, where different facial properties encode multivariate data [5].

2.4.2 Abstraction Axes and Abstraction Spaces

So far, we mainly discussed the process of abstraction from reality via data representations to visual representations. Yet we also showed that positive or negative abstraction can be *perceived* by a viewer as he or she is manipulating this abstraction chain or visualization pipeline. For better describing the latter aspect, Viola and Isenberg [19], inspired by earlier examples in visualization [11–13, 20] as well as in the arts world, proposed the notion of *axes of abstraction* which could form an *abstraction space*. With these two concepts, we can describe the abstraction that is perceived and controlled by the beholder, in contrast to the abstraction that is applied as a particular visual representation is generated (Definition 2.7).

An *abstraction axis* in this concept is the previously mentioned virtual, perceived connection between different endpoints of the previously discussed abstraction process. This connection arises for observers as they adjust the settings of the visualization pipeline. This notion, however, assumes that, for each abstraction axis, there is a clearly identifiable succession of changes to the visual representation that (a) decreases the amount of information in each step and (b) provides a meaningful generalization of the depicted content to the viewer. In fact, Viola and Isenberg [19] even state that abstraction axes do not need to be unique: for their chosen example from structural biology [20], they show that a molecular van-der-Waals surface-based molecular representation can be subjected to two alternative forms of structural abstraction (a phenomenon they call "forking" of axes)—one leading to a surface-based abstraction via different probe sizes and one leading to a second-order representation via balls-and-sticks, licorice (sticks-only), and backbone representations.

In particular for this latter form of abstraction, one could argue that condition (a) is not necessarily met: while the transition from van-der-Waals surfaces to the licorice representation in van der Zwan's [20] model certainly removes the detail of the graphical atom representations, it simultaneously also adds representations of the bonds between atoms that did not exist in the starting configuration: the representations of atoms with implicitly represented bonds are continuously replaced with representations of bonds with implicitly represented atoms. One could thus argue that in this transition no abstraction happens, only one representation is smoothly transitioned into another. This transition, however, only happens in the eyes of the beholder; at any given point, along the transition still abstraction happens from reality (source things) via data representations to visual representations.

A more recent example is the work by Miao et al. [12] who similarly constructed a progression of transitions from an atom-based representation of DNA nanostructures and the mechanical building blocks of the nanostructures to be built. Interesting in their progression of ten abstraction stages is that, while the first and last are fairly clear, the specific order of the sequence in-between is not and was created based on the discussions with and needs of their collaborating domain scientists.

Based on these two examples, we thus suggest that pure and continuous abstraction axes are rather rare. Instead, abstraction axes are typically composed of smaller building blocks where one representation is (typically) seamlessly transformed into another and as such forms a constructed sequence. In practice, we often find abstraction axes that progress from a representation with more information to a representation with less detail, thus the name abstraction axis. We can also find transitions, however, that remove one type of visual detail and replace it with another type of visual information. Abstraction axes are always constructed with a given purpose and application case in mind and are not unique. If two abstraction axes work on independent attributes of the visual representation and can thus be independently controlled, then they form an *abstraction space* (Definition 2.7).

2.5   An Information-Theoretic Analysis of Abstraction as a Process

As described above, the notion of "abstraction" encompasses a wide range of definitions in different contexts. It can be quite difficult for a single mathematical formulation to encapsulate the essences of these definitions. In this section, we examine the
characteristics of the process of abstraction using information-theoretic measures.
We show that the definitions given at the beginning of this chapter can be explained
using an information-theoretic metric, which therefore offers a potential means for
modeling and measuring visual abstraction.

Figure 2.2 shows several visualization images generated using some typical
visualization techniques. Most visualization researchers would unreservedly refer
to the first four images, (a)–(d), as results of visual abstraction, and many would
contentedly accept a suggestion that (e) and (f) are also results of visual abstraction,
but some would be hesitant to consider (g) and (h) as such abstracted representations.
Nevertheless, one can also argue that the latter four images, (e)–(h), are also results
of visual abstraction because, in comparison with the source data, some information
has been abstracted away and, in comparison with statistical abstraction of the source
data, the information presented is visual.

First, the level of willingness for people to consider a visualization image as
an abstracted visual representation does not appear to be related to the quality of
the image or the usefulness of the technique that generates the image. Second, we
can observe that both Fig. 2.2a, h feature some deformation, and deformation does
not seem to be a critical factor that influences the perception of visual abstraction
results. Similarly, from a comparison of (b) versus (h) and (c) versus (g), we can
observe that the types of data to be visualized do not have decisive influence upon
the perception of the term "visual abstraction." Third, we can also observe that an
impression of photorealism or just a perceived intention seems to bring about the
hesitation in characterizing a visualization image as the result of visual abstraction.
Meanwhile, having no or less photo-realistic effect in an image (e.g., (e) or (f)) does
not immediately imply visual abstraction either, at least to some people. Here the
adjective "photo-realistic" indicates that the rendering algorithm used was designed
to achieve a photo-realistic effect, without implying that the image resulting from
the rendering process actually resembles a photograph.

One hypothesis is that our willingness or hesitation to consider a visualization
image as resulting from visual abstraction relates to an unconsciously integrated
reasoning about two conditions of visual abstraction.

A. **A visual abstraction is a transformation from data to its visual representation
   with some information loss**—Here, data can be of any data types including
   visual data (e.g., image corpora and videos). This can be considered as a broad
   definition of visual abstraction and encapsulates the aforementioned definitions in
   cartography and shape analysis. While introducing a constraint of visual output,
   it exhibits a parallel with the definitions in relation to mathematical abstraction,
   abstract thinking, and grouping in object-oriented design. All eight images in

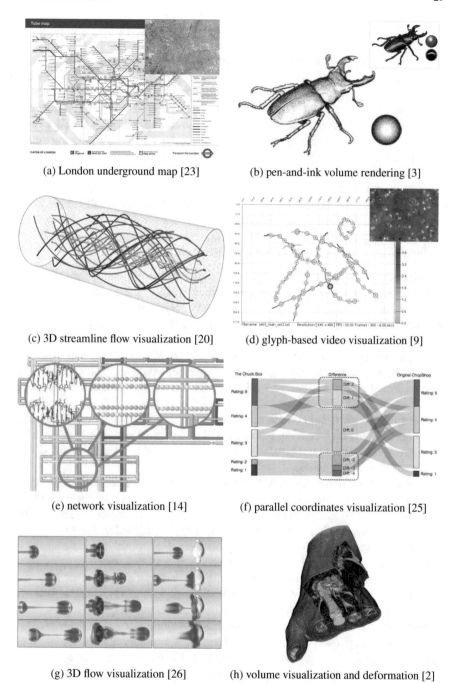

(a) London underground map [23]

(b) pen-and-ink volume rendering [3]

(c) 3D streamline flow visualization [20]

(d) glyph-based video visualization [9]

(e) network visualization [14]

(f) parallel coordinates visualization [25]

(g) 3D flow visualization [26]

(h) volume visualization and deformation [2]

Fig. 2.2 Examples of visualization images that may attract different views as to whether they are the results of visual abstraction processes. Most would agree that **a–d** are visually abstract, many would content that **e** and **f** are considered so, and some might be hesitant about **g** and **h**

Fig. 2.2 satisfy this condition in general. We will discuss information loss in detail later.

B. **A visual abstraction is a transformation from a more photo-realistic visual representation to a less photo-realistic one**—This can be considered as a narrow definition of visual abstraction and encapsulates the aforementioned definitions in art, cartography, and shape analysis. It applies to transformations with visual input as well as visual output. Considering the examples in Fig. 2.2, for images (a), (b), (c), and (d), it is relatively easy for one to imagine their photo-realistic counterparts. Although some of these images can be generated directly from source data that may not be visual, a subjective impression of a transformation that decreases photorealism is sufficient for viewers to associate these images with abstraction. Meanwhile, it is harder to imagine a photo-realistic version of (e) or (f), and therefore, this condition does not appear to be applicable to them. For images (g) and (h), it is intuitive to consider them more photo-realistic than less. They not only fail to satisfy but also negate this condition.

We can easily see that reading data using a spreadsheet or reading their statistical summary does not meet either condition. Images (a), (b), (c), and (d) in Fig. 2.2 satisfy both conditions. Images (e) and (f) satisfy condition A but not B. Images (g) and (h) satisfy condition A but negate B. Suppose that we had a numerical score 2 for condition A, score 1 for condition B, score 0 for not applicable, and score −1 for negation. Spreadsheet or statistical summary would score 0; (g) and (h) would score 1; (e) and (f) would score 2; (a), (b), (c), and (d) would score 3. Such a scoring system would reflect the level of willingness for one to characterize a visualization image as the result of visual abstraction.

We can also infer that condition A is more essential than condition B. Without A, images (e) and (f) would not be considered as results of visual abstraction at all. Without B, there would not be any hesitation about whether images (g) and (h) are results of visual abstractions.

However, condition A does not in itself meet the expectation for the minimal quality that the process of meaningful visual abstraction should possess, since arbitrarily throwing away information should not be referred to as meaningful abstraction. Below, we use several information-theoretic measures to clarify Condition A.

Let $P_{d \to v}$ be a process for transforming a dataset d to a visualization image v. Let $\mathbb{D}$ be the data space containing all possible datasets that $P_{d \to v}$ can take as its input, and $\mathbb{V}$ be the data space containing all possible visualization images that $P_{d \to v}$ can generate. In information theory, $\mathbb{D}$ and $\mathbb{V}$ are referred to as *alphabets*. The dataset d is thus a letter in the input alphabet $\mathbb{D}$, and the visualization image v is a letter of the output alphabet $\mathbb{V}$. The process $P_{d \to v}$ can thus be written as $P_{d \to v} : \mathbb{D} \longrightarrow \mathbb{V}$.

The Shannon entropy measures the amount of uncertainty or variation of an alphabet. Let $p(d)$ be the probability of a dataset d in the context of an application. The Shannon entropy of $\mathbb{D}$ is thus defined as:

$$\mathcal{H}(\mathbb{D}) = -\sum_{d \in \mathbb{D}} p(d) \log_2 p(d)$$

When all letters in $\mathbb{D}$ have the same probability, we have $\mathscr{H}(\mathbb{D}) = \log_2 \|\mathbb{D}\|$, where $\|\mathbb{D}\|$ is the number of different letters in $\mathbb{D}$. Similarly, we can measure the Shannon entropy $\mathscr{H}(\mathbb{V})$ as the amount of uncertainty or variation of $\mathbb{V}$.

Alphabet Compression [3] is the difference $\mathscr{H}(\mathbb{D}) - \mathscr{H}(\mathbb{V})$, which is a coarse indication of the amount of information loss of the visualization process $P_{d \to v}$. Consider a simple example. $\mathbb{D}$ is defined by a real variable, X, which may take valid values between 0.00 and 10,000.00 at two decimal point precision. There are thus 1,000,001 possible values. Let all values have the same probability. We have $\mathscr{H}(\mathbb{D}) \approx 20$ bits.

Meanwhile, we consider a process $P_{d \to v}$ that plots a value $d \in \mathbb{D}$ as a bar in a single-variable bar chart using a canvas with 1000H $\times$ 100W pixels. The maximum resolution available for the mapping function $P_{d \to v} : \mathbb{D} \longrightarrow \mathbb{V}$ is 1000 pixels, thus 1001 bar charts with different bar heights. We have $\mathscr{H}(\mathbb{V}) \approx 10$ bits. The alphabet compression is therefore about 10 bits. In terms of Condition A, there are about 10 bits of information loss. Therefore, any visualization process, which features many-to-one mapping from data to visual objects, typically exhibits positive alphabet compression. Only when the variation of $\mathbb{D}$ is very small, e.g., using the above canvas to plot an integer variable in the range of [0, 100], the amount of alphabet compression can be zero. In the worst scenario, the plotting function randomly depicts a bar with a height between 0 and 1000 pixels, the amount of alphabet compression would be negative.

All images in Fig. 2.2 feature many-to-one mappings. For example, the distortion in Fig. 2.2a is a kind of many-to-one mapping, since many potential track layouts would lead to the same metro map. In the image rendered with a pen-and-ink effect in Fig. 2.2b, each white pixel could be a placeholder for many differently colored pixels that have been abstracted away. Each glyph in Fig. 2.2d is a very low-resolution visual representation of some 20 values, most of which are real numbers. In the volume-rendered image in Fig. 2.2h, each pixel results from a rendering integral that transforms a few hundred voxel values to an RGB trio. Many different combinations of these voxel values could result in pixels with the same color.

Hence, a process for generating visualization images from relatively complex datasets features many-to-one mappings, which means information loss or positive alphabet compression. According Condition A, such a process is thus a process of visual abstraction.

However, what quantifies a visualization or a meaningful visual abstraction must be a process that is intended to generate "meaningful" visualization images from input datasets. The word "meaningful" implies three factors: (i) the viewer can interpret what is being depicted; (ii) the viewer's interpretation of what is depicted is reasonably correct in relation to the original data; and (iii) the viewer's interpretation errors due to information loss do not have serious impact on the viewer's task.

Consider that a viewer's interpretation is a process $Q_{d \leftarrow v} = P_{d \to v}^{-1}$ that attempts to reconstruct a dataset from a given visualization image. This process can be written as $Q_{d \leftarrow v} : \mathbb{V} \longrightarrow \mathbb{D}'$. We use $\mathbb{D}'$ to denote an alphabet that has the same set of letters as $\mathbb{D}$ but a different probability mass function from that of $\mathbb{D}$. For example, given a bar that is 499 pixels tall, a viewer may interpret it as one of these values in the original $\mathbb{D}$, $\{498.00, 498.01, \ldots, 499.99, 500.00\}$. Imagine that the interpretation is biased

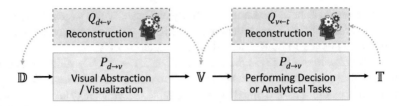

Fig. 2.3 The effectiveness of a visual abstraction process depends on the succeeding task process as well as the viewer's knowledge, biases, and cognitive capability

toward 500.00 due to the corresponding mark on the vertical axis. The probability $q(500)$ would be undesirably higher than the original probability $p(500)$.

In information theory, such errors in the interpretation can be collectively measured by the Kullback–Leibler divergence, which is defined as:

$$\mathscr{D}_{KL}(\mathbb{D}'\|\mathbb{D}) = \sum_{d\in\mathbb{D}} q(d)\log_2 \frac{q(d)}{p(d)}$$

where $p()$ and $q()$ are the probability mass functions of $\mathbb{D}$ and $\mathbb{D}'$ respectively, and $q(d)/p(d)$ is a discrete representation of the Radon–Nikodym derivative of q with respect to the original p.

In the context of visual abstraction, this measurement offers a counterbalance to the measurement alphabet compression. It is referred to as *Potential Distortion* [3]. While it is desirable to have the results of visual mapping $P_{d\to v}$ as abstract as possible, i.e., for $P_{d\to v}$ to have a high amount of alphabet compression, it is also necessary to keep the inaccuracy of the interpretation function $Q_{d\leftarrow v}$ as low as possible, i.e., for $Q_{d\leftarrow v}$ to have a low amount of potential distortion.

Since $Q_{d\leftarrow v}$ is a human-centric process, $Q_{d\leftarrow v}$ may feature inaccuracy due to perceptual errors and cognitive biases. However, $Q_{d\leftarrow v}$ can also make use of human knowledge that is not encoded in the data to help more accurate reconstruction. For example, imagine that a viewer is asked to guess what would be the original colors on the patch of white pixels between two black lines in the pen-and-ink visualization image in Fig. 2.2b. A naïve guess would be either white (as what is seen) or an arbitrary selection from various gray colors. Most viewers, especially those familiar with the depicted object or volume visualization methods, can do much better than the naïve guess. Hence the process of "knowledge-assisted guessing"—a heuristic process—has a lower amount of potential distortion than the naïve guessing. In general, it is this human knowledge that enables visual abstraction to be deployed effectively in many situations, such as those illustrated in Fig. 2.2. Whether users have the adequate ability to interpret the results of visual abstraction is thus one of the key criteria for judging if a visual abstraction process is appropriate or its results are meaningful, which reflecting the two factors (i) and (ii) described above.

Nevertheless, since $P_{d\to v}$ is usually a many-to-one mapping, and $Q_{d\leftarrow v}$ is usually a one-to-many mapping, one may wonder why we should go through such "unnecessary fuss" to apply the process $P_{d \to v}$ first to $\mathbb{D}$ and another process $Q_{d \leftarrow v}$ to reconstruct $\mathbb{D}'$. One important rationale is about the *task* succeeding $P_{d \to v}$ and $Q_{d \leftarrow v}$. The judgment about whether a visual abstraction process is appropriate or its results are meaningful thus depends on another process $P_{v \to t}$. As illustrated in Fig. 2.3, $P_{v \to t}$ takes $\mathbb{V}$ as the input, and generates another output alphabet $\mathbb{T}$ that may consist of a collection of letters, e.g., different options of a decision, different levels of an assessment, different categories of a situation, etc.

The process of visual abstraction $P_{d \to v}$ and the reconstructive interpretation $Q_{d \leftarrow v}$ can collectively affect the task process $P_{v \to t}$, especially its *Cost* $\mathbf{Ct}(P_{v \to t}, Q_{v \leftarrow t})$. Similar to $Q_{d \leftarrow v}$, here $Q_{v \leftarrow t}$ is an interpretation process for reconstructing $\mathbb{V}$ from $\mathbb{T}$. For a univariate value (e.g., 499.38), there is little merit to visualize it using a bar chart. The difference in the cost of reading the number and that of viewing a bar is negligible for most tasks. The potential distortion caused by visual abstraction can only affect the process $P_{v \to t}$ negatively. However, if the number of variables increases, e.g., 10 variables, the cognitive load for viewing and comparing 10 numbers is likely to be higher than viewing and comparing 10 bars using a bar chart. It is not difficult to imagine the merits of visualization when the number of variables increases. For the volume datasets featured in Fig. 2.2b, h, the number of variables in a dataset is typically at the scale of $256 \times 256 \times 256$ or more. It is inconceivable to perform a decision task by reading the numerical values of such a volume dataset. Hence, visual abstraction can be used to transform a volume dataset with a huge number of variables to visualization images as shown in (b) and (h), which reduces the cost $\mathbf{Ct}(P_{v \to t}, Q_{v \leftarrow t})$ significantly.

The above information-theoretic discourse on visual abstraction is based on the cost-benefit metric for data intelligence proposed by Chen and Golan [3]. For any data intelligence process P_i with an input alphabet $\mathbb{Z}_i$ and an output alphabet $\mathbb{Z}_{i+1}$, its cost-benefit ratio is defined as:

$$\frac{Benefit}{Cost} = \frac{Alphabet\ Compression - Potential\ Distortion}{Cost}$$

$$= \frac{\mathbf{AC}(P_i) - \mathbf{PD}(Q_i)}{\mathbf{Ct}(P_i, Q_i)} = \frac{\mathscr{H}(\mathbb{Z}_i) - \mathscr{H}(\mathbb{Z}_{i+1}) - \mathscr{D}_{KL}(\mathbb{Z}_i' \| \mathbb{Z}_i)}{\mathbf{Ct}(P_i, Q_i)}$$

When this metric is applied to the two processes in Fig. 2.3, we have the combined cost-benefit ratio as:

$$\frac{Benefit}{Cost}(d \to v \to t) = \frac{\mathbf{AC}(P_{d \to v}) - \mathbf{PD}(Q_{d \leftarrow v}) + \mathbf{AC}(P_{v \to t}) - \mathbf{PD}(Q_{v \leftarrow t})}{\mathbf{Ct}(P_{d \to v}, Q_{d \leftarrow v}) + \mathbf{Ct}(P_{v \to t}, Q_{v \leftarrow t})}$$

$$\text{(2.1)}$$

$$= \frac{\mathscr{H}(\mathbb{D}) - \mathscr{H}(\mathbb{V}) - \mathscr{D}_{KL}(\mathbb{D}' \| \mathbb{D}) + \mathscr{H}(\mathbb{V}) - \mathscr{H}(\mathbb{T}) - \mathscr{D}_{KL}(\mathbb{V}' \| \mathbb{V})}{\mathbf{Ct}(P_{d \to v}, Q_{d \leftarrow v}) + \mathbf{Ct}(P_{v \to t}, Q_{v \leftarrow t})}$$

$$= \frac{\mathscr{H}(\mathbb{D}) - \mathscr{H}(\mathbb{T}) - \mathscr{D}_{KL}(\mathbb{D}' \| \mathbb{D}) - \mathscr{D}_{KL}(\mathbb{V}' \| \mathbb{V})}{\mathbf{Ct}(P_{d \to v}, Q_{d \leftarrow v}) + \mathbf{Ct}(P_{v \to t}, Q_{v \leftarrow t})}$$

In comparison, if one has to perform the task by reading the data without visualization, the cost-beneficial ratio would be:

$$\frac{Benefit}{Cost}(d \to t) = \frac{\mathbf{AC}(P_{d \to t}) - \mathbf{PD}(Q_{d \leftarrow t})}{\mathbf{Ct}(P_{d \to t}, Q_{d \leftarrow t})} \quad (2.2)$$
$$= \frac{\mathscr{H}(\mathbb{D}) - \mathscr{H}(\mathbb{T}) - \mathscr{D}_{KL}(\mathbb{D}''\|\mathbb{D})}{\mathbf{Ct}(P_{d \to t}, Q_{d \leftarrow t})}$$

Note that the term $\mathscr{D}_{KL}(\mathbb{D}'\|\mathbb{D})$ in Eq. (2.1) and the term $\mathscr{D}_{KL}(\mathbb{D}''\|\mathbb{D})$ in Eq. (2.2) are of different quantities as they relate to $Q_{d \leftarrow v}$ and $Q_{d \leftarrow t}$ respectively.

When the dataset d is large and complex, we can see that the cost $\mathbf{Ct}(P_{d \to t}, Q_{d \leftarrow t})$ in Eq. (2.2) would be much higher than the combined costs in Eq. (2.1) in terms of time and cognitive load in performing the task. In other words, we have:

$$\mathbf{Ct}(P_{d \to t}, Q_{d \leftarrow t}) > \mathbf{Ct}(P_{d \to v}, Q_{d \leftarrow v}) + \mathbf{Ct}(P_{v \to t}, Q_{v \leftarrow t})$$

Although reading data might appear to be more accurate, the reconstruction process $Q_{d \leftarrow t}$ from the task alphabet $\mathbb{T}$ (e.g., the patient has a tumor or not) to the data alphabet $\mathbb{D}$ (e.g., a volume dataset) is much more error-prone than the reconstruction process via visualization. In other words, we have:

$$\mathbf{PD}(Q_{d \leftarrow t}) > \mathbf{PD}(Q_{d \leftarrow v}) + \mathbf{PD}(Q_{v \leftarrow t})$$

With $\mathbf{AC}(P_{d \to v}) + \mathbf{AC}(P_{v \to t}) = \mathbf{AC}(P_{d \to t})$, it is not difficult to conclude:

$$\frac{Benefit}{Cost}(d \to v \to t) > \frac{Benefit}{Cost}(d \to t)$$

Under Condition A, we can thus mathematically reason that, for any slightly large or complex dataset, the process from data alphabet $\mathbb{D}$ to task alphabet $\mathbb{T}$ with visual abstraction is usually more cost-beneficial than the process without.

For some very simple datasets, such as a univariate value, visual abstraction may not have an information-theoretic merit. However, this is not to say that it could not have cognitive merit in disseminative visualization. More likely, the results of visual abstraction could attract more attention from the viewers who unconsciously devote more cognitive load to the task. Although the viewers' cost-beneficial ratio increases, the presenter of the disseminative visualization benefits from the contribution of extra cognitive load from the viewers. In many ways, this is similar to scenarios of disseminative visualization, where the amount of visual abstraction is purposely reduced in order to attract viewers' attention and hence their cognitive load. Such scenarios may include, for instance, showing an animated chart, while a static chart could adequately convey the information, or showing visualization in theater-based virtual environments [4].

Similarly, we can also use the cost-benefit metric to analyze the scenarios under condition B by comparing the cost-benefit ratio of a more photo-realistic technique with a less photo-realistic technique. Similar to Condition A, the potential distortion is affected by viewers' knowledge as well as their biases. The cost is affected by the viewer's task as well as cognitive capability.

Furthermore, this metric can be applied to human-centric processes (e.g., visualization and interaction) as well as machine-centric processes (e.g., statistics and algorithms). In general, statistical abstraction and algorithmic abstraction usually result in more alphabet compression as well as more potential distortion but less cost than visual abstraction. In designing a visual analytics workflow, the metric can be used to compare the cost-benefit of a human-centric process with that of a machine-centric process by analyzing the trade-off among alphabet compression, potential distortion, and cost. The metric can also be used to guide a visualization designer in choosing different forms of visual abstraction, e.g., in reasoning about the trade-off among the amount of abstraction, the potential perceptual errors, and the cost of task performance.

In summary, as defined at the beginning of this chapter, meaningful visual abstraction depends on some points of view and some tasks. From the perspective of information theory, the **points of view** may be in either or both of the following forms:

- The factors that influence the *alphabet compression* and *cost* of the process $P_{d \rightarrow v}$ for transforming data to visualization. These factors may include the designers' wish to keep or highlight some information while removing or deemphasizing other information, their understanding of the task requirements, their appreciation of the resources available for visualization, and their awareness of the viewers' knowledge of visual representations and skills of visual analysis.
- The factors that influence the *potential distortion* and the *cost* of the process $Q_{d \leftarrow v}$ for reconstructing data from visualization. These factors may include the viewers' knowledge related to the data being depicted and the visual representations used, their understanding of the information required for performing their tasks, and their cognitive load and time constraint in executing the process $Q_{d \leftarrow v}$.

Meanwhile, **tasks** can be defined as processes that succeed the processes $P_{d \rightarrow v}$ and $Q_{d \leftarrow v}$. As long as the tasks fall broadly in the category of data intelligence tasks, the cost-benefit metric proposed by Chen and Golan [3] can also be applied to these succeeding processes. Therefore, from the information-theoretic perspective, the most meaningful visual abstraction, or the most effective visualization in general, is the process with the optimal cost-benefit measure.

2.6 Summary

In this chapter, we thus formally defined the concepts of abstraction and visual abstraction as they relate to the field of visualization and based on existing notions of the terms in related fields such as the arts and in philosophy. We argued that any visual representation is the result of multiple abstraction steps from reality, and we called the step from data representation to visual representation visual abstraction. We also showed that as users of a visualization system, we do not observe this abstraction process but instead adjust settings to transition from one visual representation to another—each being an independent result of the abstraction process from source thing to sign thing. Yet as designers of visualization systems, we can provide guided interaction such that several results of meaningful abstractions can be assembled into sequences that we call abstraction axes to better illustrate how different aspects of reality relate to each other, and several of these abstraction axes can be assembled into abstraction spaces to illustrate the interrelation of several independent aspects. So while we argue that any visual representation is the result of an abstraction process, it is still important to discuss abstraction

and visual abstraction as it teaches us about visualization as a process in general.

Acknowledgements The authors would like to thank Jos Roerdink, Helwig Hauser, Stefan Bruckner, Hans-Christian Hege, and Torsten Möller for fruitful discussion that helped shaping the chapter. Thanks to Peter Mindek for illustrating the abstraction space in Fig. 2.1. This work was funded through the ILLUSTRARE grant by both the Austrian Science Fund (FWF; I 2953-N31) and the French National Research Agency (ANR; ANR-16-CE91-0011-01).

References

1. Arnheim, R.: Visual Thinking. University of California Press, Berkeley (1969)
2. Buttenfield, B.P., McMaster, R.B. (eds.): Map Generalization: Making Rules for Knowledge Representation. Longman Scientific & Technical, Harlow (1991)
3. Chen, M., Golan, A.: What may visualization processes optimize? IEEE Trans. Vis. Comput. Graph. 22(12), 2619–2632 (2016). https://doi.org/10.1109/TVCG.2015.2513410
4. Chen, M., Gaither, K., John, N.W., McCann, B.: An information-theoretic approach to the cost-benefit analysis of visualization in virtual environments. IEEE Trans. Vis. Comput. Graph. 25(1), 32–42 (2019). https://doi.org/10.1109/TVCG.2018.2865025
5. Chernoff, H.: The use of faces to represent points in k-dimensional space graphically. J. Am. Stat. Assoc. 68(342), 361–368 (1973). https://doi.org/10.1080/01621459.1973.10482434
6. Cornea, N.D., Silver, D., Yuan, X., Balasubramanian, R.: Computing hierarchical curve-skeletons of 3D objects. Vis. Comput. 21(11), 945–955 (2005). https://doi.org/10.1007/s00371-005-0308-0
7. Frege, G.: Der Gedanke [The thought]. Beiträge zur Philosophie des deutschen Idealismus 1(2), 58–77 (1918). https://www.jstor.org/stable/2251513. English translation in Mind: Q. Rev. Psychol. Philos. 65(259), 289–311 (1956)
8. Isenberg, T.: Capturing the essence of shape of polygonal meshes. Ph.D. thesis, Otto-von-Guericke University of Magdeburg, Germany (2004). https://doi.org/10.25673/5454
9. Leppänen, M.: Towards an abstraction ontology. In: Duží, M., Jaakkola, H., Kiyoki, Y., Kangassalo, H. (eds.) Information Modelling and Knowledge Bases XVIII, Frontiers in Artificial Intelligence and Applications, vol. 154, pp. 166–185. IOS Press, Amsterdam (2007). http://ebooks.iospress.nl/volumearticle/3413
10. Lewis, D.K.: On the Plurality of Worlds. Blackwell, Oxford (1986)
11. Miao, H., De Llano, E., Isenberg, T., Gröller, M.E., Barišić, I., Viola, I.: DimSUM: dimension and scale unifying maps for visual abstraction of DNA origami structures. Comput. Graph. Forum 37(3), 403–413 (2018). https://doi.org/10.1111/cgf.13429
12. Miao, H., De Llano, E., Sorger, J., Ahmadi, Y., Kekic, T., Isenberg, T., Gröller, M.E., Barišić, I., Viola, I.: Multiscale visualization and scale-adaptive modification of DNA nanostructures.

IEEE Trans. Vis. Comput. Graph. **24**(1), 1014–1024 (2018). https://doi.org/10.1109/TVCG.2017.2743981

13. Mohammed, H., Al-Awami, A.K., Beyer, J., Cali, C., Magistretti, P., Pfister, H., Hadwiger, M.: Abstractocyte: a visual tool for exploring nanoscale astroglial cells. IEEE Trans. Vis. Comput. Graph. **24**(1), 853–861 (2018). https://doi.org/10.1109/TVCG.2017.2744278
14. Mosby: Mosby's Medical Dictionary, 9th edn. Elsevier, Amsterdam (2012)
15. Munzner, T.: Visualization Analysis and Design. A K Peters Visualization Series. CRC Press, Boca Raton (2014). https://doi.org/10.1201/b17511
16. Ogden, C.K., Richards, I.A.: The Meaning of Meaning: A Study of the Influence of Language upon Thought and of the Science of Symbolism. Kegan Paul, London (1923)
17. Stanford encyclopedia of philosophy. Web site. https://plato.stanford.edu/contents.html
18. Thompson, P.: Margaret Thatcher: a new illusion. Perception **9**(4), 483–484 (1980). https://doi.org/10.1068/p090483
19. Viola, I., Isenberg, T.: Pondering the concept of abstraction in (illustrative) visualization. IEEE Trans. Vis. Comput. Graph. **24**(9), 2573–2588 (2018). https://doi.org/10.1109/TVCG.2017.2747545
20. van der Zwan, M., Lueks, W., Bekker, H., Isenberg, T.: Illustrative molecular visualization with continuous abstraction. Comput. Graph. Forum **30**(3), 683–690 (2011). https://doi.org/10.1111/j.1467-8659.2011.01917.x

Chapter 3
Measures in Visualization Space

Fabian Bolte and Stefan Bruckner

Abstract Measurement is an integral part of modern science, providing the fundamental means for evaluation, comparison, and prediction. In the context of visualization, several different types of measures have been proposed, ranging from approaches that evaluate particular aspects of visualization techniques, their perceptual characteristics, and even economic factors. Furthermore, there are approaches that attempt to provide means for measuring general properties of the visualization process as a whole. Measures can be quantitative or qualitative, and one of the primary goals is to provide objective means for reasoning about visualizations and their effectiveness. As such, they play a central role in the development of scientific theories for visualization. In this chapter, we provide an overview of the current state of the art, survey and classify different types of visualization measures, characterize their strengths and drawbacks, and provide an outline of open challenges for future research.

3.1 Introduction

Considering the vast amounts of data involved in many scientific disciplines, it is essential to provide effective and efficient means for forming a mental model of the underlying phenomena. Visualization seeks to provide these means through interactive computer-generated graphical representations, taking advantage of the extraordinary capability of the human brain to process visual information. Specifically, the term "visualization" refers to the process of extracting meaningful information from

F. Bolte · S. Bruckner (✉)
University of Bergen, Bergen, Norway
e-mail: Stefan.Bruckner@UiB.no

F. Bolte
e-mail: Fabian.Bolte@UiB.no

© Springer Nature Switzerland AG 2020
M. Chen et al. (eds.), *Foundations of Data Visualization*,
https://doi.org/10.1007/978-3-030-34444-3_3

Fig. 3.1 240 different tree visualization techniques [48]—which one should be used?

data and constructing a visual representation of this information. This process is composed of three basic stages [26]

A. making data displayable by a computer,
B. transmitting visual representations to human viewers, and
C. forming a mental picture about the data.

Significant effort has been devoted to the formulation of taxonomies and categorizations of this general process. For instance, Shneiderman [49] introduced a task-by-data taxonomy, while Tory and Möller [53] focused on the classification of visualization algorithms. In an influential contribution, Munzner [38] proposed a nested model for designing and developing visualization pipelines, that has inspired a considerable amount of subsequent work. Wang et al. [60], for instance, proposed a two-stage framework for designing visual analytics systems, while Ren et al. [42] proposed a multi-level interaction model of goal, behavior, and operation to facilitate system development with formal descriptions. The multi-level typology of Brehmer and Munzner [7] distinguishes between the basic questions of *why*, *how*, and *what*, in order to classify abstract visualization tasks. These types of classifications are highly valuable resources for visualization practitioners and researchers to steer the design process and to compare competing approaches.

Ultimately, however, in order to assess the effectiveness of visualization, it is crucial to know whether or not the mental picture of the data established by a human viewer is consistent with the original data, and whether or not one specific visualization technique or parameter setting is more effective than another. Displaying and analyzing data is of ever-increasing importance in almost all research disciplines. Consequently, the field of visualization is constantly growing and reliable visualizations are of more and more importance for domain experts to gain authentic insights. This progress comes along with a steady growth in diversity and complexity of visualization methods, making judgment of their effectiveness and suitability for a certain task difficult. Figure 3.1, for instance, which shows 240 different techniques to visualize tree data taken from a visual bibliography on the topic [48], illustrates the challenges in selecting appropriate visualization techniques.

Traditionally, visualization techniques and their parameter settings are evaluated by carrying out user studies which measure their performance for particular sets of tasks. However, such studies require considerable effort and their design is non-trivial [38]. Their specialized nature also makes it difficult to generalize the outcomes. Furthermore, when developing new visualization techniques, frequently only a small number of initial users is available, making it difficult to obtain statistically significant results. The alternative of solely relying on the visualization creator's judgment, is also scientifically questionable because it often reflects personal preference and may include bias. Hence, it is highly desirable to support a visualization process by enabling visualization creators to conduct an evaluation using objective measures.

In principle, such quality measures could then be used to automatically select and/or parameterize a visualization from a set of choices according to these measures by using an appropriate optimization process. Moreover, measures may also inform us about the structure of the visualization space itself; i.e., they may lead us to deeper insights into how the phenomenon of visualization works and hence could be of utility beyond a descriptive or evaluative usage. Hence, questions related to visualization measures are tightly connected to the bigger effort of specifying a theory of visualization. In this paper, we survey approaches that seek to enable the systematic analysis of visualization algorithms and their properties with respect to the underlying data characteristics and their perceptual qualities. While we cover the significant body of research that has been devoted to various types of visualization measures, we also specifically look at approaches that regard the interplay between data, algorithms and their parameters, and visual perception and cognition as a phenomenon that deserves study in its own right.

In many disciplines of science, hypotheses are formulated based on empirical data, and then subsequently developed into models and complete theories of the phenomenon under investigation. The predictions of these models and theories are then continuously validated and, once they are supported by sufficient data, are generally accepted as scientific "facts".[1] Importantly, the consequences of these theories can lead to the discovery of new relationships and insights due to their predictions. Theoretical physics, for instance, heavily relies on the mathematical structure of existing well-validated theories in the development of more comprehensive models of our universe. There are many instances—for example within the standard model of particle physics—where subsequent discoveries have been predicted based on structural and mathematical aspects such as symmetries of the underlying theory. For instance, the famous Higgs mechanism and one of its important predictions, the Higgs boson, were already described in the 1960s, but strong evidence for its existence only became available in 2013.

The formulation of measures forms an important first step in the development of such theories, as they are often the fundamental building blocks from which more complex relationships can be derived. Thus, measures play a central role in the ongoing search for a more comprehensive theory of visualization.

---

[1]While a scientific theory can never be proven "true" in a mathematical sense, there are many examples of well-established theories such as evolution, quantum mechanics, and general relativity that form the basis of modern science and that are rarely questioned on a principle level.

3.2   Measurement in Science

In philosophy, the topic of measurement in science has been illuminated from many different points of view. Tal [52] gives a comprehensive account of the different schools of thought, and here, we will only briefly summarize his considerations in order to provide additional background. In principle, he distinguishes between the following perspectives:

A. **Mathematical theories of measurement** regard measurement as the mapping of qualitative empirical relations to relations among numbers or other mathematical entities. Measurement theory aims to identify the assumptions related to the use of different mathematical structures for describing aspects of the empirical world. In particular, it attempts to make statements about the adequacy and limits related to the use of these structures. One of the key insights of measurement theory is that mathematical structures used for measurement should mirror relevant relations among the real-world objects being measured. For instance, we could mistakenly assume that an object measured at a temperature of 60 °C is twice as hot as one measured at 30°. However, when expressed using the Fahrenheit scale, the temperatures of these objects are 86 and 140, respectively. This is because the zero points of these two scales are arbitrary and do not correspond to the absence of temperature.

B. **Realist views** consider measurement as the estimation of mind-independent properties and/or relations. A measurement is regarded as the empirical estimation of an objective property or relation. The term "objective", in this context, is meant to signify that these properties are independent of the conventions and beliefs of the humans conducting the measurement and of the methods used in their execution. The values of measurements are regarded as approximations of true values, and measurement itself is aimed at obtaining knowledge about properties and relations, rather than the assignment of values to objects themselves. For instance, a realist about length measurement would say that the ratio of the length of an object to the standard meter has a definite objective value, irrespective of how it is measured. The measurement itself is merely an approximation of this value.

C. **Operationalist views** are concerned with the meaning and use of quantity terms. A realist would argue that these terms refer to sets of properties that exist independently of being measured. The operationalist point of view, on the other hand, is that the meaning of quantity concepts is solely determined by the set of operations used for their measurement. They view measurement as a set of operations that shape the meaning and/or regulate the use of a quantity term. For example, length could be defined as the result of concatenating rigid rods, but it could also be defined by timing electromagnetic pulses. A strict operationalist would distinguish these two into distinct quantity concepts such as "length-1" and "length-2".

D. **Information-theoretic accounts** view measurement as the gathering and interpretation of information about a system. Measuring instruments are regarded as "information machines" that interact with an object in a given state, encode that state into a signal, and convert this signal into an output. The accuracy of a measurement is dependent on the instrument as well as the level of noise in the environment. Information-theoretic accounts of measurement were originally developed by metrologists and hence are practically oriented and tailored toward evaluating and improving the accuracy of measurement standards. As such, their connection to more philosophical considerations is less explored.

E. **Model-based accounts** view measurement as the coherent assignment of values to parameters in a theoretical and/or statistical model of a process. According to model-based views, measurement consists of two levels: (1) a process involving interactions between an object of interest, an instrument, and the environment; and (2) a theoretical and/or statistical model (i.e., an abstract representation based on simplifying assumptions) that describes this process. Hence, the central goal of measurement is to assign values to the parameters of these models such that they satisfy certain criteria such as coherence and consistency.

While these considerations are important and relevant lines of philosophical investigation, for the purposes of the discussion here we will largely gloss over these partially subtle distinctions. Nevertheless, we will see that some of these views are more prominent in the visualization domain than others. Many of the visualization quality measures are constructed in an operationalist manner, providing different means to measure the same property of a visualization. Several phenomena in visualization have been described by applying communication models from information theory, and several theoretical models try to explain, e.g., perceptual processes in the human visual system or the visualization process as a whole. Mathematical theories of measurement and realist views have received less attention in visualization research. As this topic gains more attention, we expect a more explicit exploration of the philosophical underpinnings of different approaches. In the following sections, we will describe different types of measurements in visualization and how they can be combined to build a better understanding of visualization as a research field in the future.

3.3 Types of Visualization Measures

There are numerous different aspects of the visualization process that one can set out to quantify. Partially, the boundaries between different types of measures can be fuzzy, but in the following we will attempt to characterize some principal categories of measures that have been investigated.

3.3.1 Measures of Perceptual Characteristics

The measurement of perceptual characteristics of visualizations aims to mimic low-level processing of visual stimuli in the human perceptual system. Essentially, the idea is that by—at least partially—modeling and simulating the early processing stages of the perception pipeline, we can predict how particular visual elements influence the interpretation of a particular visualization by a human observer.

Significant efforts have been devoted to understanding the effectiveness of different visual variables for encoding quantitative and qualitative data in the visualization literature. For example, Cleveland and McGill [15] ran a well-known series of graphical perception experiments to measure accuracy in comparing values and to derive the rankings of encoding variables that still form the basis for many visualization design decisions. Similar types of experiments have also been used to compare different types of charts, and their results have been employed to aid the automatic construction of visualizations [35, 36].

A major early contribution to the study of visual perception was made by the Gestalt School of Psychology. Developed in the early twentieth century, the intent was to understand the principles behind how humans acquire and maintain meaningful perceptions of the world given its complex and chaotic nature. The main idea maintains that the human perceptual system employs a notion of "Gestalt" (German for shape or form) that it uses to organize and interpret its inputs. By further investigating this basic thought, psychologists were able to establish a series of Gestalt principles of perception, which are still respected today as accurate descriptions of visual behavior. Since then, several works have set out to describe these and related observations and their effects in a more formal manner.

At the most basic level, we can look at physiologically based models which typically idealize neural behavior using mathematical functions. The response of retinal ganglion cells, which have a center-surround behavior, can be described by a difference-of-Gaussians function which contains a narrow excitatory center within a larger inhibitory surround [43]. A Gabor function, mathematically defined as a 1D sinoid within a 2D Gaussian envelope, has been shown to be a good approximation of the edge patterns which the primary visual cortex (V1) neurons are sensitive to [17]. Li [33] presented a model of contour perception in the primary visual cortex. While it does not include retinal processing or edge pattern recognition, it focuses on lateral connections in the visual cortex and how they can give rise to contour integration phenomena. Grossberg and Williamson [63] proposed a more detailed physiologically based model which includes center-surround processing and Gabor-like pattern matching of neurons. It divides the primary visual cortex into several layers associated with particular behaviors such as contour enhancement and convergence of neural activity. Pineo and Ware [39] combine aspects of the models by Li and Grossberg and Williamson. They realize a difference-of-Gaussians retinal response and a V1 Gabor response. Furthermore, their approach is specifically tailored toward the viewing of data visualizations, which—they argue—tend to be viewed in an exploratory manner. Hence, they seek to model perception in the moments after viewing, before steady-state activity is reached. This also allows them to make the computational evaluation of the model sufficiently fast to be embedded in an optimization loop. Thus, in addition to their model of low-level perception, Pineo and Ware [39] also present an application of their perceptual model for 2D flow visualization. They argue that the brain generates its high-level understanding of a visualization from the activity of low-level neurons, and erroneous low-level perception thus has a degrading effect on this high-level understanding. Based on this reasonable assumption, they propose a predictor for the perceived direction at a point in visual space from the activity of edge selective neurons that surround it. Likewise, they predict the perceived speed of flow from the activity of blue-yellow neurons (which correspond to their chosen color mapping) weighted by the distance of the receptive field to the point being predicted. These measures are then used in a hill climbing optimization process to adjust the parameters of a streaklet-based visualization.

Such perceptual measures focus on the low-level processing of visual stimuli in the human perceptual system such as preattentive processing [23]. Hence, they are primarily concerned with how basic visual encoding variables, such as position, length, area, shape, and color, and the interaction of the variables (e.g., integrable or separable), influence the efficiency of low-level perceptual features such as visual search, change detection, and magnitude estimation [3]. While physiological models taking into account neural response are scientifically attractive due to their "first-principles" nature, an obvious challenge is to scale them up to more informative aspects of higher-level perception. As is the case in many areas of science, it is far from trivial to connect multiple scales in a meaningful manner while preserving important practical aspects such as computational feasibility. For this reason, the modeling of higher-level phenomena often ignores some of the more detailed aspects. In the context of perceptual measures, the concept of saliency [24] is a prominent example for this.

In general, visual saliency models assess the features of an image to predict which areas of that image will draw a viewer's attention. While they are typically inspired by the structure and function of the human visual cortex and are designed to be "biologically plausible", most approaches make a number of simplifying assumptions. Several practical saliency models have been proposed that, while inspired by basic principles such as the center-surround mechanism, forego more detailed modeling of the neural response and instead take a more phenomenological approach. Saliency models can be categorized as models of bottom-up visual attention. Bottom-up visual attention is drawn to regions that are distinct from their surroundings with respect to their basic visual features such as contrast, color, or motion. Top-down visual attention, on the other hand, is driven by the viewer's goals, expectations, and experience. It is hence allocated voluntarily based on the viewer's task and prior knowledge [16, 40]. This makes saliency an attractive basic task-agnostic measure for investigating how viewers read a visualization in principle, and thus, saliency-based measures have garnered the interest of visualization researchers.

Kim and Varshney [28], for instance, presented a method that enhances the saliency of selected regions in volumetric data which they validated using an eye-tracking study. Lee et al. [32] applied the concept of saliency to surface meshes and showed how the measure can be used for targeted simplification as well as viewpoint selection. Jänicke and Chen [25] proposed an approach which uses a saliency-based metric to measure the mismatch between data-space feature maps and the visual representation of the data. While most types of saliency models are tailored toward natural scenes, Matzen et al. [37] developed a method specifically targeted at abstract data visualizations.

Overall, perceptual measures are a useful tool for determining and/or predicting which parts of a visualization will be most prominently seen by a user. Combined with an appropriate way to characterize relevant features in the data, they can be utilized to detect potential mismatches between the importance of regions in data space and their perceptual prominence in the final image. However, at present, only low-level perceptual processing can be feasibly taken into account and higher-level aspects or even cognition are still beyond the reach of current approaches.

3.3.2 Task-Oriented Quality Measures

In contrast to lower-level perceptual measures, the goal of quality measures is to inform about the performance of a visualization technique with respect to a particular well-defined task assumed to be important for the overall goal of the visualization. As discussed in the survey by Behrisch et al. [3], a particular characteristic of such measures is that they do not explicitly consider the user. Instead, they often attempt to heuristically quantify the presence and/or extent of an "anti-pattern", i.e., an assumed known defect or undesirable characteristic of a visualization. These types of measures are commonly referred to as "quality metrics" in the visualization literature. However, as pointed out by Behrisch et al. [3], this is a somewhat misleading term as "metric" has a precise meaning in mathematics with well-defined properties (i.e., non-negativity, identity of indiscernibles, symmetry, and the triangle inequality) which need not necessarily hold in all cases. Thus, we adopt the more neutral term "measure" which does not have these implications.

As the recent state-of-the-art report by Behrisch et al. [3] focuses on these types of measures (classified as "mid-level perceptual quality metrics" in their work), we will only briefly summarize well-known approaches and refer the reader to their comprehensive survey for further details. Given their specialized nature, it makes sense to discuss task-oriented quality measures according to the type of visualization they are designed for, as shown in Fig. 3.2. For instance, in scatterplots and scatterplot matrices "scagnostics"—based on an idea by Tukey and Tukey [56]—have been introduced as an approach to identify anomalies based on attributes of their shape and appearance. These measures themselves form a multi-dimensional space which can be explored in a scatterplot matrix in order to identify outliers in the form of unusual scatterplots. Wilkinson et al. [62] later presented graph-theoretic methods to implement the same approach using a set of measure categories (outliers, shape, trend, density, and coherence), and each composed of multiple numerical measures. For example, the shape of scattered points in a plot can be described by

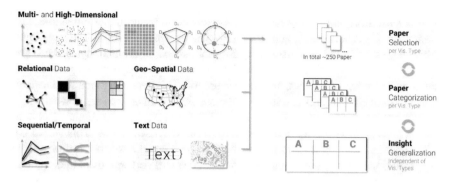

Fig. 3.2 Behrisch et al. [3] analyzed and categorized quality measures from around 250 papers in visualization. These mid-level measures are mostly specific to the underlying data, task, and visualization technique

the following measures: convexity, skinniness, stringiness, and straightness. Bertini and Santucci [4, 5] proposed a model for visual clutter in scatterplots based on an estimate of colliding points vs. available space. They subsequently derived a quality measure that aims to quantify whether the relative data density is preserved when considering the represented density in the plot. It is also common for quality measures to be defined implicitly, for example as part of a layout algorithm. For instance, Byron and Wattenberg [10] presented an approach to optimize the appearance of stacked graphs by using measures such as deviation and wiggle.

Task-oriented quality measures have arguably received the most attention in the field of visualization, as they often tend to encode—with varying degree of fidelity—known best practices or common shortcomings specific to a particular class of visualizations. In essence, they can be seen as (partial) formalizations of design recommendations, and thus tend to be quite practically oriented. Typically being grounded in well-established principles in visualization makes these types of measures semantically meaningful and expressive, providing a sound basis for optimization as well as for the comparison of different but related algorithms. A potential downside of this applied nature of task-oriented quality measures is their limited generalizability.

For instance, when rendering streamlines in flow visualizations, there exist different seeding strategies to define the starting points and number of streamlines. The overall goal is to display all features in the flow without introducing clutter. This goal introduces a trade-off between increasing the number of streamlines to cover all features, but decreasing it for better clarity. While this optimization is crucial for streamline visualizations, it is so specific that it can hardly be applied to any other type of visualization. This is true for many task-specific quality metrics. As a consequence, whenever a new visualization technique is discovered, new task-oriented quality measures need to be developed to optimize the specific aspects of this visualization. It would be desirable to define general measures that express the quality of a visualization independent of its type and allow for their comparison.

As an example, edge crossings can be optimized for multiple visualizations, i.e., graphs, parallel coordinates, and storylines. This is because all of these techniques utilize edges (or links) as a visual encoding for aspects of the underlying data. The user's ability to read a chart is influenced by the number of edge crossings as well as the angle at which they cross, and there seem to be higher-level perceptual aspects of such visual embeddings that increase the cognitive load on the user. If we manage to define these aspects instead of task-specific features, then we might be able to form a more general theory about the perception of visualizations. This would not only allow us to compare different techniques on an equal basis, but further enable the prediction of how new visualization techniques will perform given their defined visual mapping.

3.3.3   Structure-Oriented Measures

In contrast to task-oriented measures, this class attempts to quantify general structural elements of the visualization process. More specifically, structure-oriented measures aim to express in a—at least in principle—measurable form fundamental characteristics of the visualization process itself. Classical examples for these types of measures are Tufte's data-to-ink ratio, as well as his lie factor [55]. The former describes the proportion between the amount of pixels used to present data and the total amount of pixels, whereas the latter describes the ratio between the size of a data value and the size of its corresponding visual element. Both express desirable relationships between the data and its visual representation, but are not tied to any particular visual encoding. On the contrary, they aim to describe general qualities of visualizations, and thus play a particularly important role in considerations toward a theory of visualization.

Mackinlay [35] was one of the first to discuss the expressiveness and effectiveness of visualizations as general means to compare and choose different visual designs. He describes expressiveness as the ability to encode all facts of a dataset without introducing additional facts that are not in the data. Effectiveness, on the other hand, further depends on the user's capabilities to read a certain visualization. Having the user introduced as a deciding factor for the effectiveness of a visualization requires a detailed understanding of the human visual system and although a lot of research has been contributed toward this goal, we still do not possess a sufficiently complete model that would allow us to predict this on a general level. We are therefore further reliant on empirical results of user studies to describe the perceptual capabilities of visualization users.

Demiralp et al. [18] evaluated visual mappings in general by assessing how well the input data is represented by visual elements. They describe visualization as a function that maps from a domain of data points to a range of visual primitives. They further argue that the same measures that can be found in data, like symmetry and distance, should be reflected in the visual elements. In this sense, we could encode pairwise difference in data space as pairwise perceptual difference in color, shape, size, or others. One problem with this approach is that perceptual distance is not given in most visual spaces and needs to be estimated empirically. Additionally, we often utilize several visual encodings at the same time, and it is unclear how they interact and potentially interfere. The authors argue that when two visual spaces are combined, a measure for that space can be constructed from the individual measures. When acquiring perceptual measures for all kinds of visual spaces, we could then create a standard library to validate the pairwise distances between elements in all kinds of visualizations.

Inspired by these considerations, Kindlmann and Scheidegger [29] argue that distance functions and metrics have limits, since for example, partial orders are not symmetric. They instead developed an algebraic framework for describing symmetries between manipulations in data space and their resulting consequences in visualization space. From this, they derived three principles that should be true for a mapping from data to visualization, i.e., unambiguous data depiction, representation invariance, and visual-data correspondence. In short, the visual mapping should make sure that a change in the data is reflected by a corresponding change in the visualization, while changes in the data representation (e.g., the specific data structures used in the implementation) do not affect the visualization, and significant changes in data should result in noticeable changes in the visualization. Given some examples, it becomes clear that not always all of these principles can be met. The visualization designer needs to be aware of certain shortcomings and make sure that the right principle is respected given the task at hand. In this work, the authors introduced a uniform description of different design choices. They adhered to a mathematical model that describes the process of visualization based on its structural properties. They further mention that user studies can be utilized to test perceptual distinguishability and thereby complement mathematical models. The conjunction of evaluated visualizations and mathematical models can help to make statements about visualizations which are not yet evaluated through user studies. While this approach still relies on some notion of perceptual distance, it is notable as it does lead to measurable predictions that in principle can be verified without reliance on user studies. For instance, it can be tested without user involvement whether a significant change in the data leads to no change in the visualization. This opens up the door for a set of "unit tests" for visualization, which could verify at least some objective characteristics fully automatically.

Silver [50] employed the concept of object orientation to conceptualize the visualization process, arguing that the definition and abstraction of features into objects, and their interactions in local regions, allows for a better measurability of phenomena and understanding of their evolution. Abstracting the features of a scientific domain into such a concept allows for generally applicable measurements such as volume, diameter, and curvature and provides a basis for objective comparison. Jankun-Kelly et al. [27] proposed the P-Set model to describe a user's interactions as choosing a parameter based on a previous parameter set, and applying the new set to derive a transformed visualization result. As demonstrated by Liu et al. [34], distributed cognition can be utilized as a theoretical framework in visualization. Purchase et al. [41] analyzed which existing theoretical models can be applied to visualization and provided suggestions for their integration. In particular, they considered visualization under the light of data-centric predictive theory, information theory, and scientific modeling. Chen and Jänicke [12] applied information theory to describe phenomena in visualization with communication models. They argued that many problems and features in visualization can be explained by similar phenomena from information theory which can be applied to evaluate visualizations on a more general level. Xu et al. [64] followed a similar idea to evaluate visualizations by measuring the amount of information that is transported through the visual channels and applied this framework to flow visualization examples. Wang and Shen [61] complemented this work by additional principles with a particular focus on scientific visualization. Category theory and semiotics were employed by Vickers et al. [58] to facilitate an improved understanding of visualizations in practice and to describe a well-formed visualization process. The conceptual framework of visual multiplexing by Chen et al. [13] facilitates the study of different mechanisms for integrating and overlaying multiple pieces of visual information.

Based on these information-theoretic considerations, Chen and Golan [11] introduced a comprehensive cost-benefit model of visualization, defining cost as the search space for answers. They utilized the big O notation to classify tasks accordingly. Presenting a fact or piece of information has cost $O(1)$, observations as in "What happened?" require the user to read all data points, which has a complexity of $O(n)$. When looking into correlations, causes, and other complex relationships, we must consider a broader spectrum of relations, ending up at $O(n^k)$. And, finally, when we want to derive a model for visualization, taking into account all parameters and algorithmic steps, the complexity might be $O(n!)$. They further introduced a cost function, which can be derived from energy, time, or monetary measurements necessary to find the answer. They defined benefit as a gain in certainty about the information. Based on these definitions, they derived an incremental cost-benefit ratio that describes the amount of effort required to compress the information toward the point that the user's initial question can be answered and a decision can be made. Based on this formulation, it is in principle possible to use an optimization process to discover the best visualization method.

Bruckner et al. [9] proposed a model to analyze the directness of interaction techniques in visualization. They considered the different mappings involved in the visualization process, i.e., the mapping from data space via the visualization space to the output space (e.g., a monitor or a head-mounted display), as well as the subsequent perceptual and cognitive processes involved in generating the user's mental model. They then investigated the parallel process of interaction, starting from an intended action (based on the user's mental model) via the manipulation space (i.e., a physical interaction device such as a computer mouse) to the interaction space and finally back to the data space. Based on this model, they introduced a measure for the degree of indirectness of an interactive visualization setup based on how invertible the involved mappings are and demonstrated how this measure can be practically realized.

Compared to task-specific quality metrics, describing visualizations on a general level not only provides us with a better understanding of visualization as a scientific research field, but further allows us to make predictions about non-evaluated, or even not yet developed visualization techniques. For instance, when the interaction with the visualization does not coincide with gathered knowledge about interaction directness, the user is likely to experience a discrepancy between their intended and executed manipulation. One major question is how the sheer number of theoretical frameworks and models can be combined and integrated into one coherent knowledge base. Similar to other research fields like physics, where theories about electricity and magnetism have been combined into a larger theory of electromagnetism, visualization could gather greater insights by combining existing theoretical frameworks, leading to a fundamental strengthening of the research field as a whole.

3.3.4 Meta-Perceptual Process Measures

So far we have primarily examined well-established and generally accepted measures and models to evaluate visualization with the goal of optimizing task execution time, easing data exploration, or increasing the gained insight. But when visualization is utilized as a knowledge source for the general public, we can formulate other equally important goals for visualization design. In education, we might be interested in creating memorable knowledge or engage students in working with a visualization. In commercial scenarios, aspects such as aesthetics and impact, or even the profitability of a visualization can be the main goals of a specific design. We summarize these higher-level aspects as meta-perceptual process measures that aim to characterize additional qualities that go beyond what are typically considered to be primary desired properties of a visualization in the research community. In some sense, such measures aim to capture the attributes of a visualization from the point of view of other domains, such as art or economics.

For instance, Healey et al. [22] conducted experiments to evaluate how hue and orientation allow users to accurately estimate features in visualizations through preattentive processing. The question was if a short glimpse at a visualization can convey the general message, and if it can, which factors influence this capability. While Skau et al. [51] showed that even small visual embellishments increase the error rate when reading bar charts, Bateman et al. [2] found that visually embellished visualizations are more memorable than plain charts. Figure 3.3 shows two visual mappings for bar

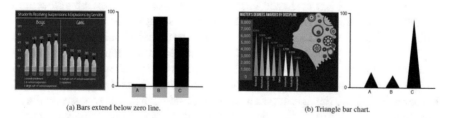

(a) Bars extend below zero line.   (b) Triangle bar chart.

Fig. 3.3 Comparison of plain and embellished bar charts [51]. Even small visual modifications, like using triangular instead of rectangular bar charts, will increase the error rate. Embellished representations can increase the memorability of the visualization

charts, as well as their embellished counterparts. Borkin et al. [6] investigated which elements of visualizations make them memorable. They showed, for example, that color, human recognizable objects, high visual density, and unique design improve the ability of humans to remember a visualization. Furthermore, memorability was independent of subjects' context and biases.

Aesthetics of a visualization are hard to measure and in most cases subjective. Tractinsky et al. [54] found a strong correlation between aesthetics and usability, which suggests it as an important factor for designing and evaluating visualizations. Lau and Moere [31] proposed a model for aesthetics in information visualization, seeing aesthetics as the degree of artistic influence on the data mapping, rather than as a measure of appeal. Filonik and Baur [20] summarized several possible measures of aesthetics for information visualization from the literature and concluded that many aspects of this phenomenon remain unexplored. Harrison et al. [21] ran a user study and found correlations between certain measurable visual features and visually appealing aesthetics. They found that colorfulness and visual complexity have a positive correlation to perceived aesthetics, but depend on gender, age, and level of education.

Saket et al. [45] summarized and reviewed several of these meta-perceptual criteria in the field of visualization. They described engagement as the amount of time spent with the visualization, proposed a model for measuring enjoyment [46], and found that pictorial representations and embellished visualizations increase enjoyment [47]. Their work concluded that memorability, engagement, and enjoyment are complex aspects of visualizations that are hard to quantify, and require further study. It is, for example, not yet clear how interactions affect these measures, and many more factors that influence a user's experience might exist.

A somewhat different class of measures is related to non-cognitive aspects of visualization. For instance, Van Wijk [57] proposed a model to measure the "profitability" of a visualization in an economic sense. In this model, the cost of a visualization (e.g., development cost and users' time to understand the visualization) is considered in relation to the return on investment in the form of knowledge gain. The value of a visualization can thus be increased if many people use it regularly, obtain valuable knowledge, and spend less time or money to make a decision. Unfortunately, knowledge gain is a rather broad and vague concept, so more precise notions are needed to quantify this aspect more accurately.

Compared to previously discussed approaches, meta-perceptual process measures have so far mostly been evaluated in rather narrow scenarios, providing guidelines for visualization design. More quantitative measures that would allow for the comparison of different visualizations with respect to the outlined qualities have not been explored extensively. Furthermore, the fact that some qualities like aesthetics are not necessarily directly related to common visualization goals such as the generation of insight, may have led some visualization researchers to discard them as irrelevant. However, we believe that it is important to also consider the impact of visualization in a broader context, and hence find that the measurement of such properties is an important and worthwhile endeavor. Parallels may be drawn to other fields—for instance, organizational performance was once mostly viewed in terms of its economic characteristics, but organizational psychology has shown that measures of occupational health and well-being such as job satisfaction can be important predictors for the financial success of a company.

3.4 Toward a "Bigger Picture"

As can be seen from previous examples, there is still a long way to go toward quantitative statements about visualizations in general. Figure 3.4 provides a high-level overview of the discussed measure categories with respect to their practicality as well as their ability to describe general phenomena. Many of the presented quality measures are specific to a certain type of visualization, like wiggle in streamgraphs, or scagnostics for scatterplots. Counting the number of edge crossings in a visualization is an example that can be applied to several different visualization techniques, like graphs, streamgraphs, and parallel coordinate plots, but is still specific to visualizations that utilize visual links for their layout. It could be argued, that a meter can measure width, height, and length in the real world, because every object has to have these properties, given their underlying molecular structure. Visualizations, on the other hand, utilize a number of visual properties to encode varying information, even encoding semantically similar information with different visual encodings. From this point of view, it is no surprise that different subareas in visualization have developed vastly varying quality measures. Kosara [30] looks at many of the best practices followed in visualization and encourages researchers to build a better, well-justified basis for knowledge about visualizations.

Some properties, like clutter, empty space, and overplotting are more general and can be used to characterize visualizations on a more fundamental level. But their effect on the users' perception varies and is therefore often evaluated through user studies. Several examples, as for instance discussed by Harrison et al. [21], have shown that measures which by themselves do not make a statement about quality (e.g., colorfulness and visual complexity) can be transformed into quality measures, when evaluated with a user study. The users' perception can be measured or quantified and thereby operate as an indicator for quality. The fact that this type of evaluation can be performed for visualizations in general means that there might be a common ground that allows for comparability. While Behrisch et al. [3] provided an excellent summary of task-dependent quality measures, a survey of existing studies would be able to provide an overview of what these studies have in common and on how specific they are to their individual task and visualization types.

In order to compare not only different visual encodings, but visualization types, we would require a standardized way of evaluating common properties. For instance, different visualizations might apply the same color map and when asking users of different visualizations the same questions we would acquire comparable answers. One major problem of this approach is that visualizations are, among other things, data-, task-, and user-dependent. While a given dataset might create clutter in one visualization, it might not in another, and the opposite can be true for yet another dataset.

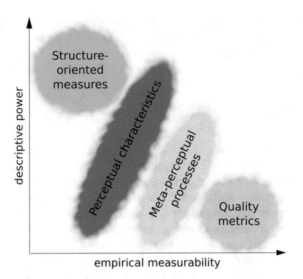

Fig. 3.4 Overview of visualization measures regarding their ease of being measured and their capability of describing visualizations as a whole. While quality metrics are easy to measure, they are in most cases too specific to find applicability in generalized observations. Meta-perceptual measurements try to capture more general, higher-level phenomena, but require user studies to be quantified. Although they are applicable to a large range of visualization techniques, their generated insight follows specific intents (like making a visualization memorable). Perceptual studies try to understand the human visual system and could, if fully understood, explain many phenomena in the analysis of visualizations. At present, however, where only the low-level visual processing is well understood, their applicability is limited to rankings of visual channels and encodings for rather isolated situations. Theories about visualization are among the most general and descriptive approaches for describing visualizations. Although some of them propose varying measures for the quality of a visualization and allow for their comparison, they are in many cases still too abstract to be applied in practical use cases

Some visualizations are better in giving an overview, while others provide detailed insights, and the questions asked in user studies are often task-dependent to investigate exactly these specific strengths or weaknesses. When asking task-independent questions in order to keep the results comparable, insight on these specific differences might get lost. Lastly, visualizations can be targeted toward a certain audience, being more specific for experts, or more intuitive for the broader public. For this reason, participants of a user study are often chosen from the specific audience, introducing a bias toward the background and knowledge the participants have. If the goal is to create comparable results, the distribution of participants would need to be as general as possible, introducing additional problems like participants not having the background knowledge required to, e.g., benefit from a visualization in the medical domain. Independent of the mentioned shortcomings, we might be able to come up with some general statements that provide insight into the users' mental model and opinion about the visualization given their data and purpose they operate on, similar to the System Usability Score introduced by Brooke [8]. The interpretation of such a score, as in this case demonstrated by Bangor et al. [1], can lead to a description of the general performance of a visualization, and bring us closer to a common basis for comparability across fields.

Perceptual studies in particular provide more general means by analyzing the human vision and ranking different visual channels based on their capability of presenting information. They build a fundamental understanding of the basic principles of visualization and are applicable to all kinds of visualization types. So far, we merely understand low-level perceptual processes. This fact limits the applicability of perceptual studies to make general statements about visualizations and predict their usability. Meta-perceptual metrics, on the other hand, try to evaluate higher-level features independent of the specific visual encoding. Aesthetics, engagement, and enjoyment have a major impact on the way users interact with the visualization and on how the gained knowledge is memorized. Despite several efforts taken in this direction, these measures have mostly been explored in information visualization and require further research in other fields like, e.g., scientific visualization. When we have a better understanding of how these phenomena behave in different visualization types, we can build a more general theory and learn from the insights gained. In addition to already mentioned measures, proxy measures can be used to quantify properties that are otherwise hard to observe. The idea is to find a measurable property that strongly correlates with the phenomenon we want to analyze.

As a reflection of a discussion panel on how to pursue theoretical research in visualization, Chen et al. [14] described different evaluation approaches and how they can contribute to a theoretical foundation. Taxonomies classify objects of interest, such as data types, visual encodings, user tasks, or interaction techniques into groups and subgroups. Ontologies then describe additional relationships between these different groups and entities, providing a more detailed picture of the underlying interactions. Guidelines describe the quality of a certain approach and make statements about which practices should or should not be used in order to achieve a desired outcome. The authors argue that guidelines need to be evaluated and refined over time, as well as transformed into quantitative laws when applicable. VisGuides [19, 59] provides a platform to openly discuss guidelines in visualization and allow for their continuous refinement. When a guideline has shown to be useful over the years, it can be established as a principle. Conceptual models describe a general idea or understanding of how certain processes or systems work in order to reason about their structure and functioning. For example, a perceptual model describes how we think the human visual system works, which allows us to derive conclusions and best practices, although we have not fully understood this system yet. Such models can further be supported by mathematical frameworks, like information theory. In our opinion, the combination of quantitative measures and a mathematical framework can form the basis of more general models of visualization. These can then be used to reason about causal relationships and make testable predictions. We believe, that the main goal of our community should be to unify existing approaches into larger theories about visualization that incorporate acquired knowledge into a more general understanding of the subject itself. Sacha et al. [44] demonstrated how perceptual and theoretical frameworks, as well as guidelines, can be combined into a model for understanding the process of knowledge generation. We should continue this line of thought to further integrate quality and meta-perceptual measures into theoretical frameworks and to create general models of the visualization process. By continuously verifying and refining these models, we can continuously advance visualization theory and strengthen the research field for greater accomplishments to come.

3.5   Conclusion

Other research fields have shown how incremental refinement and verification of theoretical models can lead to major leaps in knowledge and understanding. In visualization, we have seen several promising attempts toward a theoretical foundation, as well as greater acknowledgement and presence of theoretical papers. We can learn from other scientific disciplines and bear in mind that the formulation of a theory and definition of measures in visualization do not need to be perfect from the very beginning. Practical barriers, like not being able to compute a measure due to technical limitations, should not prevent us from suggesting and formulating such concepts. Many important milestones in scientific history, like Einstein's general relativity or Feynman's quantum electrodynamics, have been postulated much earlier than they could be verified. Similarly, Fermat's Last Theorem took 358 years from its proposition to a mathematical proof. Such theories allow us to state our assumptions, formulate predictions, and develop technological advances, even if they are not well-verified or "proven" yet. Evaluation efforts can be made not only to assess specific visualization techniques or applications, but to empirically test theories. Based on such continuously validated and refined theories, we are optimistic that we will eventually be able to evaluate and compare visualization techniques on a more general level, predict how users will perceive and interact with the visualization, and develop new visualization techniques for better decision making.

Acknowledgements The authors would like to thank all participants of the Dagstuhl Seminar for constructive and fruitful discussions. This work was supported by the MetaVis project (#250133) funded by the Research Council of Norway.

References

1. Bangor, A., Kortum, P.T., Miller, J.T.: An empirical evaluation of the system usability scale. Int. J. Hum.-Comput. Interact. **24**(6), 574–594 (2008). https://doi.org/10.1080/10447310802205776
2. Bateman, S., Mandryk, R.L., Gutwin, C., Genest, A., McDine, D., Brooks, C.: Useful junk?: The effects of visual embellishment on comprehension and memorability of charts. In: Proceedings of ACM CHI, pp. 2573–2582 (2010). https://doi.org/10.1109/tvcg.2013.234
3. Behrisch, M., Blumenschein, M., Kim, N.W., Shao, L., El-Assady, M., Fuchs, J., Seebacher, D., Diehl, A., Brandes, U., Pfister, H., Schreck, T., Weiskopf, D., Keim, D.A.: Quality metrics

for information visualization. Comput. Graph. Forum **37**(3), 625–662 (2018). https://doi.org/10.1111/cgf.13446

4. Bertini, E., Santucci, G.: Improving 2D scatterplots effectiveness through sampling, displacement, and user perception. In: Proceedings of the International Conference on Information Visualisation, pp. 826–834 (2005). https://doi.org/10.1109/IV.2005.62

5. Bertini, E., Santucci, G.: Quality metrics for 2D scatterplot graphics: automatically reducing visual clutter. In: Proceedings of Smart Graphics, pp. 77–89 (2004). https://doi.org/10.1007/978-3-540-24678-7_8

6. Borkin, M.A., Vo, A.A., Bylinskii, Z., Isola, P., Sunkavalli, S., Oliva, A., Pfister, H.: What makes a visualization memorable? IEEE Trans. Vis. Comput. Graph. **19**(12), 2306–2315 (2013). https://doi.org/10.1109/tvcg.2013.234

7. Brehmer, M., Munzner, T.: A multi-level typology of abstract visualization tasks. IEEE Trans. Vis. Comput. Graph. **19**(12), 2376–2385 (2013). https://doi.org/10.1109/TVCG.2013.124

8. Brooke, J.: SUS - a quick and dirty usability scale. Usability Eval. Ind. **189**(194), 4–7 (1996)

9. Bruckner, S., Isenberg, T., Ropinski, T., Wiebel, A.: A model of spatial directness in interactive visualization. IEEE Trans. Vis. Comput. Graph. **25**(8), 2514–2528 (2019). https://doi.org/10.1109/TVCG.2018.2848906

10. Byron, L., Wattenberg, M.: Stacked graphs - geometry and aesthetics. IEEE Trans. Vis. Comput. Graph. **14**(6), 1245–1252 (2008). https://doi.org/10.1109/TVCG.2008.166

11. Chen, M., Golan, A.: What may visualization processes optimize? IEEE Trans. Vis. Comput. Graph. **22**(12), 2619–2632 (2015). https://doi.org/10.1109/tvcg.2015.2513410

12. Chen, M., Jaenicke, H.: An information-theoretic framework for visualization. IEEE Trans. Vis. Comput. Graph. **16**(6), 1206–1215 (2010). https://doi.org/10.1109/tvcg.2010.132

13. Chen, M., Walton, S., Berger, K., Thiyagalingam, J., Duffy, B., Fang, H., Holloway, C., Trefethen, A.E.: Visual multiplexing. Comput. Graph. Forum **33**(3), 241–250 (2014). https://doi.org/10.1111/cgf.12380

14. Chen, M., Grinstein, G., Johnson, C.R., Kennedy, J., Tory, M.: Pathways for theoretical advances in visualization. IEEE Comput. Graph. Appl. **37**(4), 103–112 (2017). https://doi.org/10.1109/mcg.2017.3271463

15. Cleveland, W.S., McGill, R.: Graphical perception: theory, experimentation, and application to the development of graphical methods. J. Am. Stat. Assoc. **79**(387), 531–554 (1984). https://doi.org/10.2307/2288400

16. Connor, C.E., Egeth, H.E., Yantis, S.: Visual attention: bottom-up versus top-down. Curr. Biol. **14**(19), R850–R852 (2004). https://doi.org/10.1016/j.cub.2004.09.041

17. Daugman, J.G.: Uncertainty relation for resolution in space, spatial frequency, and orientation optimized by two-dimensional visual cortical filters. J. Opt. Soc. Am. A **2**(7), 1160–1169 (1985). https://doi.org/10.1364/JOSAA.2.001160

18. Demiralp, Ç., Scheidegger, C.E., Kindlmann, G.L., Laidlaw, D.H., Heer, J.: Visual embedding: a model for visualization. IEEE Comput. Graph. Appl. **34**(1), 10–15 (2014). https://doi.org/10.1109/mcg.2014.18

19. Diehl, A., Abdul-Rahman, A., El-Assady, M., Bach, B., Keim, D., Chen, M.: Visguides: a forum for discussing visualization guidelines. In: Proceedings of EuroVis (Short Papers), pp. 61–65 (2018). https://doi.org/10.2312/eurovisshort.20181079

20. Filonik, D., Baur, D.: Measuring aesthetics for information visualization. In: Proceedings of the International Conference on Information Visualisation, pp. 579–584 (2009). https://doi.org/10.1109/iv.2009.94

21. Harrison, L., Reinecke, K., Chang, R.: Infographic aesthetics: designing for the first impression. In: Proceedings of ACM CHI, pp. 1187–1190 (2015). https://doi.org/10.1145/2702123.2702545

22. Healey, C.G., Booth, K.S., Enns, J.T.: High-speed visual estimation using preattentive processing. ACM Trans. Comput.-Hum. Interact. **3**(2), 107–135 (1996). https://doi.org/10.1145/230562.230563

23. Healey, C., Enns, J.: Attention and visual memory in visualization and computer graphics. IEEE Trans. Vis. Comput. Graph. **18**(7), 1170–1188 (2012). https://doi.org/10.1109/TVCG.2011.127

24. Itti, L., Koch, C., Niebur, E.: A model of saliency-based visual attention for rapid scene analysis. IEEE Trans. Pattern Anal. Mach. Intell. **20**(11), 1254–1259 (1998). https://doi.org/10.1109/34.730558
25. Jaenicke, H., Chen, M.: A salience-based quality metric for visualization. Comput. Graph. Forum **29**(3), 1183–1192 (2010). https://doi.org/10.1111/j.1467-8659.2009.01667.x
26. Jänicke, H., Weidner, T., Chung, D., Laramee, R.S., Townsend, P., Chen, M.: Visual recon-structability as a quality metric for flow visualization. Comput. Graph. Forum **30**(3), 781–790 (2011). https://doi.org/10.1111/j.1467-8659.2011.01927.x
27. Jankun-Kelly, T.J., Ma, K.L., Gertz, M.: A model and framework for visualization exploration. IEEE Trans. Vis. Comput. Graph. **13**(2), 357–369 (2007). https://doi.org/10.1109/TVCG.2007.28
28. Kim, Y., Varshney, A.: Saliency-guided enhancement for volume visualization. IEEE Trans. Vis. Comput. Graph. **12**(5), 925–932 (2006). https://doi.org/10.1109/TVCG.2006.174
29. Kindlmann, G., Scheidegger, C.: An algebraic process for visualization design. IEEE Trans. Vis. Comput. Graph. **20**(12), 2181–2190 (2014). https://doi.org/10.1109/TVCG.2014.2346325
30. Kosara, R.: An empire built on sand: Reexamining what we think we know about visualization. In: Proceedings of the Workshop on Beyond Time and Errors on Novel Evaluation Methods for Visualization, pp. 162–168 (2016). https://doi.org/10.1145/2993901.2993909
31. Lau, A., Moere, A.V.: Towards a model of information aesthetics in information visualization. In: Proceedings of the International Conference on Information Visualisation, pp. 87–92 (2007). https://doi.org/10.1109/iv.2007.114
32. Lee, C.H., Varshney, A., Jacobs, D.W.: Mesh saliency. ACM Trans. Graph. **24**(3), 659–666 (2005). https://doi.org/10.1145/1073204.1073244
33. Li, Z.: A neural model of contour integration in the primary visual cortex. Neural Comput. **10**(4), 903–940 (1998). https://doi.org/10.1162/089976698300017557
34. Liu, Z., Nersessian, N., Stasko, J.: Distributed cognition as a theoretical framework for infor-mation visualization. IEEE Trans. Vis. Comput. Graph. **14**(6), 1173–1180 (2008). https://doi.org/10.1109/tvcg.2008.121
35. Mackinlay, J.: Automating the design of graphical presentations of relational information. ACM Trans. Graph. **5**(2), 110–141 (1986). https://doi.org/10.1145/22949.22950
36. Mackinlay, J., Hanrahan, P., Stolte, C.: Show me: automatic presentation for visual analysis. IEEE Trans. Vis. Comput. Graph. **13**(6), 1137–1144 (2007). https://doi.org/10.1109/TVCG.2007.70594
37. Matzen, L.E., Haass, M.J., Divis, K.M., Wang, Z., Wilson, A.T.: Data visualization saliency model: a tool for evaluating abstract data visualizations. IEEE Trans. Vis. Comput. Graph. **24**(1), 563–573 (2018). https://doi.org/10.1109/TVCG.2017.2743939
38. Munzner, T.: A nested model for visualization design and validation. IEEE Trans. Vis. Comput. Graph. **15**(6), 921–928 (2009). https://doi.org/10.1109/TVCG.2009.111
39. Pineo, D., Ware, C.: Data visualization optimization via computational modeling of perception. IEEE Trans. Vis. Comput. Graph. **18**(2), 309–320 (2012). https://doi.org/10.1109/TVCG.2011.52
40. Pinto, Y., van der Leij, A.R., Sligte, I.G., Lamme, V.A.F., Scholte, H.S.: Bottom-up and top-down attention are independent. J. Vis. **13**(3), 16 (2013). https://doi.org/10.1167/13.3.16
41. Purchase, H.C., Andrienko, N., Jankun-Kelly, T., Ward, M.: Theoretical foundations of infor-mation visualization. In: Information Visualization, pp. 46–64. Springer, Berlin (2008). https://doi.org/10.1007/978-3-540-70956-5_3
42. Ren, L., Cui, J., Du, Y., Dai, G.: Multilevel interaction model for hierarchical tasks in infor-mation visualization. In: Proceedings of the International Symposium on Visual Information Communication and Interaction, pp. 11–16 (2013). https://doi.org/10.1145/2493102.2493104
43. Rodieck, R.: Quantitative analysis of cat retinal ganglion cell response to visual stimuli. Vision. Res. **5**(12), 583–601 (1965). https://doi.org/10.1016/0042-6989(65)90033-7
44. Sacha, D., Stoffel, A., Stoffel, F., Kwon, B.C., Ellis, G., Keim, D.A.: Knowledge generation model for visual analytics. IEEE Trans. Visual Comput. Graph. **20**(12), 1604–1613 (2014). https://doi.org/10.1109/TVCG.2014.2346481

45. Saket, B., Endert, A., Stasko, J.: Beyond usability and performance: a review of user experience-focused evaluations in visualization. In: Proceedings of the Workshop on Beyond Time and Errors on Novel Evaluation Methods for Visualization, pp. 133–142 (2016). https://doi.org/10.1145/2993901.2993903
46. Saket, B., Scheidegger, C., Kobourov, S.G.: Towards understanding enjoyment and flow in information visualization. In: Proceedings of EuroVis (Short Papers) (2015). https://doi.org/10.2312/eurovisshort.20151134
47. Saket, B., Scheidegger, C., Kobourov, S.: Comparing node-link and node-link-group visualizations from an enjoyment perspective. Comput. Graph. Forum **35**(3), 41–50 (2016). https://doi.org/10.1111/cgf.12880
48. Schulz, H.: Treevis.net: a tree visualization reference. IEEE Comput. Graph. Appl. **31**(6), 11–15 (2011). https://doi.org/10.1109/MCG.2011.103
49. Shneiderman, B.: The eyes have it: A task by data type taxonomy for information visualizations. In: Proceedings of IEEE Symposium on Visual Languages, pp. 336–343 (1996). https://doi.org/10.1109/VL.1996.545307
50. Silver, D.: Object-oriented visualization. IEEE Comput. Graph. Appl. **15**(3), 54–62 (1995). https://doi.org/10.1109/38.376613
51. Skau, D., Harrison, L., Kosara, R.: An evaluation of the impact of visual embellishments in bar charts. Comput. Graph. Forum **34**(3), 221–230 (2015)
52. Tal, E.: Measurement in science. In: Zalta, E.N. (ed.) The Stanford Encyclopedia of Philosophy, fall 2017th edn. Metaphysics Research Lab, Stanford University (2017)
53. Tory, M., Moller, T.: Rethinking visualization: A high-level taxonomy. In: IEEE Symposium on Information Visualization, pp. 151–158 (2004). https://doi.org/10.1109/INFVIS.2004.59
54. Tractinsky, N., Katz, A.S., Ikar, D.: What is beautiful is usable. Interact. Comput. **13**(2), 127–145 (2000). https://doi.org/10.1016/S0953-5438(00)00031-X
55. Tufte, E.R.: The Visual Display of Quantitative Information, vol. 2. Graphics Press, Cheshire (2001)
56. Tukey, J.W., Tukey, P.A.: Computer graphics and exploratory data analysis: an introduction. In: Proceedings of Computer Graphics, pp. 773–785 (1985). https://doi.org/10.1007/978-3-319-43742-2_15
57. Van Wijk, J.J.: The value of visualization. In: Proceedings of IEEE VIS, pp. 79–86 (2005). https://doi.org/10.1109/VISUAL.2005.1532781
58. Vickers, P., Faith, J., Rossiter, N.: Understanding visualization: a formal approach using category theory and semiotics. IEEE Trans. Vis. Comput. Graph. **19**(6), 1048–1061 (2012). https://doi.org/10.1109/TVCG.2012.294
59. VisGuides. http://visguides.org/. Accessed 27 Aug 2019
60. Wang, X., Dou, W., Butkiewicz, T., Bier, E.A., Ribarsky, W.: A two-stage framework for designing visual analytics system in organizational environments. In: IEEE Conference on Visual Analytics Science and Technology, pp. 251–260 (2011)
61. Wang, C., Shen, H.W.: Information theory in scientific visualization. Entropy **13**(1), 254–273 (2011). https://doi.org/10.3390/e13010254
62. Wilkinson, L., Anand, A., Grossman, R.: Graph-theoretic scagnostics. In: Proceedings of IEEE InfoVis, pp. 157–164 (2005). https://doi.org/10.1109/INFVIS.2005.1532142
63. Williamson, J.R., Grossberg, S.: A neural model of how horizontal and interlaminar connections of visual cortex develop into adult circuits that carry out perceptual grouping and learning. Cereb. Cortex **11**(1), 37–58 (2001). https://doi.org/10.1093/cercor/11.1.37
64. Xu, L., Lee, T.Y., Shen, H.W.: An information-theoretic framework for flow visualization. IEEE Trans. Vis. Comput. Graph. **16**(6), 1216–1224 (2010). https://doi.org/10.1109/tvcg.2010.131

Chapter 4
Knowledge-Assisted Visualization and Guidance

Silvia Miksch, Heike Leitte and Min Chen

Abstract Visualization envisions to intertwine the strengths of humans and computers for effective interactive visual and analytic data analysis and exploration. To this end, humans' tacit/implicit knowledge from prior experience is an important asset that can be leveraged by both human and computer to improve the visual and analytic exploration processes. However, acquiring, structuring, formalizing, storing, and utilizing implicit and explicit knowledge within the whole visualization process are provocative and widely-discussed research challenge. This chapter elaborates on (1) knowledge-assisted visualization, which aims to incorporate implicit and explicit knowledge as well as information-theoretical considerations into the visualization process to support users for decision making and (2) guidance, which is a computer-assisted process that aims to actively resolve a knowledge gap encountered by users during an interactive visualization session. This chapter ends with critical reflections about applicability, usability, and utility of the proposed knowledge enhanced visualization processes.

4.1 Introduction

Analytical reasoning for real-world decision making involves volumes of uncertain, complex, and often conflicting data that analysts need to make sense of. In addition

S. Miksch (✉)
TU Wien, Wein, Austria
e-mail: silvia.miksch@tuwien.ac.at

H. Leitte
TU Kaiserslautern, Kaiserslautern, Germany
e-mail: leitte@cs.uni-kl.de

M. Chen
University of Oxford, Oxford, UK
e-mail: min.chen@oerc.ox.ac.uk

© Springer Nature Switzerland AG 2020
M. Chen et al. (eds.), *Foundations of Data Visualization*,
https://doi.org/10.1007/978-3-030-34444-3_4

to sophisticated analysis methods, knowledge about the data, the domain, and prior experience is beneficial in this endeavor. Ideally, visualization and/or visual analytics (VA) environment would leverage this knowledge to better support domain users, their data, and their analytical tasks in an application context.

Visual Analytics (VA), "the science of analytical reasoning facilitated by interactive visual interfaces" [52, p.28], provides means to obtain information, derive insights, and build new knowledge from data. It is a multidisciplinary approach, integrating aspects of data mining and knowledge discovery, information visualization, human–computer interaction, and cognitive science. It has also been defined as a combination of "**automated analysis** techniques with **interactive visualizations**" [32, p.7]. Indeed, VA leverages, the specific strengths of computers and humans for the best possible outcome: on the one hand, computers are better at managing and processing large amounts of data by exploiting their enormous computational power; on the other hand, humans have better perceptual and cognitive means, which enable them to visually perceive unexpected patterns and to interpret data.

Without disputing the clear and well-established advantages of this human-in-the-loop approach, we envision that the overall analytic process would benefit if the **prior knowledge** that can be properly represented in computers could be effectively utilized.

In this chapter, we will investigate the feasibility of a **knowledge-assisted visualization** (KAV), where explicit prior knowledge can be exploited by the machine to support interactive visualization and data mining. We will also discuss the information-theoretical considerations about the knowledge-assisted visualization processes for supporting users' decision-making tasks (from both ontological and information-theoretic perspectives). This will be followed by a related discourse where we will elaborate on **guidance** seen as a mixed-initiative process that aims to actively resolve a knowledge gap encountered by users during an interactive visualization session.

In Sect. 4.2 the foundations of knowledge-assisted visualization are detailed, including definitions, models for KAV, and examples. KAV strongly hinges on the integration of humans and human knowledge in the data analysis process. Human involvement, however, is often criticized as error-prone and biased. In Sect. 4.3, we make an information-theoretic argument for the integration of the user in (knowledge-assisted) visualization systems and data intelligence processes and detail the benefits. To use human knowledge on a larger scale, the gained knowledge has to be formalized and made accessible to machines. This can be achieved through ontologies which are detailed in Sect. 4.4. Section 4.5 continues with integration of guidance in KAV-systems to resolve knowledge gaps in the data analysis process. The chapter closes with critical reflections and conclusions in Sect. 4.6.

4.2 Knowledge-Assisted Visualization: Definitions and Models

Sharing prior knowledge has been identified as one of the top 10 unsolved problems in visualization [10]. It ensures that all users have a common ground for the exchange of information and can profit from previous experiences, and is consistent with the user-centered design tradition in human–computer interaction. However, before we can discuss ways to represent, store, and share prior knowledge, we need to understand what knowledge is. The three terms "data", "information", and "knowledge" often occur together in computer science and are commonly not well distinguished and even occasionally used interchangeably. In this section, we will first summarize definitions and variants of knowledge, continue with an overview of models that integrate knowledge in the visualization process, and close with three examples.

4.2.1 Definitions

An introduction to information and knowledge in visualization was given in a paper by Chen et al. [11]. They observed that many competing definitions for the three terms existed in various areas of research such as computer science, engineering, psychology, or management. They also concluded that the concepts behind the terms data/information/knowledge were neither identical nor mutually disjoint and that none of them was a subset of another. A classic model for describing the relationships among these concepts is the data–information–knowledge–wisdom (DIKW) hierarchy [1], which provides a means for classifying human understanding. Russell Ackoff therein distinguished the three terms by the type of questions that could be answered [1]:

- **Data**: are just a collection of symbols.
- **Information**: is data that are processed to be useful, providing answers to "who", "what", "where", and "when" questions.
- **Knowledge**: is the application of data and information, providing answers to "how" questions.

Translating these definitions to visualization, an information-assisted system is, for example, one that provides generic filters that help people to find data points that are relevant for their application. When automatically suggesting dates for an appointment, such a filter may select and highlight empty time slots of appropriate length in the calendar. Taking additional optimization criteria into account, such as preferred time of day, transition times between tasks and overall time management would turn this system into a knowledge-assisted system as it encodes humans' knowledge about good temporal planning and automatically provides this knowledge to other users in answering their "how-to" questions, such as "you may choose this time-slot because it is optimal in terms of X, Y, and Z."

An alternative example of information-assisted visualization for spatial data is isosurface extraction for scalar fields that automatically extracts some structural information that is relevant and useful in many applications. If additional knowledge-based supports were provided to select appropriate isovalues or classify structures in a medical setting based on prior or domain knowledge, this system would turn into a knowledge-assisted system.

Both examples have in common that the knowledge-assisted variant integrates digital representations of knowledge. Chen et al. [11] defined **digital representations of knowledge** as data that represents the results of a computer-simulated cognitive process (such as perception, learning, association, and reasoning) or the transcripts of some knowledge acquired by human beings (such as ontologies, best practices, or guidelines). Wang et al. [57] further elaborated on this discussion of locations of knowledge. Following the ideas of Nonaka and Takeuchi [40], they distinguished between **explicit knowledge** and **tacit knowledge**. Explicit knowledge is knowledge that can be processed by a computer, transmitted electronically, or stored in a database. Tacit knowledge, on the other hand, is personal, specialized, and can only be extracted by a human. In their paper, Wang et al. [57] explored four **conversion processes** between these two types of knowledge (internalization, externalization, collaboration, and combination), paving ways to make personal tacit knowledge available in knowledge-assisted visualization systems.

Knowledge-assisted visualization systems require digital representations of knowledge for two types of prior knowledge: **operational knowledge** (how to interact with the visualization system), and **domain knowledge** (how to interpret the content). Both of them are vital to understand the intended message in the visualization. Integrated operational knowledge can help the user to select an appropriate algorithm for their data, define good default values for parameters, or define sorting strategies for small multiple techniques. Chen [10] stated that while a focus on usability and a perception- and cognition-aware design could alleviate the need for operational knowledge, the domain knowledge could not easily be replaced. Thus, the research on the problem of operational knowledge in visualization has focused on the science of interaction. For example, Pike et al. [43] identified the design of knowledge-based interfaces as an open challenge, stating that the ability of visual analysis tools to represent and reason with human knowledge is underdeveloped.

4.2.2 Models of Knowledge-Assisted Visualization

In the knowledge-assisted visualization paradigm [11], a solution to alleviate the problem of missing expert knowledge is to rely on explicit machine-readable knowledge to provide guidance to the user through the interactive exploration process, or to automate the process partially. Examples of techniques applying this paradigm are APT [36], SemViz [25], and Show Me [37]. Integrating operational and domain knowledge in the analysis process can improve not only the user-controlled analysis processes but also the automatic ones. The fundamental role of prior knowledge in the knowledge discovery process (KDD) has already been acknowledged two decades ago [20]. Intelligent data analysis, the applications of artificial intelligence (AI) techniques in data analysis, aims at automatically extracting information from data by exploiting explicit domain knowledge (sometimes called background knowledge in this context) [27]. Knowledge-based systems enable the integration of background knowledge into the reasoning process, so that it is easy to model exceptional rules, which, for example, can prevent the system to reason over abnormal conditions [42]. A central questions that still remains is how to acquire, structure, formalize, store, and utilize prior knowledge. Several models have already been proposed to augment existing pipelines and process models as well as incorporate explicit and implicit knowledge.

Chen et al. [11] has extended the classical interactive visualization pipeline with an additional knowledge-based system that stores knowledge retrieved from data and the user. Knowledge from data can be obtained by simulating cognitive processes, for example, through rule-based reasoning. Expert knowledge is typically the tacit knowledge concerning the operational aspects of visualization processes (e.g., how to use a complex visualization system) and that concerning the application domain (e.g., what is the suitable transfer function for visualizing a particular type of volume data). Both types of knowledge need to be digitized in order to be actionable. They also observed that directly retrieving tacit knowledge from the user could be difficult and inconvenient. Often it is not clear which a priori knowledge to collect and how to create a comprehensive knowledge base. Additionally externalizing tacit knowledge, as already observed by Chen [10], is a challenging task. Hence, they proposed an alternative pipeline that progressively learns relevant information such as frequently used parameters and design options, or analysis routines.

Wang et al. [57] extended the analytical expression of visualization proposed by van Wijk [58]. In their mode, explicit knowledge is stored in a knowledge base and tacit knowledge remains with the users. In their paper, they discussed four strategies how the tacit knowledge could be turned into explicit knowledge and captured in the knowledge base.

While the importance of prior knowledge has been recognized separately for both interactive visualization and data mining, the close examination of its role in VA processes only started a few years ago. Sacha et al. [48] recently introduced the knowledge generation model for VA, which is an extension of the VA process model by Keim et al. [32] in order to capture the different stages of the human reasoning process for obtaining new knowledge from data. Nevertheless, the direct utilization of explicit prior knowledge to assist, guide, or automate the different processes of the VA pipeline has not been adequately investigated. The published work generally focuses on the objective of capturing and reusing operational knowledge, but rarely on domain knowledge. Some recent examples of this can be found in Flöring and Appelrath [24] and Wagner [54].

Federico et al. [23] proposed a model of knowledge-assisted VA, extending van Wijk's model of visualization [58] to a knowledge-assisted VA scenario by incorporating an explicit knowledge store and several knowledge-related processes. The conceptual model is depicted in Fig. 4.1. The outer boxes enclose processes related to machines and humans, respectively. Explicit knowledge can be stored in a digital form and used to steer the visualization and the analysis process. Tacit knowledge is extended in the cognition process and influences reasoning about perceived data and the exploration strategy. Tacit knowledge can be made available to the machine through externalization. Edges encode the potential mutual influences of data/knowledge stores and processes.

4.2.3 Examples

To further illustrate the concept of knowledge-assisted visualization in practical applications, this subsection details three application scenarios of knowledge-assisted visual analytics: Gnaeus integrates clinical practice guidelines, a form of explicit knowledge, to improve patient treatment in health care. KAMAS is a malware analysis system with a knowledge base to apply and externalize the identified patterns in execution traces. KAVAGait is a system for clinical gait analysis that supports the interactive exchange between explicit and tacit knowledge.

4.2.3.1 Gnaeus: Guideline-Based Health Care for Cohorts

Gnaeus [22] is a guideline-based knowledge-assisted visualization of electronic health records for cohorts (as shown in Fig. 4.2). Evidence-based clinical practice guidelines are sets of statements and recommendations used to improve health care by providing a trustworthy comparison of treatment options in terms of risks and benefits according to a patient's status; they condense the complex domain knowledge underlying the clinical practice in a narrative form. Gnaeus utilizes their formalization as computer-interpretable guidelines (CIGs).

In Gnaeus, both the **declarative** knowledge and the **procedural** knowledge are exploited to drive two analytical components: the temporal mediator and the com-

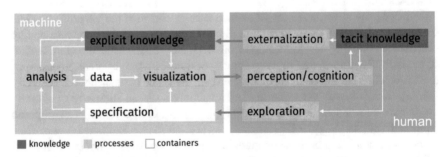

Fig. 4.1 A model for knowledge-assisted Visual Analytics: The model by Federico et al. [23] encodes the generation, transformation, and exploitation of knowledge in a visual analytics scenario

pliance analyzer. The declarative knowledge, specified as guideline intentions, is exploited to process the raw input, time-stamped data, such as blood glucose (BG) values at particular times, to produce a set of clinically meaningful summarizations and interpretations. The "BG monthly good pattern", for example, is defined as a month when the patient had up to one abnormal value of BG per week and no more than four abnormal values per month, while the BG abnormal values are defined in the context of pregnant diabetic patients according to taking insulin medication and fetus size.

Several chronic conditions can be managed with a combination of the right amount of physical activity, appropriate diet, and drugs. Thus, it is particularly important to assess not only the general efficacy of treatments but also the compliance of patients and caregivers with the clinical guidelines for the management of these diseases. An executed treatment is compliant if the recommendations that the patient was eligible for were fulfilled by performing the corresponding actions within the suggested response time windows. In Gnaeus, a rule-based reasoning engine ingests the procedural knowledge of CIGs, patient data, and treatment data, and computes compliance [3].

The CIGs are also directly visualized. In particular, the hierarchical structure of the guideline is visualized as a tree diagram with a top-down layered layout, whose nodes represent treatment plans and leaves represent clinical actions; the logical structure of a treatment plan is shown as a node-link diagram of a hierarchical task network. Gnaeus also features **knowledge-assisted interactions** to support user exploration.

Since an EHR can contain a large amount of multivariate time-oriented data for each patient, the guideline can be used as an index to browse the EHR data both across the different variables and along the time axis. When the users select a subplan in the guideline views, only relevant data is shown in the temporal views, identified through the plan-parameter dependency specified in the CIG declarative knowledge.

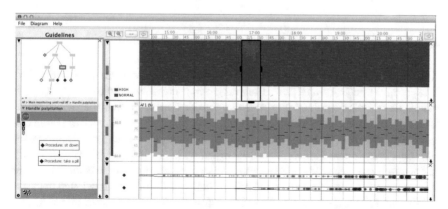

Fig. 4.2 Gnaeus, a guideline-based knowledge-assisted electronic health records visualization for cohorts [22] (Figure taken from [23] © 2017 IEEE. Used with permission)

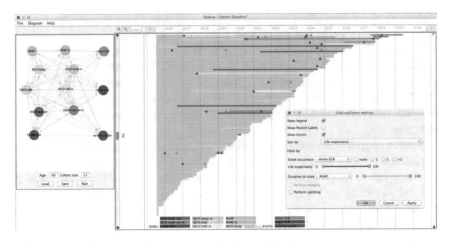

Fig. 4.3 Scipio, a plugin of Gnaeus [22] for simulating patient cohorts (Figure taken from [23] ©
2017 IEEE. Used with permission)

Moreover, the user can switch from absolute to relative time, thus data of all patients
are aligned according to the execution time of the selected subplan.

The Scipio plugin of Gnaeus (see Fig. 4.3) supports shared decision making by
interactive visualization of patient-level microsimulation [46]. The evidence-based
knowledge about probability of critical event occurrence as well as transition probabilities between conditions of increasing severity are modeled as Markov models.
Since these models might be too complex to be communicated to the patient as
such, Scipio utilizes microsimulation to generate data of a synthetic cohort of virtual
patients with similar conditions (age, disease, treatment); this data is then visualized
for an easier understanding of treatment consequences.

4.2.3.2   KAMAS: Behavior-Based Malware Analysis

KAMAS [55] is a knowledge-assisted malware analysis system (as shown in Fig. 4.4).
It supports IT-security analysts in learning about previously unknown samples of
malicious software (malware) or malware families based on their behavior. Therefore, the analysts need to identify and categorize suspicious patterns from large collections of execution traces. With KAMAS, the analysts can explore preprocessed
call sequences (rules) in their sequential order, which include system and API calls,
to find out if the observed samples are malicious or not. If a sample is malicious, the
system can be used to determine the related malware family. A knowledge database
(KDB) **storing explicit knowledge** in the form of rules is integrated into KAMAS
to ease the analysis process and to share it among colleagues. Based on the explicit
knowledge, **automated data analysis methods** are used to compare the rules featured in the loaded execution traces based on the specification with the stored explicit

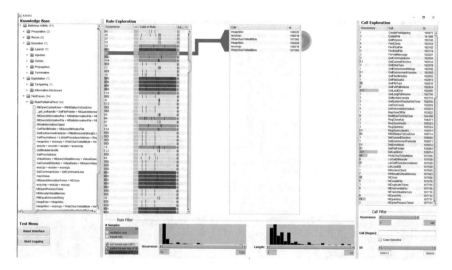

Fig. 4.4 KAMAS, a knowledge-assisted malware analysis system [55], supporting IT-security experts during behavior-based malware analysis (Figure taken from [23] © 2017 IEEE. Used with permission)

knowledge. Thereby, the specification gets adapted to highlight known rules. Additionally, the explicit knowledge can be turned on and off partially or completely by interaction.

If the analyst loads execution traces into the system, the featured rules are visualized based on the systems specification. If there is no specification prepared in the first visualization cycle (e.g., zooming, filtering, sorting), all read-in data are visualized and compared with the KDB. The image, which is generated by the visualization process, is perceived by the analyst for **gaining new tacit knowledge**, which may influence the user's perception in the future operations. Depending on the gained tacit knowledge, the analyst has now the ability to interactively explore the visualized malware data (rules) by the system-provided methods (e.g., zooming, filtering, sorting), which are affecting the specification. During this interactive process, the analyst gains new tacit knowledge based on the adjusted visualization. For the **integration of new knowledge** into the KDB, the analyst can, on the one hand, add a whole set of rules and, on the other hand, the analyst can add a selection of interesting calls that represent an extraction resulting from his/her tacit knowledge. Moreover, KAMAS allows the analyst to visualize directly the whole collection of explicit knowledge stored in the KDB.

4.2.3.3 KAVAGait: Clinical Gait Analysis

KAVAGait [56] is a knowledge-assisted VA system for clinical gait analysis (as shown in Fig. 4.5) that supports analysts during diagnosis and clinical decision mak-

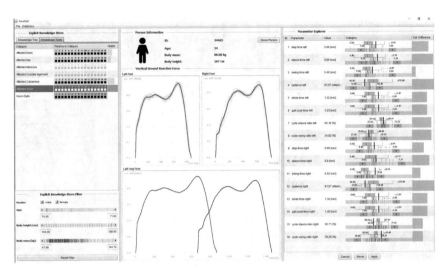

Fig. 4.5 KAVAGait, a knowledge-assisted clinical gait analysis system, supporting analysts during clinical decision making (Figure taken from [23] © 2017 IEEE. Used with permission)

ing. Users can load patient gait data containing *ground reaction forces* (GRF) measurements. These collected GRF data are visualized as waveforms in the center of the interface, representing a separated view for the left (red) and the right (blue) foot as well as providing a combined visualization. Additionally, 16 *spatio-temporal parameters* (STP) (e.g., step time, stance time, cadence) are calculated, visualized, and used for automated patient comparison and categorization.

Since one primary goal during clinical gait analysis is to assess whether a recorded gait measurement displays *normal gait* behavior or if not, which specific *gait abnormality* are present. Thus, the system's internal *explicit knowledge store* (EKS) contains several categories of gait abnormalities (e.g., knee, hip, ankle) as well as a category including healthy gait pattern data. Each category is defined by a set of parameter ranges $[min, max]$ of the 16 calculated STPs. All EKS entries are used for analysis and comparison by default. However, analysts can apply their expertise (tacit knowledge) as specification to filter entries by patient data (e.g., age, height, weight).

Automated data analysis of newly loaded patient data is provided for categories (e.g., automatically calculated category matching) influencing the systems specification. The EKS stores **explicit knowledge** and the automated data analysis methods are strongly intertwined with the visual data analysis system in KAVAGait. Thus, the combined analysis and visualization pipeline consist of the following process chain, and support the analysts during their interactive data exploration. Based on the visualization, the generated image is perceived by the analyst, gaining **tacit knowledge**, which also influences the analysts perception. As data **exploration and analysis** is an iterative process, the analyst gains further tacit knowledge based on the adjusted visualization and driven by the specification. To generate explicit knowledge, the analyst can include the STPs of analyzed patients based on his/her clinical decisions to the EKS, which can be described as the extraction of tacit knowledge.

Moreover, KAVAGait provides the ability to **interactively explore and adjust the systems EKS**, whereby the explicit knowledge can be visualized in a separated view. Two different options (one for a single patient and one for a category) are provided in KAVAGait for the adjustment of the stored explicit knowledge by the analysts' tacit knowledge.

4.3   The Importance of Knowledge in Data Analysis: An Information-Theoretic Perspective

In the previous section, we saw KAV models and examples of KAV-systems in VA that benefit strongly from the integration of tacit and explicit knowledge. However, while humans' knowledge may be appreciated in the field of visualization, there are also wide-spread doubts about the relative merits of humans' involvement in data intelligence workflows. Not only are humans considered to be slow, costly, and incapable of handling "big data" but they are also considered to be a liability of errors and biases (e.g., [30]). Such doubts have stimulated much concern among visualization researchers. For example, among the 13 chapters in the volume on Cognitive Biases in Visualization [19], the majority of the chapters there focus on addressing biases as a serious problem in visualization. Only a few chapters, such as [44, 50], argued that we should see biases as the side-effect of heuristics and knowledge application, and should not overlook the potential biases of statistical inference and algorithmic decisions. In this section, we make an information-theoretic argument for the importance of human involvement not only in visualization but also in data intelligence workflows in general.

In the field of visualization, it has been widely appreciated that the effectiveness of a visual representation depends on data, users, and tasks [38]. The dependence on users implicitly suggests the importance of **knowledge** in visualization processes. Perhaps, the first theoretical rationale was given by Chen and Jänicke when they outlined an information-theoretic framework for visualization [14] and noticed that interactive visualization violated the Markov chain condition of **data processing inequality** (DPI), which is one of the most important theorems in information theory.

Consider a fully automated workflow that consists of a sequence of processes for data processing and analysis, $P_1, P_2, \ldots, P_i, \ldots, P_n$, where each process P_i transforms from an input data space $\mathbb{Z}_i$ to an output data space $\mathbb{Z}_{i+1}$. Intuitively, an input data space encompasses all possible input datasets that a process may receive, and an output data space encompasses all possible output datasets that a process may generate. In information theory, these data spaces are also referred to *alphabets* [17]. The DPI states that if a sequence of alphabets forms a Markov chain $\mathbb{Z}_1 \longrightarrow \mathbb{Z}_2 \longrightarrow$ $, \ldots, \longrightarrow \mathbb{Z}_{n+1}$, the mutual information in the workflow can only decrease, such that:

$$\mathscr{I}(\mathbb{Z}_1, \mathbb{Z}_2) \geq \mathscr{I}(\mathbb{Z}_1, \mathbb{Z}_3) \geq \ldots \geq \mathscr{I}(\mathbb{Z}_1, \mathbb{Z}_n) \geq \mathscr{I}(\mathbb{Z}_1, \mathbb{Z}_{n+1})$$

When an automated workflow exhibits a huge amount of information loss, such as:

$$\mathscr{I}(\mathbb{Z}_1, \mathbb{Z}_2) \ggg \mathscr{I}(\mathbb{Z}_1, \mathbb{Z}_{n+1})$$

the integrity of the decisions as the final output $\mathbb{Z}_{n+1}$ could be undermined by the significant loss of information during the process. Indeed, in their most influential book on information theory [17], Cover and Thomas concluded *"no clever manipulation of the data can improve the inferences that can be made from the data."*

As Chen and Jänicke [14] noticed, interactive visualization may break the condition of Markov chain in two ways. First, when a user views the interim data between two processes (e.g., $P_1 \longrightarrow \mathbb{Z}_2 \longrightarrow P_2$), and uses the acquired knowledge to interact with a later process (e.g., changing a view parameter in P_5), the output to P_5 no longer depends only on the output of P_4. Second, if a user interacts with any process under the influence of additional knowledge that is not in the initial input data space $\mathbb{Z}_1$, this also breaks the condition of a Markov chain. Since DPI highlights a potentially critical risk of any fully automated workflow, breaking the condition of DPI is a good thing. In other words, if a "clever manipulation" that does not satisfies the Markov chain condition, such as interactive visualization, it is possible to "improve the inferences that can be made from the data."

In many scientific studies (e.g., [30]), the biases of human decisions are often measured by assuming that the decisions made using statistics or algorithms are the ground truth in conjunction with an experimental design where humans' decisions cannot benefit from any additional information that is not in the data. Unjustifiably, this approach to measuring biases was in favor of the third-party metric (i.e., a statistical measure or an algorithmic decision) that served both as a "judge" (assumed to be correct) and a "contestant" (to be compared with humans).

Chen and Golan proposed an objective measurement without a third-party metric [13]. For any decision process P_i that transforms its input alphabet $\mathbb{Z}_i$ to its output alphabet $\mathbb{Z}_{i+1}$, one can imagine that there is an inverse process $Q_i = P_i^{-1}$ that transforms the output back to the input. For example, consider an examination taken by a class of $n > 0$ students and the examination marks are of an integer value between 0 and 100. Let the input alphabet $\mathbb{Z}_i$ contain all possible combinations of n marks for the n students. We can imagine the following four scenarios for someone to perform an inverse process Q_i to reconstruct the n marks:

A. P_i is a statistical process for computing the mean μ of the n values. One has to guess the n marks with the given output μ.
B. P_i is an algorithmic process for dividing the n numbers into $k < n$ ordered groups and output k integers representing the numbers of students in individual groups (e.g., 5 students in group A, 22 students in group B, and so on). One has to guess the n marks based on these k numbers and the definitions of the groups.

C. P_i is a visualization process for plotting the n numbers in a bar chart such that each bar is of 4-pixel wide and up to 20-pixel high. One has to guess the n marks using the bar chart.

D. P_i is an interactive query process, with which one can perform $k < n$ queries and each time, one enters a student's name and obtains the corresponding mark. One has to guess the n marks based on k marks obtained from the interaction.

None of these processes is prefect because all of them cause some forms of information loss. Their merits depend on the values of n and k as well as the tasks to be performed using the output in each case. Hence, the inverse process in any of the four cases is not accurate. We can define the deviation of the guessing results from the actual n marks using an information-theoretic measure referred to as *Potential Distortion*. The measure is one of the three fundamental measures in a cost–benefit metric. For the detailed mathematical definition and explanation of the metric, see [13]. From the perspective of this chapter, the important consideration here is the knowledge of the person who has to make the guess. For a teacher who taught the n students, the potential distortion will be much lower than for one knows nothing about the n students or the subject being examined. The knowledge of the teacher can bring benefit to all four scenarios, which exemplify four types of processes in VA—statistics, algorithms, visualization, and interaction [15].

The knowledge that is used in or is available to a VA process can potentially be measured or estimated. Tam et al. conducted an observational study [51], and analyzed two case studies where visualization-assisted machine learning (ML) processes produced better classification models than fully automated ML processes. Using information-theoretic analysis, they estimated the amount of knowledge that was available to the visualization-assisted ML processes, and discovered that the amount of entropy (or information) in such knowledge is considerably larger than that of a small or skewed training dataset. Evidently, the use of such knowledge was the reason for the better models resulting from the visualization-assisted ML processes. This finding was further confirmed by a manipulation study by Kijmongkolchai et al. [34], where they experimentally confirmed that participants used their knowledge in visualization and made better decisions than the would-be decisions without such knowledge (i.e., chances). Because their stimuli were designed to measure individual pieces of knowledge in bits, they were able to transform the traditional accuracy measure to an information-theoretic measure for quantifying human knowledge used in the visualization tasks in their empirical study.

4.4 Ontologies as External Knowledge Bases

As discussed in the previous section, domain knowledge can alleviate the challenges of large and complex data analysis. In many practical applications, we can observe the impact of human knowledge upon visualization processes. For example, in the current discussion on climate change, the general public is mainly presented with highly condensed information like the average temperature of the earth's climate system when talking about global warming or expected rising of the sea level. Choropleth maps may be used to illustrate regions that are affected most severely. Climate scientists routinely produce various forms of numerical data and visualization images by computing, aggregating, curating, and selecting relevant pieces of information from a large set of simulated and measured multivariate time-series ensemble data. Making sense of this data requires specialized expertise and skills. The knowledge gained through training is only partly covered in books, documented analysis routines, and climate data analysis libraries, and a good portion of it is tacit knowledge that is usually gained through experience and often remains with the individual scientist.

Jänicke et al. and Kappe et al. reported joint work with climate experts [29, 31], where the visualization scientists encountered several of these analysis procedures to extract critical weather and climate events and procedures to aggregate information. Some of the procedures were formalized as routines for analyzing El Niño events [53], while others require interventions and decisions by a domain expert. In the projects, some knowledge of feature extraction were formalized for supporting rapid data analysis for both experts and layman users and enabling them to filter large volumes of data for relevant time frames rapidly. This is particularly helpful for users with a little climate background to focus their attention. While these projects have confirmed the benefits of formalized knowledge, they have also highlighted the need for some effective and efficient mechanisms to assist domain experts in externalizing their knowledge and assist data scientists in formalizing such knowledge.

Apart from mathematical definitions and computational procedures for formalizing knowledge, ontologies are another variant to externalize tacit knowledge. In computer science, an ontology is a form of knowledge representation. The most basic data structure for an ontology is a graph, where nodes represent concepts and edges represent the relationships among the concepts. There are a number of specification languages for defining ontologies, including popular ones such as the Web Ontology Language (OWL), the knowledge interchange format (KIF), Resource Description Framework Schema (RDFS), and The DARPA Agent Markup Language + the Ontology Inference Layer (DAML+OIL). These ontology languages have enabled many disciplines and communities to acquire, preserve, and share various domain-specific knowledge. The uses of such stored knowledge include search engines, data integration, and text analysis.

Carpendale et al. discussed the potential uses of ontologies in biological data visualization [4]. In addition to the technical challenges that large ontologies present to network visualization, they pointed out that ontologies could be used to aid (i) text mining and text and document visualization, (ii) the automated or semi-automated creation of visualization, and (iii) information retrieval in interactive visualization.

Gilson et al. made the first attempt to use ontologies to automate the visual mapping from a dataset to a visual representation [26]. Their knowledge-assisted method requires the construction of three types of ontologies:

- a *Domain Ontology* (DO), which stores the semantics of a specific subject domain (e.g., music charts),

- one or more *Visual Representation Ontologies* (VROs), each of which captures the semantics of a visual representation (e.g., bar charts, tree maps) and stores the mapping from each visual representation to its implementation mechanism in an external visualization toolkit.
- a *Semantic Bridging Ontology* (SBO), which specifies the relationships between the concepts in the DO and those in the VROs.

After receiving an input dataset, an automated system first analyzes the dataset with the aid of the DO, identifying semantics of various data labels such as column headings. It then searches the SBO to identify optional visual representations (e.g., bar charts, tree maps, etc.), and optional visual objects (e.g., vertical bars, horizontal bars, treemap nodes, etc.) with their corresponding visual channels. It compares various optional mappings and selects the most highly ranked mapping. Finally, it uses the VRO of the selected visual representation to retrieve the implementation mechanism for the visual representation. This completes the visual mapping process from a dataset to a visual representation.

Khan et al. developed a visualization-assisted search engine for supporting a team of users who routinely search for hundreds and thousands of files in the context of building industry [33]. A file ontology serves as the backbone of the search engine and is dynamically updated according to users' interaction. The users' interaction is thus a form of knowledge externalization, while the ontology is a form of knowledge formalization. In addition, the ontology enables users to visualize the provenance of the past search activities (i.e., captured facts or simple knowledge) in order to minimize the repeated search attempts with the same search criteria.

Recently, Sacha et al. constructed an ontology capturing major processes in machine learning (ML) workflows, including those processes in the four iterative stages: prepare data, prepare learning, model learning, and evaluate model [47]. They surveyed a number of papers on visualization-assisted ML, extracted the knowledge about the different places where visualization was used, and recorded these places in the ontology. They believe that the knowledge captured in this ontology can encourage researchers to develop advanced visualization techniques for aiding ML, and they hope that with more and more knowledge captured, this ontology may be used to aid the design of visualization-assisted ML workflows in the future.

More recently, Chen and Ebert outlined an ontological framework that can be used as the core for building a comprehensive ontology to support the design, evaluation, and improvement of visual analytics (VA) systems [12]. It provides 24 abstract entities for reasoning about the causal relations among the symptoms, causes, remedies, and side-effects in the life-cycles of VA systems, providing an early step toward building a more comprehensive knowledge base (in the form of an ontology) for the VA community to record a large collection of concrete instances of problems and solutions that VA researchers and practitioners have experienced in real-world applications.

4.5   Providing Guidance in KAV-Systems

Ontologies are one way to externalize tacit knowledge and to make it applicable in digital analysis systems. The overarching goal is to store knowledge and integrate it in the analysis pipeline to ease work or support users in their decision-making processes. This is particularly true in the design of VA solutions, where the design space is a very-high-dimensional manifold and there is a plethora of potential interactive visualizations and analytical methods available as well as parameters and sequences for fine tuning thereof (e.g., to set parameters, but suitable values are unknown upfront). Often it is not clear to the user(s), which methods or techniques will eventually lead to the intended results [35]. To this end, the users need more support in the sense of guidance. Guidance requires the storage, integration, and application of operational and domain knowledge in KAV-systems to help the users close the knowledge gap. In the following subsections, we will define guidance and its application in VA systems (Sects. 4.5.1 and 4.5.2), propose a model for its integration in VA systems (Sect. 4.5.3), and demonstrate its application using an example from cyclical pattern detection (Sect. 4.5.4).

4.5.1   Guidance

Guidance is seen as a process that gradually narrows the gap that hinders effective continuation of the data analysis. It provides prospective assistance so that users can make sense of the data on their own. For example, an anti-fraud manager needs guidance to detect harmful behaviors of some customers using highlighting techniques, whereas an investigator needs guidance to fine-tune the fraud transaction monitor providing various alternative options. Furthermore, the quality of VA-supported workflow can be significantly improved by guidance-oriented solutions [6]. While there are already some approaches that offer guidance to VA users (e.g., [8, 16]), there is only limited knowledge about the general guidance mechanisms and underlying structures of guidance. Furthermore, existing guidance approaches do not cover the entire VA process. New techniques are needed to offer guidance on all intertwined phases of the VA process (e.g., how to wrangle the data, how to read and interact with the visual means in a way that the user can derive appropriate information for decision making). This major challenge was also identified in the roadmap of VA [52].

In the context of VA, guidance has been described as a strategy to assist in data exploration and analysis. We characterized guidance from the conceptual perspective as follows Ceneda et al. [6, p.112]:

> **Guidance** is a computer-assisted *process* that aims to actively resolve a *knowledge gap* encountered by users during an *interactive* visual analytics session.

The three important aspects of this definition are emphasized in italics. First, guidance is a dynamic process that runs alongside the regular data analysis activities

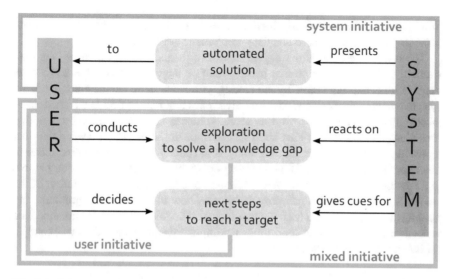

Fig. 4.6 Guidance is a mixed-initiative process. On the one hand, the user explicitly or implicitly expresses his/her analysis target and a possible knowledge gap that hinders progression, by interacting with the system. On the other hand, the system reacts to the user's actions and gives cues that help to decide which steps to take to reach the target [7] (Figure adapted from [7] with permission of Davide Ceneda)

of the user. Second, there is a knowledge gap that causes the data analysis to stall. The user does not know how to proceed. The goal of guidance is to narrow the knowledge gap. Finally, the definition of guidance describes an interactive scenario. That is, guidance assumes the existence of a human in the loop.

Guidance provides one or multiple *suggestions* to the user. Suggestions can be considered or ignored by the user. Suggestions are to help users in forming decisions. Making the decisions remains the responsibility of the user. Guidance does not aim to close the knowledge gap automatically with a definite or exact answer. Typically, this is not even possible due to ill-defined or highly complex problems. If guidance were able to compute a precise answer, we could neglect VA at all, compute the answer, and provide it to the user as an instruction. But this would contradict with the idea of the human in the loop.

4.5.2   Guidance-Enriched Visual Analytics

Guidance has its roots in human–computer interaction (HCI) [18, 41, 45] and can be seen as a mixed-initiative process [28] (as shown in Fig. 4.6). A mixed-initiative process is an approach where both users and systems can "take the initiative" and both contribute to the process. The central questions are the degree and type of involvements. Guidance is a dialogue between users and systems, in which users provide, implicitly or explicitly, their own needs and issues as input, and the system provides possible answers to alleviate problematic situations [7].

The central question is the degree of involvements. On the one hand, the user explicitly or implicitly expresses her/his analysis target and a possible knowledge gap that hinders progression, by interacting with the system. On the other hand, the system reacts to the user's actions and gives cues that help to decide which steps to take to reach the target [7]. On this score, guidance is (1) a dynamic process that runs alongside the regular data analysis activities of the users; (2) needed due to a knowledge gap that causes the data analysis to stall, and (3) an interactive scenario assuming the existence of a human in the loop. In this sense, guidance is comparable to a mentor helping a student. While the mentor does not know the solution of the student's problem, she/he can provide hints as how to approach the problem, and guiding the student toward finding the solution on her/his own. To this end, guidance is not merely an additional algorithm that computes results but is indeed as a catalyst for human–computer cooperation [5].

4.5.3 A Model for Guidance-Enriched Visual Analytics

We build upon the initial characterization of guidance by Schulz et al. [49], but we focused on VA and adapt it with respect to the knowledge gap of users, the input and the output of a guidance generation process, as well as a refined characterization of the degree to which guidance is provided (as shown in Fig. 4.7).

Ceneda et al. extended van Wijk's model of visualization [58] to cover the guidance generation process as well as the VA process [6]. Here, they made a slight modification by replacing the term *visualization* with *visual analytics*.The model thus covers both visual and analytical methods. The components of the model are shown with gray outlines in Fig. 4.8. Boxes represent artifacts, such as data or images, while circles represent functions that process some input and generate some output. Visual and analytical means [V] transform data [D] into images [I] based on some specifications [S]. The images are then perceived (P) to generate some knowledge

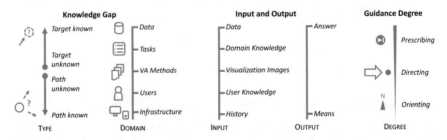

Fig. 4.7 Aspects of guidance: knowledge gap, input and output, and guidance degree [6] (Figure taken from [6] © 2017 IEEE. Used with permission)

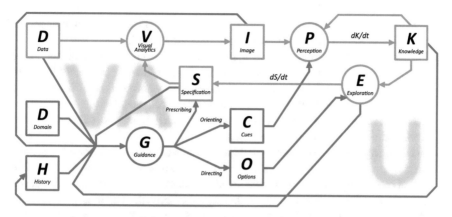

Fig. 4.8 Guidance and VA (in blue) extend van Wijk's model [58] (in gray). Aspects of visual analytics (VA) are shown to the left, while user aspects (U) are on the right. Guidance considers the user's knowledge (or lack thereof) and may build upon various inputs, including data, interaction history, domain conventions, VA specifications, and visualization images. Different degrees of guidance are possible. Orienting uses visual cues to enhance perception. Directing supports exploration by providing alternative options. Prescribing directly operates on the specification [5, 6] (Figure taken from [5]. Used with permission)

[K]. Based on their accumulated knowledge, users can interactively explore (E) the data by adjusting the specifications (e.g., choose a different clustering algorithm or change the perspective on the data). As such, van Wijk's model effectively conveys the iterative and dynamic nature of knowledge generation mediated through VA. This makes it perfectly suited to be expanded to a model of guided VA.

Ceneda et al. attached new guidance-related components to the model, shown with blue outlines in Fig. 4.8. A central position is taken by the guidance generation process (G). It is hooked up first and foremost with the user's knowledge [K]. The reason is that before one can take any measures of guidance, one needs to know what the particular problem of the user is. Similar to the worldview gap [2], they coin the term *knowledge gap* to capture the actual deficit that hinders continuation of the data analysis. The guidance generation process (G) is further connected to sources of information based on which guidance can be generated. These sources include the original data [D], visualization images [I], VA specifications [S], interaction history or provenance [H], and domain conventions or models [D]. Taken together, these components represent the *input* to the guidance generation process.

On the *output* side, results of a guidance generation process can be delivered in various ways. Figure 4.8 illustrates three different scenarios: (1) *Orienting*, (2) *Directing*, and (3) *Prescribing*:

- **Orienting guidance** is at the low end of the guidance degree. It provides basic guidance through visual cues ([C] in Fig. 4.8). The main goal is to build or maintain the user's mental map [39]. Such a map may contain potential targets and paths as well as relations among them. Providing visual cues hinting at these targets and paths is a common strategy for implementing orienting. Visual overview techniques may provide some kinds of orientation as well.

- **Directing guidance** represents a medium degree of guidance. It offers useful options or alternatives ([O] in Fig. 4.8) that the user may or may not choose to follow. The suggestions may differ in terms of quality and costs for different paths leading to the same result or, in terms of interest for paths, leading to similar or new results. Directing can benefit from preview techniques that help users make informed decisions for one or the other option.
- **Prescribing guidance** captures the highest degree of guidance. It directly operates on the specification ([S] in Fig. 4.8) in order to automatically generate suitable visual results [36, 59]). It implements a largely automated process, which proceeds toward a specified target. In a sense, this degree of guidance can be compared to an interactive presentation. A user may interrupt the presentation and ask for details, or rewind/reverse it to revisit a nugget of information that has been found earlier. Depending on the degree of automation, the user can recover control for a while and direct the presentation to another path or even another target.

The main goal of guidance is to create and maintain an environment in which users are able to make progress and perform their tasks effectively. This dynamic progressive procedure is well expressed by the knowledge change (dK/dt) occurring as a consequence of the guided visual analysis and the interactive adjustment (dS/dt) of the specification. A critical concern is that knowledge is acquired through perception and cognition (P). So the leverage point of guidance is to facilitate perception and cognition at different *degrees*.

4.5.4 Example of Guidance-Enriched VA: Cyclical Patterns in Univariate, Evenly Spaced Time-Series

To illustrate our guidance approach, Ceneda et al. recently proposed a data-driven guidance technique [9] to support the visual exploration of cyclical patterns in univariate, evenly spaced time-series (as shown in Fig. 4.9). A classical spiral plot is enhanced with data-driven guidance mechanisms to support the identification of patterns and push the exploration forward. They statistically determine cycle lengths that reveal strong patterns and visually indicate these interesting cycle lengths while the user interacts with the slider. It provides an orienting support via visual cues that aim at solving an unknown target problem for which the knowledge gap is a data problem. In a qualitative user-study, they showed that guidance could enhance the data exploration. The participants developed a deeper understanding of the data and had an increased confidence in the analysis outcome.

4.6   Conclusions and Critical Reflection

The main contribution of this chapter is the investigation of the theoretical and practical underpinnings of visualization and VA in order to (1) incorporate the function
and role of tacit (implicit) and explicit knowledge in the analytical reasoning process,
including discourses from the information-theoretical and ontological perspectives,
and (2) explore a conceptual characteristics of guidance in visualization and VA
solutions. In the following, we reflect critically on these concepts and methods.

We have recalled Ceneda et al.'s conceptual model that generalizes existing
approaches of knowledge-assisted VA [6]. It is based on the well-known visualization model of van Wijk [58] and allows for modeling a broad range of analytics
systems (both with and without explicit knowledge as well as automated data analysis). Hence, it connects seamlessly to existing theoretical foundations while extending
their descriptive, evaluative, and generative power. The proposed model contains the
essential components, processes, and connections needed in a knowledge-assisted
VA system, i.e., (1) tacit knowledge extraction, (2) automated data analysis methods,
(3) explicit knowledge-based specification, (4) explicit knowledge visualization, and
(5) tacit knowledge generation.

However, this work represents an early step in this conceptual and practical area, a
number of opportunities for future research arise. One major issue is the necessity for
novel evaluation methods that can measure knowledge flows in a VA pipeline in order

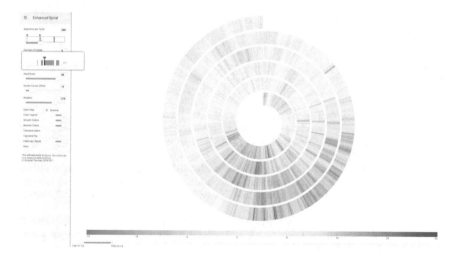

Fig. 4.9 Ceneda et al. [9] proposed a guided solution for the exploration of cyclical patterns in time-
series. A classical spiral visualization (center) is enhanced with data-driven guidance mechanisms
to support the identification of patterns. The user interface with the slider (left) allows to configure
the spiral plot. The slider is used to modify the cycle length displayed in the spiral plot. They
statistically determine cycle lengths that reveal strong patterns and visually indicate these interesting
cycle lengths while the user interacts with the slider (Figure adapted from [9] with permission of
Davide Ceneda)

to assess the effectiveness of different components in the VA pipeline. Such methods can be based on explicit knowledge as conceptualized in Ceneda et al.'s model. For example, the **nested workflow model** [21] points in this direction, enabling the description of VA processes at different design levels, also in terms of data and knowledge flows, as well as collaboration between users. Further areas of future research may include validation methods for extracted explicit knowledge, extracting knowledge indirectly via user interactions, or more specific support for collaboration and multi-user systems.

Within the characterization of guidance in VA by Ceneda et al. [6], where various aspects of guidance have been identified, the guidance generation process remains to be a black box and it is desirable to detail this process in relation to the whole VA process beyond the extension of van Wijk's model [58]. On the one hand, the operationalization of these aspects needs to be incorporated within the guidance generation process. For example, the understanding of users' knowledge gaps is limited and most existing approaches implicitly infer the knowledge gap. On the other hand, we need a better understanding about the internals of guidance and how guidance is actually generated ("opening the black box"). To this end, we could utilize the HCI model by Norman [41] for clarification (using his "why", "what", and "how" layers).

Acknowledgements The authors would like to thank Davide Ceneda, Theresia Gschwandtner, Thorsten May, Hans-Jörg Schulz, Marc Streit, Christian Tominski, and all participants of the WG group at the Dagstuhl Seminar for constructive and fruitful discussions and thank the reviewers for their invaluable feedback. This work was funded by the "Austrian Science Fund (FWF), grant P31419-N31".

References

1. Ackoff, R.L.: From data to wisdom. J. Appl. Syst. Anal. **16**(1), 3–9 (1989)
2. Amar, R., Stasko, J.: Knowledge precepts for design and evaluation of information visualizations. IEEE Trans. Vis. Comput. Graph. (TVCG) **11**(4), 432–442 (2005). https://doi.org/10.1109/TVCG.2005.63
3. Bodesinsky, P., Federico, P., Miksch, S.: Visual analysis of compliance with clinical guidelines. In: Proceedings of the 13th International Conference on Knowledge Management and Knowledge Technologies, i-Know '13, pp. 12:1–12:8. ACM, New York, NY, USA (2013). https://doi.org/10.1145/2494188.2494202
4. Carpendale, S., Chen, M., Evanko, D., Gehlenborg, N., Goerg, C., Hunter, L., Rowland, F., Storey, M.A., Strobelt, H.: Ontologies in biological data visualization. IEEE Comput. Graph. Appl. **34**(2), 8–15 (2014)
5. Ceneda, D., Gschwandtner, T., May, T., Miksch, S., Schulz, H., Streit, M., Tominski, C.: Amending the Characterization of Guidance in Visual Analytics (2017). arXiv:1710.06615 [cs.HC]
6. Ceneda, D., Gschwandtner, T., May, T., Miksch, S., Schulz Schulz, H., Streit, M., Tominski, C.: Characterizing guidance in visual analytics. IEEE Trans. Vis. Comput. Graph. (TVCG) **23**(1), 111–120 (2017)
7. Ceneda, D., Gschwandtner, T., May, T., Miksch, S., Streit, M., Tominski, C.: Guidance or no guidance? a decision tree can help. In: Proceedings of the EuroVis Workshop on Visual

Analytic (EuroVA), pp. 19–23. Eurographics Digital Library (2018). https://doi.org/10.2312/eurova.20181107

8. Ceneda, D., Gschwandtner, T., Miksch, S.: A review of guidance approaches in visual data analysis: a multifocal perspective. Comput. Graph. Forum (CGF) **38**(3), 861–879 (2019)

9. Ceneda, D., Gschwandtner, T., Miksch, S., Tominski, C.: Guided visual exploration of cyclical patterns in time-series. In: Proceedings of the IEEE Symposium on Visualization in Data Science (VDS). IEEE Computer Society (2018)

10. Chen, C.: Top 10 unsolved information visualization problems. IEEE Comput. Graph. Appl. Mag. **25**(4), 12–16 (2005). https://doi.org/10.1109/MCG.2005.91

11. Chen, M., Ebert, D., Hagen, H., Laramee, R., van Liere, R., Ma, K.L., Ribarsky, W., Scheuermann, G., Silver, D.: Data, information, and knowledge in visualization. IEEE Comput. Graph. Appl. Mag. **29**(1), 12–19 (2009). https://doi.org/10.1109/MCG.2009.6

12. Chen, M., Ebert, D.S.: An ontological framework for supporting the design and evaluation of visual analytics systems. Comput. Graph. Forum **36**(3), 131–144 (2019)

13. Chen, M., Golan, A.: What may visualization processes optimize? IEEE Trans. Vis. Comput. Graph. (TVCG) **22**(12), 2619–2632 (2016). https://doi.org/10.1109/TVCG.2015.2513410

14. Chen, M., Jänicke, H.: An information-theoretic framework for visualization. IEEE Trans. Vis. Comput. Graph. **16**(6), 1206–1215 (2010)

15. Chen, M., Trefethen, A., Banares-Alcantara, R., Jirotka, M., Coecke, B., Ertl, T., Schmidt, A.: From data analysis and visualization to causality discovery. IEEE Comput. **44**(10), 84–87 (2011)

16. Collins, C., Andrienko, N., Schreck, T., Yang, J., Choo, J., Engelke, U., Jena, A., Dwyer, T.: Guidance in the human-machine analytics process. Vis. Inf. **2**(3), 166–180 (2018)

17. Cover, T.M., Thomas, J.A.: Elements of Information Theory. Wiley, New York (2006)

18. Dix, A., Finlay, J., Abowd, G., Beale, R.: Human-Computer Interaction, 3rd edn. Pearson Education, London (2004)

19. Ellis, G. (ed.): Cognitive Biases in Visualizations. Springer, Berlin (2018)

20. Fayyad, U., Piatetsky-Shapiro, G., Smyth, P.: From data mining to knowledge discovery in databases. AI Mag. **17**(3), 37 (1996)

21. Federico, P., Amor-Amorós, A., Miksch, S.: A nested workflow model for visual analytics design and validation. In: Proceedings of the Workshop on Beyond Time And Errors (BELIV), pp. 104–111. ACM, New York, NY, USA (2016). https://doi.org/10.1145/2993901.2993915

22. Federico, P., Unger, J., Amor-Amorós, A., Sacchi, L., Klimov, D., Miksch, S.: Gnaeus: utilizing clinical guidelines for knowledge-assisted visualisation of EHR cohorts. In: E. Bertini, J.C. Roberts (eds.) Proceedings of the EuroVis Workshop on Visual Analytic (EuroVA). The Eurographics Association (2015). https://doi.org/10.2312/eurova.20151108

23. Federico, P., Wagner, M., Rind, A., Amor-Amorós, A., Miksch, S., Aigner, W.: The role of explicit knowledge: a conceptual model of knowledge-assisted visual analytics. In: Proceedings of IEEE Conference on Visual Analytics Science and Technology (VAST) (2017)

24. Flöring, S., Appelrath, H.J.: KnoVA: introducing a reference model for knowledge-based visual analytics. In: Proceedings of the International Conference on Imaging Theory and Applications and International Conference on Information Visualization Theory and Applications (IVAPP), pp. 230–235 (2011)

25. Gilson, O., Silva, N., Grant, P., Chen, M.: From web data to visualization via ontology mapping. Comput. Graph. Forum (CGF) **27**(3), 959–966 (2008). https://doi.org/10.1111/j.1467-8659.2008.01230.x

26. Gilson, O., Silva, N., Grant, P., Chen, M.: From web data to visualization via ontology mapping. Comput. Graph. Forum **27**(3), 959–966 (2008)

27. Hand, D.J.: Intelligent data analysis: issues & opportunities. Intell. Data Anal. **2**(1–4), 67–79 (1998). https://doi.org/10.1016/S1088-467X(99)80001-8

28. Horvitz, E.: Principles of mixed-initiative user interfaces. In: Proceedings of the SIGCHI conference on Human Factors in Computing Systems, pp. 159–166. ACM (1999)

29. Jänicke, H., Böttinger, M., Mikolajewicz, U., Scheuermann, G.: Visual exploration of climate variability changes using wavelet analysis. IEEE Trans. Vis. Comput. Graph. **15**(6), 1375–1382 (2009)

30. Kahneman, D.: Thinking, Fast and Slow Paperback. Penguin (2012)
31. Kappe, C.P., Böttinger, M., Leitte, H.: Exploring variability within ensembles of decadal climate predictions. IEEE Trans. Vis. Comput. Graph. **25**(3), 1499–1512 (2019)
32. Keim, D., Kohlhammer, J., Ellis, G., Mansmann, F.: Mastering the Information Age: Solving problems with Visual Analytics. Eurographics (2010)
33. Khan, S., Kanturska, U., Waters, T., Eaton, J., Banares-Alcantara, R., M.Chen: Ontology-assisted provenance visualization for supporting enterprise search of engineering and business files. Adv. Eng. Inf. **30**(2), 244–257 (2016)
34. Kijmongkolchai, N., Abdul-Rahman, A., Chen, M.: Empirically measuring soft knowledge in visualization. Comput. Graph. Forum **36**(3), 73–85 (2017)
35. Kriglstein, S., Pohl, M., Smuc, M.: Pep Up Your Time Machine: Recommendations for the Design of Information Visualizations of Time-Dependent Data, pp. 203–225. Springer, New York (2014)
36. Mackinlay, J.: Automating the design of graphical presentations of relational information. ACM Trans. Graph. (TOG) **5**(2), 110–141 (1986). https://doi.org/10.1145/22949.22950
37. Mackinlay, J., Hanrahan, P., Stolte, C.: Show me: automatic presentation for visual analysis. IEEE Trans. Vis. Comput. Graph. (TVCG) **13**(6), 1137–1144 (2007). https://doi.org/10.1109/TVCG.2007.70594
38. Miksch, S., Aigner, W.: A matter of time: applying a data-users-tasks design triangle to visual analytics of time-oriented data. Comput. Graph. Spec. Sect. Vis. Anal. **38**, 286–290 (2014). https://doi.org/10.1016/j.cag.2013.11.002
39. Misue, K., Eades, P., Lai, W., Sugiyama, K.: Layout adjustment and the mental map. J. Vis. Lang. Comput. **6**(2), 183–210 (1995). https://doi.org/10.1006/jvlc.1995.1010
40. Nonaka, I., Takeuchi, H.: The knowledge-creating company. Harvard business review **85**(7/8), 162 (2007)
41. Norman, D.A.: The Design of Everyday Things. Basic Books Inc, New York (2013)
42. Perner, P.: Intelligent data analysis in medicine-recent advances. Artif. Intell. Med. **37**(1), 1–5 (2006). https://doi.org/10.1016/j.artmed.2005.10.003
43. Pike, W.A., Stasko, J., Chang, R., O'Connell, T.A.: Inform. Vis. The science of interaction. **8**(4), 263–274 (2009). https://doi.org/10.1057/ivs.2009.22
44. Pohl, M.: Cognitive biases in visual analytics – a critical reflection. In: Ellis, G. (ed.) Cognitive Biases in Visualizations, pp. 177–184. Springer, Berlin (2018)
45. Preece, J., Sharp, H., Rogers, Y.: Interaction Design: Beyond Human-Computer Interaction, 4th edn. Wiley, New York (2015)
46. Rubrichi, S., Rognoni, C., Sacchi, L., Parimbelli, E., Napolitano, C., Mazzanti, A., Quaglini, S.: Graphical representation of life paths to better convey results of decision models to patients. Med. Decis. Mak. **35**(3), 398–402 (2015). https://doi.org/10.1177/0272989X14565822
47. Sacha, D., Kraus, M., Keim, D.A., Chen, M.: VIS4ML: an ontology for visual analytics assisted machine learning. IEEE Trans. Vis. Comput. Graph. **25**(1), 385–395 (2019)
48. Sacha, D., Stoffel, A., Stoffel, F., Kwon, B.C., Ellis, G., Keim, D.A.: Knowledge generation model for visual analytics. IEEE Trans. Vis. Comput. Graph. **20**(12), 1604–1613 (2014)
49. Schulz, H.J., Streit, M., May, T., Tominski, C.: Towards a characterization of guidance in visualization. In: Poster at IEEE Conference on Information Visualization (InfoVis) (2013)
50. Streeb, D., Chen, M., Keim, D.A.: The biases of thinking fast and thinking slow. In: Ellis, G. (ed.) Cognitive Biases in Visualizations, pp. 97–107. Springer, Berlin (2018)
51. Tam, G.K.L., Kothari, V., Chen, M.: An analysis of machine- and human-analytics in classification. IEEE Trans. Vis. Comput. Graph. **23**(1), 71–80 (2017)
52. Thomas, J.J., Cook, K.A.: Illuminating the Path: The Research and Development Agenda for Visual Analytics. IEEE Computer Society (2005)
53. Trenberth, K.E.: The definition of el nino. Bull. Am. Meteorol. Soc. **78**(12), 2771–2778 (1997)
54. Wagner, M.: Integrating explicit knowledge in the visual analytics process. Technical report, TU Wien, Ph.D. Thesis (2017)
55. Wagner, M., Rind, A., Thür, N., Aigner, W.: A knowledge-assisted visual malware analysis system: design, validation, and reflection of KAMAS. Comput. Secur. **67**, 1–15 (2017). https://doi.org/10.1016/j.cose.2017.02.003

56. Wagner, M., Slijepcevic, D., Horsak, B., Rind, A., Zeppelzauer, M., Aigner, W.: KAVAGait: Knowledge-assisted visual analytics for clinical gait analysis. IEEE Trans. Vis. Comput. Graph. (TVCG) **25**(3), 1528–1542 (2018). https://doi.org/10.1109/TVCG.2017.2785271
57. Wang, X., Jeong, D.H., Dou, W., Lee, S.W., et al.: Defining and applying knowledge conversion processes to a visual analytics system. Comput. Graph. **33**(5), 616–623 (2009). https://doi.org/10.1016/j.cag.2009.06.004
58. van Wijk, J.: Views on visualization. IEEE Trans. Vis. Comput. Graph. (TVCG) **12**(4), 421–433 (2006). https://doi.org/10.1109/TVCG.2006.80
59. Wongsuphasawat, K., Moritz, D., Anand, A., Mackinlay, J., Howe, B., Heer, J.: Voyager: exploratory analysis via faceted browsing of visualization recommendations. IEEE Trans. Vis. Comput. Graph. (TVCG) **22**(1), 649–658 (2016). https://doi.org/10.1109/TVCG.2015.2467191

Chapter 5
Mathematical Foundations in Visualization

Ingrid Hotz, Roxana Bujack, Christoph Garth and Bei Wang

Abstract Mathematical concepts and tools have shaped the field of visualization in fundamental ways and played a key role in the development of a large variety of visualization techniques. In this chapter, we sample the visualization literature to provide a taxonomy of the usage of mathematics in visualization and to identify a fundamental set of mathematics that should be taught to students as part of an introduction to contemporary visualization research. Within the scope of this chapter, we are unable to provide a full review of all mathematical foundations of visualization; rather, we identify a number of concepts that are useful in visualization, explain their significance, and provide references for further reading. We assume the reader has basic knowledge of linear algebra [90], multivariate calculus [89], statistics, combinatorics, and stochastics [39]. Other topics not covered in this chapter, such as image analysis [88], computer graphics [86], signal processing [41], computational geometry [2], geometric modeling, mesh generation, computer-aided geometric design [35, 106], and numerics [76] can be found in well-established textbooks. More advanced topics such as information theory, dimension reduction, and kernel methods are discussed in other parts of the book.

I. Hotz (✉)
Linköping University, Norrköping, Sweden
e-mail: ingrid.hotz@liu.se

R. Bujack
Los Alamos National Laboratory, Santa Fe, NM, USA
e-mail: bujack@lanl.gov

C. Garth
Technische Universität Kaiserslautern, Kaiserslautern, Germany
e-mail: garth@cs.uni-kl.de

B. Wang
University of Utah, Salt Lake City, UT, USA
e-mail: beiwang@sci.utah.edu

© Springer Nature Switzerland AG 2020
M. Chen et al. (eds.), *Foundations of Data Visualization*,
https://doi.org/10.1007/978-3-030-34444-3_5

Fig. 5.1 Visualization pipeline. All steps in the pipeline involve the use of mathematical concepts and tools. We cover various aspects of data analysis, filtering, and mapping

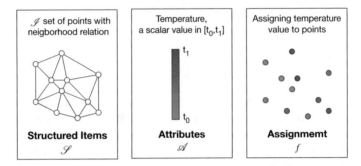

Fig. 5.2 Example of a data set: $\mathscr{S}$ consists of a set of points with a neighborhood relation. Attributes in $\mathscr{A}$ are elements of the interval $[t_0, t_1]$. f assigns temperature values to the points

5.1 Data and Basic Terminology

> You can have data without information, but you cannot have information without data.
>
> *Daniel Keys Moran, programmer and science fiction writer*

Data are at the center of every visualization task and every step of the visualization pipeline; see Fig. 5.1. The input to the visualization pipeline, the raw data, can be any collection of information in any form. In this chapter, we define a data set as a triplet $\mathscr{D} = (\mathscr{S}, \mathscr{A}, f)$ consisting of a set of structured items $\mathscr{S}$, a set of attributes $\mathscr{A}$, and a function that assigns attributes to the items. $\mathscr{S}$ consists of a set of items, continuous, or discrete, together with a structure (such as a metric for a continuous domain or neighborhood relations for networks); see Fig. 5.2 for an example. The tools used for the analysis and visualization of data sets depend on the nature of $\mathscr{S}$ and $\mathscr{A}$. The most important distinctions are continuous versus discrete structures and quantitative versus categorical attributes; see Table 5.1. In this section, we emphasize continuous structures and quantitative attributes. A more detailed classification of data sets concerning types, structures, and organizations can be found in Munzner [68]. An introduction to data representations from a scientific visualization perspective can be found in Telea [92].

Table 5.1 Examples for possible structures $\mathscr{S}$ and attribute spaces $\mathscr{A}$

Structures $\mathscr{S}$	Attributes $\mathscr{A}$
Continuous domains equipped with metrics	Ordered, ordinal, quantitative
Meshes, simplicial complexes	Scalars, vectors, tensors
Graphs, networks, trees	Categorical

5.1.1 The Structure $\mathscr{S}$

The structure $\mathscr{S}$ can vary from discrete points to continuous domains. In general, $\mathscr{S}$ consists of a set of items and some relation between the items. We describe two of the most frequently used structures in more detail.

Graphs, Networks, and Trees. *Graphs* or *networks* are structures that are frequently used for non-spatial, relational data representations. The terms graph and network are sometimes used interchangeably. Mathematically, a graph G is a pair (V, E) consisting of a set of items V, called vertices or nodes, and a set of relationships between these items expressed as a set of edges $E \subseteq V \times V$. Edges can be directed or undirected. For directed graphs, (v, w), and $(w, v) \in E$ represent different relations. If the edges are assigned a numeric attribute, the graph is weighted.

A possible representation of a finite graph is an adjacency matrix, which is a square matrix of size $|V| \times |V|$. For a simple graph, the adjacency matrix is a $(0,1)$-matrix with zeros on its diagonal and ones for each edge. If the graph is undirected, the matrix is symmetric. Typically, graphs are displayed using a set of points for the vertices, which are joined by lines for the edges. A general introduction to graphs and networks can be found in [56].

When analyzing graphs, characteristics as cycles, planarity, sparseness, and hierarchical representations are of interest.

Continuous Domains. A continuous *domain* D is a subset of $\mathbb{R}^n$ equipped with a *metric*. A metric supports measurements and determines distances in the domain. A common metric is the Euclidean distance. Other metrics include Manhattan distances and polar distances. More generally, when the domain is a *parameterized manifold*, the choice of a metric has an impact on many calculations such as derivatives; see Sect. 5.2.

A continuous domain can be represented by a finite set of discrete samples associated with an interpolation scheme. In this case, $\mathscr{S}$ consists of a set of points $\{p_i \in D \mid i = 1, \ldots, k\}$, equipped with a neighborhood structure, e.g., the points are organized as a *regular grid* (associated with piecewise multilinear interpolation) or a simplicial complex (corresponding to piecewise linear interpolation).

5.1.2   The Attribute Space $\mathscr{A}$

An attribute is a specific property assigned to data items that arise from measurement, observation, or computation. Attributes can be continuous and quantitative, e.g., temperature; discrete and ordered, e.g., the number of people in a class; as well as categorical, e.g., various types of tree species. The set of possible attributes span the *attribute space*.

The most common *continuous quantitative attributes* can be subsumed under the term *tensor*. A *tensor of order r* is defined as a multilinear mapping acting on r copies of a n-dim vector space V over $\mathbb{R}$ into the space of real numbers,

$$\mathbf{T} : \underbrace{V \times \ldots \times V}_{r} \to \mathbb{R}. \tag{5.1}$$

Sometimes, *rank*, *degree*, and *order* are used interchangeably. A tensor of order 0 corresponds to a scalar $\alpha \in \mathbb{R}$, and a tensor of order 1 is a vector $\mathbf{v} \in V$.

$$
\begin{aligned}
\alpha : \mathbb{R} \to \mathbb{R}, &\qquad \alpha(x) = \alpha x &\qquad \text{0th-order tensor or } scalar, \text{ e.g., temperature;} \\
\mathbf{w} : V \to \mathbb{R}, &\qquad \mathbf{w}(\mathbf{v}) = \mathbf{w} \cdot \mathbf{v} &\qquad \text{1st-order tensor or } vector, \text{ e.g., velocity;} \\
\mathbf{T} : V \times V \to \mathbb{R}, &\quad \mathbf{T}(\mathbf{v}, \mathbf{v}') = \mathbf{v} \cdot \mathbf{T} \cdot \mathbf{v}' &\quad \text{2nd-order } tensor, \text{ e.g., strain tensor.}
\end{aligned}
$$

Tensors of higher order, especially 3rd- and 4th-order tensors, can also be found in a few visualization applications. In the visualization literature, the term tensor often refers to 2nd-order tensors. With respect to a specific basis $\{\mathbf{e}_1, \ldots, \mathbf{e}_n\}$ of the vector space V, a tensor is fully specified by its action on the basis elements resulting in the typical component representations. For a vector, this is $w = (w_1, \ldots, w_n)^T$ and for a 2nd-order tensor, and this is a matrix

$$\mathbf{T} = \begin{pmatrix} t_{1,1} & \cdots & t_{1,n} \\ \vdots & \ddots & \vdots \\ t_{n,1} & \cdots & t_{n,n} \end{pmatrix}.$$

For a basic introduction to the use of tensors in visualization, we refer to the state-of-the-art report by Kratz et al. [59].

Enriched Attribute Space $\mathscr{A}^*$. In-depth data analysis often requires some modifications of the attribute space. The most common examples are *filtering*, e.g., removing noise, or *enrichment* of the original attributes by derived quantities, e.g., the field gradient or local histograms. Other modifications are changes of the representation or parameterization of the attribute space to emphasize data symmetries useful for feature or pattern definitions; see also Sect. 5.4. Examples include scaling, rotation in attribute space, and expressing a 2nd-order tensor by its eigenvalues and eigenvectors.

5.1.3 Fields as Example Data Sets

Field data are very common in scientific applications where they express physical quantities defined over continuous domains, for instance, temperatures in a room, or wind velocities in the atmosphere. Such data are often the results of numerical simulations or measurements from experiments. A *field* is defined as a mapping from a continuous *domain* $D \subseteq \mathbb{R}^n$ into an *attribute space* $\mathscr{A} \subseteq \mathbb{R}^m$ (similar notions include *range* and *co-domain*), given as

$$f : \mathbb{R}^n \supseteq D \rightarrow \mathscr{A} \subseteq \mathbb{R}^m. \tag{5.2}$$

Typically, the domain can be considered in a spatiotemporal context, for example, $D = D_s \times I_t \subseteq \mathbb{R}^4$, where $D_s \subseteq \mathbb{R}^3$ is the spatial domain and $I_t \subseteq \mathbb{R}$ is a time interval. Depending on the attribute space, we distinguish a *scalar field* $S : D \rightarrow \mathscr{A} \subseteq \mathbb{R}$, a *vector field* $V : D \rightarrow \mathscr{A} \subseteq \mathbb{R}^2$, a *tensor fields* T, and more generally, a combination of such fields, resulting in a *multifield* with an attribute space spanned by the individual fields.

Ensembles of Fields. Fields are often associated with a set of parameters, which typically play a different role than the domain dimensions. Parameters are often used to create collections of data sets, referred to as *ensembles* [96].

$$\{f_1, \ldots, f_k\} : D \times \{P_1, \ldots, P_k\} \rightarrow \mathscr{A} \subset \mathbb{R}^m, \tag{5.3}$$

where each P_i (for $i = 1 \ldots k$) is a parameter tuple. An example of an ensemble is the data set generated from a computer simulation with different initial conditions (described by different parameters). Each $f_i : D \times P_i \rightarrow \mathscr{A}$ is an *ensemble member* or a *realization*. Ensemble members often have internal correlations or follow certain distributions, making them especially hard to analyze. Ensemble data arise in many applications and are an important theme in visualization research [46].

5.2 Differential Structures

> Science is a differential equation. Religion is a boundary condition.
>
> *Alan Turing, mathematician and computer scientist*

Whereas real data and computations are mostly based on discrete domains and attributes, many of the concepts for their analysis are founded on continuous settings. The machinery of differential arithmetics and differential structures provides powerful analysis tools. *Differential operators* [16, 87] play a crucial part in visualization. They allow the definition and categorization of many features, including *extrema*, *ridges*, *valleys*, *saddles*, and *vortices*. *Differential equations*, for example,

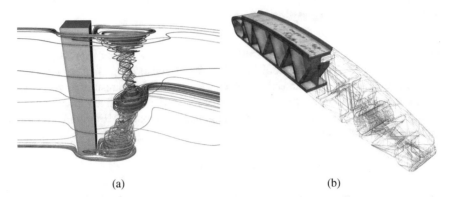

(a)   (b)

Fig. 5.3 Interplay between discrete data and continuous concepts. **a** Numerically computed stream-lines of the flow behind a cylinder approximate the solutions of an ordinary differential equation (Image courtesy Wito Engelke). **b** A discrete mesh approximates the shape of a mechanical part, where a continuous color map highlights the extremal values of the load of the material (Image courtesy Martin Falk)

are the basis for the definition of streamlines, a fundamental method in flow visual-ization; see Fig. 5.3a.

Finally, *differential geometry* provides mathematical tools to characterize curves and surfaces and plays an important role in visualization; see Fig. 5.3b. In this chapter, we summarize the most fundamental concepts of discrete structures that are fre-quently encountered in visualization research.

5.2.1 Differential Operators

Differential Operators in Euclidean Spaces. *Differential operators* [16, 87] map functions (e.g., fields) to their derivatives and thus allow us to study the rates at which continuous attributes change. They can be applied to scalar, vector, and tensor fields. They give rise to definitions of features, such as *extrema*, *ridges*, *valleys*, *saddles*, and *normals* of iso-surfaces. We describe differential operators for scalar fields $f : \mathbb{R}^n \rightarrow \mathbb{R}$ and vector fields $v : \mathbb{R}^n \rightarrow \mathbb{R}^n$. The explicit expression of the operators depends on the inherent metric of the space; here, we assume the Euclidean metric. We often use the operator

$$\nabla = \begin{pmatrix} \frac{\partial}{\partial x_1} \\ \dots \\ \frac{\partial}{\partial x_n} \end{pmatrix} \tag{5.4}$$

to simplify the notations. The *gradient* of a scalar field

$$\nabla f = \begin{pmatrix} \frac{\partial f}{\partial x_1} \\ \cdots \\ \frac{\partial f}{\partial x_n} \end{pmatrix}, \tag{5.5}$$

is a vector that indicates the direction of the steepest ascent. Locations where gradient vanishes ($\nabla f = 0$) are associated with *critical points* of the scalar field, such as maxima, minima, and saddles; see also Sect. 5.7. *Hessian matrices* consisting of 2nd-order partial derivatives are used to classify the critical points,

$$H = \begin{pmatrix} \frac{\partial^2 f}{\partial x_1^2} & \cdots & \frac{\partial f}{\partial x_1 \partial x_n} \\ \cdots & & \cdots \\ \frac{\partial^2 f}{\partial x_n \partial x_1} & \cdots & \frac{\partial^2 f}{\partial x_n^2} \end{pmatrix}. \tag{5.6}$$

The eigenvalues of the Hessian H can be interpreted as the principal curvatures and the eigenvectors as principal directions; therefore, H is often used to define ridges and valley lines in scalar fields. For example, a *topographic ridge* is defined as the set of points where the slope is minimal on the scalar field restricted to a contour line. This means that one eigenvector of H is aligned with the elevation gradient [72].

The *Jacobian* $J \in \mathbb{R}^{n \times n}$ is a matrix that generalizes the concept of a gradient for a vector field v,

$$J = \nabla v = \begin{pmatrix} \frac{\partial v_1}{\partial x_1} & \cdots & \frac{\partial v_n}{\partial x_1} \\ \cdots & & \cdots \\ \frac{\partial v_1}{\partial x_n} & \cdots & \frac{\partial v_n}{\partial x_n} \end{pmatrix}. \tag{5.7}$$

The eigenvalues of the Jacobian can be used to categorize the types of 1st-order *critical points* in vector fields, i.e., positive for sources, negative for sinks, differently signed for saddles, and complex for center points; see Fig. 5.4.

Other important differential operators are the *Laplace operator* $\Delta f = \nabla^2 f = \frac{\partial^2 f}{\partial x^2} + \frac{\partial^2 f}{\partial y^2} + \frac{\partial^2 f}{\partial z^2}$, the *divergence* $\operatorname{div} v = \nabla \cdot v$, and the *curl* $\operatorname{curl} v = \nabla \times v$ of a vector field. In an infinitesimal neighborhood, the divergence is a measure of how much the flow converges toward or repels from a point, and the curl indicates of how much the flow swirls or rotates.

Differential Operators for Field Approximations. Differential operators also play an important role in the approximation of fields as they represent the components in the *Taylor expansion*. A scalar field in the vicinity of a point $P \in \mathbb{R}^n$ can be approximated as $f(P+x) = f(P) + \nabla f(P) \cdot x + \frac{1}{2} x^T H(P) x + O(\|x\|^3)$. For vector fields, the linear approximation is given as $v(P+x) = v(P) + J(P) \cdot x + O(\|x\|^2)$.

Differential Operators in Non-Euclidean Spaces. For non-Euclidean spaces, differential operators are more complex. Consider, for example, spherical coordinates: The divergence of a vector $(v_r, v_\theta, v_\varphi)$ (where r is the radius, θ is the polar angle, and φ is the azimuthal angle) is then given as

$$\operatorname{div} v = \frac{1}{r^2} \frac{\partial\left(r^2 v_r\right)}{\partial r} + \frac{1}{r \sin \theta} \frac{\partial}{\partial \theta} \left(v_\theta \sin \theta\right) + \frac{1}{r \sin \theta} \frac{\partial v_\varphi}{\partial \varphi}. \qquad (5.8)$$

The differential operators for cylinder and spherical coordinates can be found in most textbooks.

5.2.2 Differential Equations

A *differential equation* [1, 80] is a mathematical equation that relates a function with its derivatives. Differential equations are categorized into ordinary differential

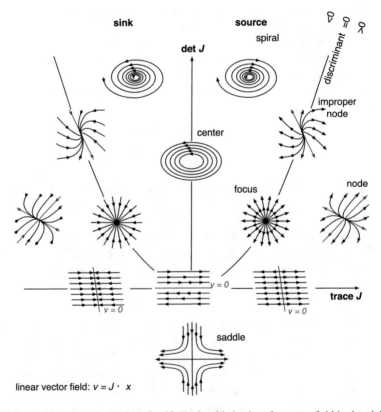

Fig. 5.4 Jacobian J can be used to classify the local behavior of a vector field in the vicinity of a critical point. Locally, the field can be approximated up to 1st order via the Taylor expansion as $v(x) = v_0 + J \cdot x$. If $v_0 = 0$, the point x is critical. The critical point can be classified based on the determinant and the trace of the Jacobian. The sign of the discriminant $\Delta = \operatorname{tr}^2(J) - 4 \det(J)$ separates the area of real and complex eigenvalues of the Jacobian. Complex eigenvalues are associated with swirling motions

equations (containing one independent variable) and partial differential equations (involving two or more independent variables).

One of the most common examples of an ordinary differential equation in visualization is given through the relation of a vector field and its trajectories (that are everywhere tangential to the field); see Fig. 5.3a. A flow can be represented either as a time-dependent vector field $\mathbb{R}^d \times \mathbb{R} \to \mathbb{R}^d$, $(x, t) \mapsto v(x, t)$ or through its *flow map*,

$$\mathbb{R} \times \mathbb{R} \times \mathbb{R}^d \to \mathbb{R}^d, \quad t \times t_0 \times x_0 \mapsto F_{t_0}^t(x_0), \tag{5.9}$$

with $F_{t_0}^{t_0}(x_0) = x_0$, and $\quad F_{t_1}^{t_2}(F_{t_0}^{t_1}(x_0)) = F_{t_0}^{t_2}(x_0)$.

The flow map describes how a flow parcel at (x_0, t_0) moves to $F_{t_0}^{t_1}(x_0)$ in the time interval $[t_0, t_1]$. The two representations of the vector field are related to the initial value problem [15],

$$\dot{F}_{t_0}^t(x_0) = v(x(t), t), \quad F_{t_0}^{t_0}(x_0) = x_0, \tag{5.10}$$

where $\dot{F}$ refers to the temporal derivative of F and inversely through integration,

$$x_0 + \int_{t_0}^t v(x(t), t)dt = F_{t_0}^t(x_0). \tag{5.11}$$

Partial differential equations are more complex than ordinary differential equations and, depending on the initial and boundary conditions [36], may not have a unique solution or a solution at all. As a popular example, we can look at the heat equation,

$$\frac{du(x, t)}{dt} - \alpha \nabla^2 u(x, t) = 0, \tag{5.12}$$

where $\alpha \in \mathbb{R}$ is called the *thermal diffusivity*. The solution to the above heat equation is a Gaussian. It describes the physical problem of heat transfer or diffusion and is used in various visualization applications, for instance, in diffusion-based smoothing, or to define a continuous scale space.

Even if solutions of differential equations exist, for visualization applications, it is rarely possible to derive them analytically, but only numerically [11, 66], due to the reliance on empirical data for coefficients, initial conditions, and boundary conditions. The most popular solvers for ordinary differential equations are the Euler and Runge–Kutta methods. For partial differential equations, the families of finite element methods (FEM), finite volume schemes, and finite differences methods are frequently used, depending on the choice of discretization.

5.2.3   Differential Geometry

We review elements from differential geometry [61] that are most relevant to visualization, including parametrized curves and surfaces, lengths, areas, and curvature. Some of these concepts can be generalized from three-dimensional to higher-dimensional spaces dealing with general manifolds, which are topics in Riemannian geometry [3].

Parametric Curves. In differential geometry, curves are defined in a parametrized form, and their geometric properties, including arc length, curvature, and torsion, are expressed using integrals and derivatives. A *parametric curve*

$$\gamma : [a, b] \subset \mathbb{R} \to \mathbb{R}^d \tag{5.13}$$

is a vector-valued function defined over a non-empty interval. Curves can be distinguished depending on how often they are differentiable. In the continuous case, we will assume the curve to be sufficiently smooth.

The fundamental theorem of differential geometry of curves guarantees that up to transformations of the Euclidean space (rotations, reflections, and translations), a three-dimensional curve can be uniquely defined by its *velocity*, *curvature*, and *torsion*. These three concepts describe changes of the *Frenet–Serret frame*, which is a local coordinate system that moves with the curve. A Frenet–Serret frame is spanned by the unit *tangent* vector $T(t)$, *normal* vector $N(t)$, and *binormal* vector $B(t)$, which are defined via derivatives of the curve $\gamma(t)$ with respect to the parameter $t \in [a, b]$,

$$T(t) = \frac{\gamma'(t)}{\|\gamma'(t)\|},$$
$$N(t) = \frac{\gamma''(t) - \left(\gamma''(t) \cdot T(t)\right) T(t)}{\|\gamma''(t) - (\gamma''(t) \cdot T(t))\|},$$
$$B(t) = T(t) \times N(t).$$

Consequently, commonly used curve descriptors include *velocity* $v(t) = \|\gamma'(t)\|$, *curvature* $\kappa(t) = \|T'(t) \cdot N(t)\| / \|T(t)\|$, and *torsion* $\tau(t) = \|N'(t) \cdot B(t)\| / \|T(t)\|$. Other useful measures are the *arclength* $l(t) = \int_a^t \|\gamma'(s)\| ds$ and the *acceleration* $a(t) = \gamma''(t)$.

Parametric Surfaces. Similar to curves, surfaces can be parametrized; see Fig. 5.5. A parametric surface

$$S : \Omega \subset \mathbb{R}^2 \to \mathbb{R}^n \tag{5.14}$$

is a vector-valued function of a non-empty area. We assume the surface to be sufficiently smooth.

The *tangent plane* of a surface at a point $S(p) \in \mathbb{R}^n$ with $p \in \mathbb{R}^2$ is the union of all tangent vectors of all curves through $S(p)$. The plane is spanned by the two partial derivatives $S_u(p) = \partial S/\partial u$ and $S_v(p) = \partial S/\partial v$. The *surface normal*, perpendicular to the tangent plane, is given by the cross product of the partial derivatives,

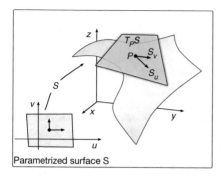

Parametrized surface S

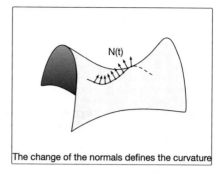

The change of the normals defines the curvature

Fig. 5.5 Left: parametrized surface. Right: The changes of the normals in a certain direction define the normal curvature of the surface

$$N(p) = \frac{S_u(p) \times S_v(p)}{\|S_u(p) \times S_v(p)\|}. \tag{5.15}$$

Measurements on Surfaces. The calculation of the length of a curve on a surface or the surface area can be easily formulated using the *first fundamental form* $\mathrm{I}(p)$: $\mathbb{R}^2 \to \mathbb{R}^{2 \times 2}$. $\mathrm{I}(p)$ defines a natural local metric induced by the Euclidean metric in $\mathbb{R}^n$. For notational simplicity, we omit the dependence of the location $p \in \mathbb{R}^2$. Its components g_{uv} are defined as the scalar product of the tangent vectors $S_u \cdot S_v$. In matrix form, the first fundamental form is given as,

$$\mathrm{I} = \begin{pmatrix} g_{uu} & g_{uv} \\ g_{vu} & g_{vv} \end{pmatrix} = \begin{pmatrix} E & F \\ F & G \end{pmatrix}. \tag{5.16}$$

Using the first fundamental form, a line element ds on the surface is expressed as $ds^2 = E\,du^2 + 2F\,du\,dv + G\,dv^2$ and an area element as $dA = \|S_u \times S_v\|\,du\,dv = \sqrt{EG - F^2}\,du\,dv$. The arclength of a curve on the surface results from integrating the line element $l = \int_a^b ds$, and the area of a surface patch results from integrating the area element.

Surface Curvature. Many different curvature measures are available. Loosely speaking, curvature is a concept that measures the amount by which a surface deviates from a plane or the variation of the surface normal. Central to the concept of curvature is the *Gauss map*, which maps the surface normals to the unit sphere $N : S \to S^2$. The differential of the Gauss map in a certain direction is a measurement of curvature in that direction. Mathematically, the curvature is summarized in the *second fundamental form*, denoted as II. In matrix form, it is given as,

$$\mathrm{II} = \begin{pmatrix} S_{uu} \cdot N & S_{uv} \cdot N \\ S_{vu} \cdot N & S_{vv} \cdot N \end{pmatrix} = \begin{pmatrix} e & f \\ f & g \end{pmatrix}, \tag{5.17}$$

where S_{uv}, S_{uu}, S_{vv} are the respective second derivatives of the surface parametrization. The *shape operator* expresses the curvature in *local coordinates*,

$$S = \frac{1}{EG - F^2} \begin{pmatrix} eG - fF & fG - gF \\ fE - eF & gE - fF \end{pmatrix}. \tag{5.18}$$

Its eigenvalues (k_1 and k_2) are called the *principal curvatures* at a given point; its eigenvectors are called the *principal directions*. The *Gaussian curvature K* is equal to the product of the principal curvatures. It can also be calculated as the ratio of the determinants of the second and first fundamental forms. The *mean curvature H* is defined as the average of the principal curvatures:

$$K = k_1 \cdot k_2 = \frac{eg - f^2}{EG - F^2}, \qquad H = \frac{1}{2}(k_1 + k_2) = \frac{1}{2} \frac{eG + gE - 2fF}{EG - F^2}.$$

Points on the surface can be categorized as *elliptic* ($K > 0$), *parabolic* ($K = 0$, $H \neq 0$), *hyperbolic* ($K < 0$), and *flat* ($K = H = 0$) using the Gaussian and mean curvatures.

The curvature κ of a surface curve γ can be decomposed into its *normal curvature* k_n normal to the surface and its *geodesic curvature* k_g, which measures the deviation of a curve from being a geodesic $\kappa^2 = k_n^2 + k_g^2$. The extrema of the normal curvature over all curves through a point correspond to the principal curvatures k_1 and k_2 of the surface. A curve where the geodesic curvature is equal to zero is called a *geodesic*, which is a generalization of a straight line on arbitrary surfaces, as the straightest and locally shortest curve.

5.2.4 Manifolds

Roughly speaking, an *n-manifold M* embedded in $\mathbb{R}^m$ is a space that is locally similar to Euclidean space $\mathbb{R}^n$. Formally, each point p of the manifold M has an open neighborhood $U_p \subset M$ that is homeomorphic to an open subset V of the Euclidean space described by a chart or local frame $\varphi : U_p \subset M \rightarrow V_p \subset \mathbb{R}^n$. The entire manifold can be described by a collection of compatible charts, which together form an atlas.

A well-known example is a sphere, which is a 2-manifold embedded in $\mathbb{R}^3$, defined by the condition $x^2 + y^2 + z^2 = R$ (R being the radius). There are many ways to define charts on the sphere. It is also possible to cover the whole sphere excluding one point with a chart, which requires at least two charts to complete the atlas. Covering a sphere with one chart, however, is not possible.

Similar to surfaces, one can define a tangent space $T_p M$ attached to every point in M. $T_p M$ has the same dimension as the manifold. The tangent space defines a local basis on the manifold and plays an important role since many fields (e.g., vector fields) live in the tangent space of the domain; see Fig. 5.6.

5.3 Sampled Data and Discrete Methods

> The world is continuous, but the mind is discrete.

> *David Bryant Mumford, mathematician*

Fields are defined over continuous domains in theory; however, they are described at discretely sampled locations in practice. Typical analysis and visualization methods rely on a *reconstruction* of the continuous fields. Two different approaches are commonly used to deal with this issue. First, the discrete data are interpolated to fill the entire domain. Second, the analysis techniques are transferred to the discrete setting.

5.3.1 Data Representation

Sampled data come in many different forms and representations depending on their origin. For measurement data, one often deals with *unstructured point clouds* resulting from practical constraints, e.g., possible placements for sensors. Data coming from simulations are mostly based on grid structures, ranging from *uniform grids* to *unstructured* and *hybrid grids*. Therefore, the attributes are assigned to either the grid vertices, the grid cells, or distinguished points inside the cells, e.g., *Gauss* or *integration points* coming from finite element simulations; see Fig. 5.7. An overview of common data representations can be found in [92].

A grid is built from a set of vertices V and neighborhood relations, defining edges, faces, and cells. The neighborhood relations can be given explicitly for unstructured grids or implicitly encoded in an index structure. An example is a quad mesh where the vertices are identified by three indices $V = \{v_{ijk} \mid 1 \leq i, j, l \leq n\}$ and edges $E = \{(v_{ijk}, v_{i+1jk}), (v_{ijk}, v_{ij+1k}), (v_{ijk}, v_{ijk+1}) \mid \forall v_{ijk} \in V\}$. The most common

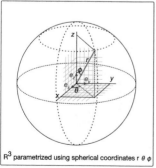 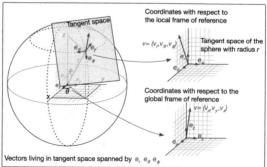

Fig. 5.6 Vector field defined on a sphere given in spherical coordinates. Left: A parametrization of the sphere with spherical coordinates, Right: The vector field can be expressed in a local reference frame, which depends on the spherical coordinates

two-dimensional cells are triangles and rectangles; *three*-dimensional cells include quads, tetrahedra, and prisms.

5.3.2 Simplicial Complexes

Simplicial complexes are data structures that are particularly useful for combinatorial algorithms (see Sect. 5.7). They can be considered as a formal generalization of triangulations to higher dimensions. A k-simplex is defined as the convex hull of $k + 1$ affinely independent points $p_i \in \mathbb{R}^k$; the convex hull of any non-empty subset of the $k + 1$ points is a *face* of the simplex. 0-, 1-, 2-, and 3-simplices are vertices, edges, triangles, and tetrahedra, respectively.

A *simplicial complex K* is a set of simplices such that every face of a simplex from K is also in K, and the intersections of two simplices in K are either empty or a face of both simplices; see Fig. 5.8. A more detailed discussion can be found in [22, 67]. A simplicial complex is a type of *cell complex* in which the cells are simplices. There are several different ways to formalize and instantiate the notion of a cell complex, including CW complex, Δ-complex, cube complex, polytopal complex, etc.; see Hatcher [47] for an introduction.

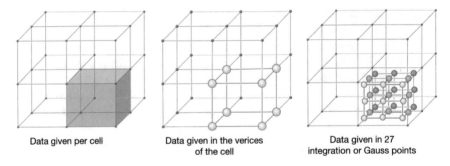

| Data given per cell | Data given in the verices of the cell | Data given in 27 integration or Gauss points |

Fig. 5.7 Data can be assigned to a regular cubic grid in many different ways

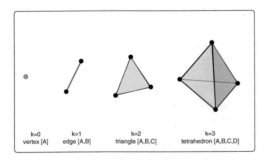 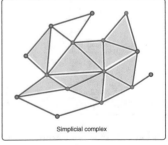

| k=0 vertex [A] | k=1 edge [A,B] | k=2 triangle [A,B,C] | k=3 tetrahedron [A,B,C,D] | Simplicial complex |

Fig. 5.8 Left: 0-, 1-, 2-, and 3-simplex, respectively. Right: A simplicial complex embedded in $\mathbb{R}^2$

5.3.3 Neighborhood Graphs

Neighborhood graphs impose combinatorial structures on point clouds that capture a certain notion of proximity. Such structures give rise to the use of grid-based analysis methods but are also of interest for clustering algorithms and many discrete theories. The most fundamental neighborhood structure is the Delaunay triangulation of a point cloud. Given a finite set of points $P = \{p_i\}_{i=1}^m \subset \mathbb{R}^n$, the *Voronoi diagram* is defined as a decomposition of the domain in regions V_i assigned to each point $p_i \in P$. V_i contains all points in $\mathbb{R}^n$ that are at least as close to p_i as to any other point in P. The dual structure of the Voronoi diagram in the plane is the *Delaunay triangulation* and in three dimensions the *Delaunay tetrahedralization*. The Delaunay triangulation maximizes the minimum angle in a triangulation and gives rise to a reasonably nice triangulation. The concept extends to higher dimensions, but its computation becomes very costly. Many other neighborhood graphs have been studied with respect to geometric properties and robustness. Examples include the Gabriel graph [38] and the k-nearest neighbors graph. A more detailed discussion about such graphs can be found in textbooks on computational geometry [2]. Neighborhood graphs in the context of high-dimensional and sparse data in visualization applications are also discussed in [18]. There is a large body of work related to meshing that is also relevant to this context [106].

5.3.4 Reconstruction and Interpolation

The goal of a *reconstruction* is to recover an approximate version of a continuous function from a sampled data set. A reconstruction that matches the values in the sampled points exactly is called *interpolation*.

Given a set of points (vertices or nodes) $P = \{p_i\}_{i=1}^m$ with $p_i \neq p_j$ for $i \neq j$ and a set of associated values $\{f_i \in \mathbb{R}\}_{i=1}^m$, a function $f : \mathbb{R}^n \to \mathbb{R}$ is called *interpolating function* for the set of points if it fulfills the interpolation condition $f(p_i) = f_i$, for $1 \leq i \leq m$.

Infinitely, many possibilities are available to interpolate a set of points. The choice of a specific interpolation is often guided by simplicity and efficiency. It is important to be aware that different interpolation schemes may have significant impact on the computation and visualization results. The most common interpolation methods for gridded data are piecewise linear, bilinear, and trilinear interpolations. For scattered data, one typically constructs a grid or uses radial basis functions [7].

5.3.5   Discrete Theories

Discrete theories typically inherit structural properties from the smooth setting and come with theoretical understandings about the preservation of relevant invariants. In general, they satisfy a subset of properties from the smooth setting, resulting in a large diversity of discrete theories [97]. For example, in the discrete setting, a geodesic defined as a locally shortest connection is not equivalent to the straightest connection, as in the continuous setting [73].

In visualization, the most important examples arise from combinatorial differential topology and geometry. For instance, discrete exterior calculus provides discrete differential operators [19]; discrete differential geometry introduces concepts for curvatures and geodesics [20]. A very useful and popular discrete theory is discrete Morse theory [37], which forms the base of many current algorithms for the extraction of the Morse–Smale complex; see also Sect. 5.7.

5.4   Symmetries, Invariances, and Features

> Symmetry is a vast subject, significant in art and nature. Mathematics lies at its root, and it would be hard to find a better one on which to demonstrate the working of the mathematical intellect.
>
> *Hermann Weyl, mathematician and theoretical physicist [98]*

Symmetries, invariances, and conserved quantities are closely related concepts that play an important role in many mathematical and physical theories, for instance, Noether's theorem links symmetries of physical spaces with conservation properties [84]. Invariants are properties of an object (a system or a data set) that remain unchanged when certain transformations (such as rotations or permutations) are applied to the object. In visualization, invariants play a central role for feature definition and pattern recognition. For example, the number of legs of a three-dimensional animal model is invariant with respect to changes due to animal movement or shape morphing. Another example is the Galilean invariance for flow features, e.g., vorticity does not change under certain changes of the reference frame even though the flow components change [71]. There are also topological invariants which characterize spaces with respect to smooth deformations [47]. A formal analysis of the symmetries that arise from group actions, with a strong emphasis on the geometry, Lie groups, and Lie algebra can be found in textbooks dealing with representation theory and invariant theory [42].

5.4.1 Features, Traits, and Properties

According to the Cambridge Dictionary, a feature is "a typical quality or an important part of something". In the visualization literature, the term feature is not well-defined and oftentimes an overloaded concept. Features often represent structures in a data set that are meaningful within some domain-specific context. They can be used as the basis for abstract visualization. Here, we define a feature $F(\mathscr{D}) \subset \mathscr{I}$ of a data set $\mathscr{D}$ as a subset of data items having a specific property; see Sect. 5.1. For field data, features are typically defined as certain subsets in the spatial domain. Typical features of a scalar field $s : D \rightarrow \mathbb{R}$ are iso-surfaces $s^{-1}(a)$ (for $a \in \mathbb{R}$) and the set of critical points of s.

In many cases, features can be locally defined by *traits* $T \subset \mathscr{A}^*$, subsets of the enriched attribute space $\mathscr{A}^*$ containing the data attributes and possibly derived quantities. Specifically, given a field $f : D \rightarrow \mathscr{A}^*$ that maps a domain D into an enriched attribute space $\mathscr{A}^*$, a *trait-induced feature* is defined to be $F_T(\mathscr{D}) = f^{-1}(T) = \{x \in \mathscr{I} | f(x) \in T\}$, for some $T \subset \mathscr{A}^*$ [53]. A point trait $T = \{p\} \in \mathscr{A}^* = \mathbb{R}$ gives rise to a trait-induced feature known as an *iso-surface*. A point trait is also referred to as a *feature descriptor*. If $\mathscr{A}^*$ encodes the derivatives of f, then the set of critical points is a trait-induced feature given by all points where the derivative of the scalar function is equal to 0. A line trait is a line in $\mathscr{A}^* = \mathbb{R}^2$ spanned by the scalar values and its derivatives. It is desirable for a descriptor to be invariant with respect to changes (e.g., rotations and scalings) to the data representation.

Other types of features based on structures of the data, such as cycles in a graph, may not be described by traits naturally. Such features are referred to as *structure-induced features*. In general, features can be defined by any combination of attribute and structural constraints.

5.4.2 Transformations, Symmetries, and Invariances

Invariants are directly linked to transformations T describing an inherent *symmetry* of the system. A *transformation* is a function that maps a set X to itself, i.e., $T : X \rightarrow X$. In the context of visualization, a transformation concerning the structure $\mathscr{S}$ is called the *inner transformation*; a transformation of the attribute space $\mathscr{A}^*$ is called the *outer transformation*. A transformation can be both an inner and an outer transformation. The notion of invariance and transformation can also be extended to changes in the model used to create the visualization, or the image itself [58].

When talking about invariants, we are interested not only in one specific transformation but also in certain classes of transformation described as transformation groups [42]. A transformation group acting on a set X is defined as a group G with neutral element e and an action

$$T : G \times X \rightarrow X,$$

where each group element $g \in G$ defines a transformation T_g as $T_g(x) \equiv T(g, x)$ with the following properties: for all $x \in X$ and all $g, h \in G$, $T_e(x) = x$, and $T_g(T_h(x)) = T_{gh}(x)$.

A *symmetry group* is a group that conserves a certain structure, property, or feature. It gives a unique relation between symmetries and invariants. Formally, let $T : \mathscr{D} \to \mathscr{D}$ be a transformation (short for $T : \mathscr{I} \times \mathscr{A}^* \to \mathscr{I} \times \mathscr{A}^*$), and $F(\mathscr{D})$ be a feature of a data set $\mathscr{D}$. Then, we say that T is a *symmetry* of $\mathscr{D}$ if $F(\mathscr{D})$ commutes with the transformation T

$$T(F(\mathscr{D})) = F(T(\mathscr{D})).$$

Typical transformations for field data are rotations in three-dimensional Euclidean space that form the group SO_3 acting on $\mathbb{R}^3$. An application is the definition of invariant moments as descriptors of flow patterns [8]. An example that plays an important role in flow visualization is the Galilean transformation, which transforms coordinates between two reference frames that differ only by constant relative motion [57]. Domain-specific invariants like shear stress or anisotropy also play a central role in tensor field visualization [60]. An example of discrete data is the permutation group $Sym(M)$ whose elements are permutations of a set M.

5.5 Cluster Analysis

> The Milky Way is nothing else but a mass of innumerable stars planted together in clusters.
>
> ———————————————————
> *Galileo Galilei, astronomer, physicist and engineer*

A frequently employed approach in visualization and exploratory analysis is cluster analysis or clustering, i.e., to assign a set of objects to groups in a manner such that objects in the same group are more similar to each other in some manner than to those in other groups. In other words, data are decomposed into a set of classes that in some sense reflect the distribution of the data.

To achieve this general goal, a very large variety of algorithms have been presented for specific problems or data modalities [33, 51]; they differ significantly in how they define and identify clusters. Clustering results are typically subject to various parameters, and it is often necessary to modify (e.g., transform) input data and choose parameters to obtain a result with desired properties. We describe four clustering techniques that are frequently applied in data analysis and visualization and illustrate how they have been used to address various visualization problems.

k-**Means Clustering**. Given a set of data $(x_1, \ldots, x_n)$ where each x_i is a d-dimensional real vector, k-means clustering (also called *Lloyd's algorithm*) seeks to partition the data into $k \leq n$ disjoint sets $C = \{C_1, \ldots, C_k\}$ (with a fixed k) such that the variance within each cluster is minimized, i.e., to find

$$\arg\min_{C} \sum_{i=1}^{k} \sum_{x \in C_i} \|x - m_i\|^2$$

where m_i is the mean of data in C_i. The result depends centrally on the chosen metric, for which the Euclidean norm is often selected. Algorithmically, C_i can be found iteratively in a manner similar to computing a centroidal Voronoi tessellation [21]: Given an initial set of cluster centers m_i, assign to each cluster the data points that are closer to m_i than to all other cluster centers. Compute a new set of means as cluster centers from the assigned points, and repeat the process until convergence. Initially, the data centers can either be chosen randomly or according to heuristics [13].

k-means clustering was used in visualization, for example, by Woodring and Shen [102], who employed it to automatically generate transfer functions for volume rendering temporal data. They achieved this by identifying clusters of data points that behave similarly over time. k-means clustering is relatively easy to understand and utilize. However, a major drawback of this approach is that the number of classes or clusters k must be specified a priori.

Spectral Clustering. Clustering is not directly applied on the data, but rather on the *similarity matrix* S (where $S_{ij} = \|x_i - x_j\|$) that contains pairwise distances between individual data items. Clustering is then performed on the eigenvectors of S. Intuitively, S can be viewed as describing a mass–spring system. Masses coupled through tight springs will largely move together relative to the equilibrium of such a system, and thus eigenvectors of small eigenvalues of S can be seen to form a suitable partition of the data.

As with clustering in general, many incarnations of this basic idea have been given. The *normalized cuts* technique is a nonparametric clustering approach often used in image segmentation [85]. For visualization purposes, it was utilized by Ip et al. to explore feature segmentation of three-dimensional intensity fields [50], and by Brun et al. to visualize white matter fiber traces in DT-MRI data [6].

Density-Based Clustering. The DBSCAN (density-based spatial clustering of applications with noise) algorithm is a widely used general-purpose clustering scheme [32, 83]. It considers the density of data points in their embedding space and subdivides them into three types. A point x_i is a *core points* if at least m points lie within a distance of δ from x_i; these points are called *directly reachable* from x_i. Both m and δ are parameters. An arbitrary point x_j is *reachable* from x_i if there is a path $x_i, x_k, \ldots, x_j$ such that each point in the path is directly reachable from its predecessor. Points that are not reachable from any core point are called *outliers*. Clusters are formed by core points and the points that are reachable from them. (There may be multiple core points in a cluster.) Due to the non-symmetric reachability relations, DBSCAN uses the notion of density connectedness for a pair x_i and x_j. That is, points x_i and x_j are connected if there is a third point x_l from which both x_i and x_j are reachable.

DBSCAN is relatively easy to implement and has a good runtime properties, but many variants of the basic technique exist that differ in various details [83, 91]. Wu et al. used DBSCAN to provide level-of-detail in visualization and exploration of academic career path [103].

Mean Shift. A *mean shift* procedure is a variant of density-based clustering; it is applied to identify the maxima (or *modes*) of a density function from discrete samples. Fixing a kernel function $K(x_i - x)$ (typically flat or Gaussian) and a point x in the embedding space, the weighted mean in a window around x is

$$m(x) = \frac{\sum_{x_i} K(x_i - x) x_i}{\sum_{x_i} K(x_i - x)}.$$

The mean shift $m(x) - x$ is then minimized by setting $x \leftarrow m(x)$ and iterating until convergence. Data points x_j are grouped into clusters according to the mode to which the mean shift converges if initialized with x_j. This process yields a general-purpose clustering technique that does not incorporate assumptions about the data and relies on a single parameter, the kernel bandwidth. In visualization, a good example of the usefulness of this algorithm is given by Böttger et al. [4], who use mean shift clustering to achieve edge bundling in brain functional connectivity graphs.

5.6 Statistics for Visualization

> If the statistics are boring, you've got the wrong numbers.
>
> *Edward R. Tufte, statistician [93]*

Statistics deals with the collection, description, analysis, and interpretation of (data) *populations*. *Descriptive statistics* are used to summarize population data. *Moments*, also called *summary statistics*, are a statistical notion to describe the shape of a function (distribution). Mathematically, the n-th *central moment* of a real-valued continuous function $f(x)$ of a real variable is given by

$$\mu_n = \int_{-\infty}^{\infty} (x - c)^n f(x) dx,$$

where c is the mean of $f(x)$. The first moment corresponds to the mean, and a usual assumption considers $c = 0$. These moments give rise to the usual statistical descriptors of a distribution such as variance ($n = 2$), skewness ($n = 3$), and kurtosis ($n = 4$). Potter et al. provide guidance on the visualization of functions via their summary statistics [75]. For multiple variables, the concept of moments can be generalized to *mixed moments*. Applications in visualization include pattern matching for feature extraction [9].

A frequent problem in comparative visualization is comparing distributions. Here, the *covariance* of two distributions

$$\text{cov}(f, g) = E[f - E[f]] \, E[g - E[g]]$$

signifies their joint variability. In the multivariate case, covariance can be generalized to the *covariance matrix*. Covariance matrices have been frequently used in
visualization, for example in glyph-based [74] or feature-based visualization [101].

Furthermore, *correlation* of functions may be used for comparison. In the broadest
sense, correlation is any statistical association between data populations; in practice,
correlation is usually used to indicate a linear relationship between functions. An
commonly used concept is the *Pearson's correlation coefficient*,

$$\rho_{f,g} = \text{corr}(f, g) = \frac{\text{cov}(f, g)}{\sigma_f \sigma_g},$$

where σ_f and σ_g refer to the standard deviation of f and g, respectively. $\rho_{f,g} \in
(0, 1]$ if f and g are positively correlated; $\rho_{f,g} \in [-1, 0)$ if f and g are negatively
correlated; $\rho_{f,g} = 0$ if f and g have no linear correlation. Finding correlations among
data is one of the most essential tasks in many scientific problems, and visualization
can be very helpful during such a process [14, 43].

Order statistics, on the other hand, characterizes a population in terms of ordering
and allows us to make statistical statements about the distribution of its values. For
example, the q-percentile ($0 \leq q \leq 100$) denotes the value below which q percent
of the samples are located. Order statistics can be easily combined with descriptive
statistics in the univariate case [75]. Higher-dimensional variants of these notions are
also available and used to represent data visually [77]. An interesting generalization
of order statistics to a widely used topological structure is the contour boxplot [99].

5.7   Topological Data Analysis

> If you can put it on a necklace, it has a one-dimensional hole.
> If you can fill it with toothpaste, it has a two-dimensional
> hole. For holes of higher dimensions, you are on your own.
>
> *Evelyn Lamb, math and science writer [62]*

For topology in visualization, two key developments from computational topology
play an essential role in connecting mathematical theories to practice: first, separat-
ing features from noise using *persistent homology*; second, abstracting topological
summaries of data using *topological structures* such as Reeb graphs, Morse–Small
complexes, Jacobi sets, and their variants.

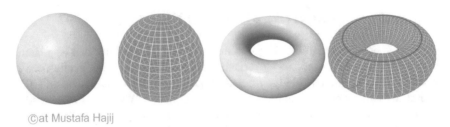

©at Mustafa Hajij

Fig. 5.9 Betti numbers for the sphere and the torus. $\beta_0 = 1$, $\beta_1 = 0$, and $\beta_2 = 1$ for the sphere (left) and $\beta_0 = 1$, $\beta_1 = 2$, and $\beta_2 = 1$ for the torus (right). Image courtesy of Mustafa Hajij

5.7.1 Topology, Homology, and Betti Numbers

Topology has been one of the most exciting research fields in modern mathematics [52]. It is concerned with the properties of space that are preserved under continuous deformations, such as stretching, crumpling, and bending, but not tearing or gluing [100].

The beginning of topology was arguably marked by Leonhard Euler, who published a paper in 1736 that solved the now famous Königsberg bridge problem. In the paper, titled *"The Solution of a Problem Relating to the Geometry of Position"*, Euler was dealing with "a different type of geometry where distance was not relevant" [70]. Johann Benedict Listing was credited as the first to use the word "topology" in print based on his 1847 work titled *"Introductory Studies in Topology"*; although many of Listing's topological ideas were borrowed from Carl Friedrich Gauss [70]. Both Listing and Bernhard Riemann studied the *components* and *connectivity* of surfaces. Listing examined connectivity in *three*-dimensional Euclidean space, and Enrico Betti extended the idea to n dimensions. Henri Poincaré then gave a rigorous basis to the idea of connectivity in a series of papers *"Analysis situs"* in 1895. He introduced the concept of *homology* and improved upon the precise definition of Betti numbers of a space [70]. In other words, it was Poincaré who "gave topology wings" [52] via the notion of homology.

The original motivation to define homology was that it can be used to tell two objects (a.k.a. topological spaces) apart by examining their holes. This process associates a topological space $\mathbb{X}$ with a sequence of abelian groups called homology groups $H(\mathbb{X})$, which, roughly speaking, count and collate *holes* in a space [40]. Informally, homology groups generalize a commonsense notion of connectivity. They detect and describe the connected components (*zero*-dimensional holes), tunnels (*one*-dimensional holes), voids (*two*-dimensional holes), and holes of higher dimensions in the space. The p-th Betti number β_p is the rank of the p-th homology group of $\mathbb{X}$, $H_p(\mathbb{X})$, and captures the number of p-dimensional holes of a topological space. For instance, a sphere contains no tunnels but a void, and a torus contains two tunnels (see Fig. 5.9).

5.7.2   From Homology to Persistent Homology

For simplicity, we work with data represented by simplicial complexes denoted by $\mathbb{X}$. In algebraic terms, the construction of homology groups begins with a chain complex $\mathsf{C}(\mathbb{X})$ that encodes information about $\mathbb{X}$, which is a sequence of abelian groups $\mathsf{C}_0(\mathbb{X}), \mathsf{C}_1(\mathbb{X}), \ldots$ connected by homomorphisms known as the boundary operators $\partial_p : \mathsf{C}_p(\mathbb{X}) \to \mathsf{C}_{p-1}(\mathbb{X})$. The *p-th homology group* is defined as $\mathsf{H}_p(\mathbb{X}) = \ker(\partial_p)/\mathrm{im}(\partial_{p+1})$. The *p-th Betti number* is the rank of this group, $\beta_p = \mathrm{rank}\ \mathsf{H}_p$, see [67] for an introduction.

Persistent homology transforms the algebraic concept of homology into a multi-scale notion by constructing an extended series of homology groups. In its simplest form, persistent homology applies a homology functor to a sequence of topological spaces connected by inclusions, called a *filtration*. Consider a finite sequence of simplicial complexes connected by inclusions $f_p^{i,j} : \mathbb{X}_i \hookrightarrow \mathbb{X}_j$,

$$\emptyset = \mathbb{X}_0 \hookrightarrow \mathbb{X}_1 \hookrightarrow \cdots \hookrightarrow \mathbb{X}_n = \mathbb{X}.$$

Applying *p*-th homology to this sequence results in a sequence of homology groups connected from left to right by homomorphisms induced by the inclusions,

$$0 = \mathsf{H}_p(\mathbb{X}_0) \to \mathsf{H}_p(\mathbb{X}_1) \to \cdots \to \mathsf{H}_p(\mathbb{X}_n) = \mathsf{H}_p(\mathbb{X})$$

for each dimension p. The *p-th persistent homology group* is the image of the homomorphism induced by inclusion, $\mathsf{H}_p^{i,j} = \mathrm{im}\, f_p^{i,j}$ for $0 \le i \le j \le n$. The corresponding *p-th persistent Betti number* is the rank of this group, $\beta_p^{i,j} = \mathrm{rank}\ \mathsf{H}_p^{i,j}$ [24, Page 151]. As the index increases, the rank of the homology groups changes. When the rank increases (i.e., $f_p^{i-1,i}$ is not surjective), we call this a *birth* event at $\mathbb{X}_i$; when the rank decreases (i.e., $f_p^{j-1,j}$ is not injective), we call this a *death* event at $\mathbb{X}_j$. Persistent homology pairs the birth and the death events as a multi-set of points in the plane called the *persistence diagrams* [29]; see [30, 31] for a comprehensive mathematical introduction. A celebrated theorem of persistent homology is the *stability* of persistence diagrams [17], that is, small changes in the data lead to small changes in the corresponding diagrams, making it suitable for robust data analysis. See Fig. 5.10 for an example in $\mathbb{R}^2$. Given a set of points in $\mathbb{R}^2$, we compute its persistent homology by studying the union of balls centered around the points as the radius increases. Here, a green component is born at time 0 and dies when it merges with a red component at time 2.5, resulting a point $(0, 2.5)$ in the persistence diagram. A tunnel is born at time 4.2 and dies at time 5.6, giving rise to a point $(4.2, 5.6)$ in the persistence diagram.

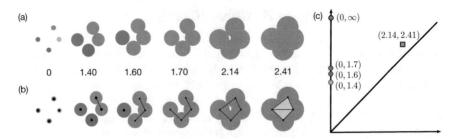

Fig. 5.10 Computing persistent homology of a point cloud in $\mathbb{R}^2$. **a** A nested sequence of topological spaces formed by unions of balls at increasing parameter values. **b** A filtration of simplicial complexes that capture the same topological information as in **b**. **c** 0- (circles) and 1-dimensional (squares) features in a persistence diagram

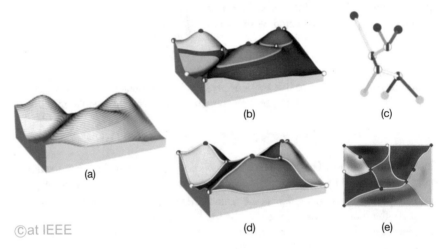

Fig. 5.11 Contour-based (**c**) and gradient-based (**e**) topological structures of a *two*-dimensional scalar function (**a**)

5.7.3 Topological Structures

Several techniques in topological data analysis and visualization construct topological structures from well-behaved functions on point clouds as summaries of data. On the one hand, the well-behaveness is formalized with the Morse theory. On the other hand, such topological structures can be roughly classified into two types: contour-based (Reeb graphs [79], Reeb spaces [27], contour trees [12], and merge trees) and gradient-based topological structures (Morse–Smale complexes [25, 28] and Jacobi sets [23]); see Fig. 5.11. All such topological structures provide meaningful abstractions of (potentially high-dimensional) data, reduce the amount of data needed to be processed or stored, utilize sophisticated hierarchical representations that capture features at multiple scales, and enable progressive simplifications [63].

Morse Function. Let $\mathbb{M}$ be a smooth, compact, and orientable d-manifold without boundary ($d \geq 2$). Suppose $\mathbb{M}$ is equipped with a Riemannian metric so that gradients are well-defined. Given a smooth function $f : \mathbb{M} \to \mathbb{R}$, a point $x \in \mathbb{M}$ is called a *critical point* if the gradient of f at x equals zero, that is, $\nabla f(x) = 0$, and the value of f at x is called a *critical value*. All other points are *regular points* with their function values being *regular values*. A critical point is *non-degenerate* if the Hessian, i.e., the matrix of second partial derivatives at the point, is invertible. A smooth function f is a *Morse function* if (a) all its critical points are non-degenerate; (b) all its critical values are distinct [24, Page 128]. A pair of two Morse functions is *generic* if their critical points do not overlap.

Morse–Smale Complexes. Given a Morse function $f : \mathbb{M} \to \mathbb{R}$, at any regular point x the gradient is well-defined and integrating it in both directions traces out an integral line, $\gamma : \mathbb{R} \to \mathbb{M}$, which is a maximal path whose tangent vectors agree with the gradient [28]. Each integral line begins and ends at critical points of f. The *ascending/descending manifolds* of a critical point x are defined as all the points whose integral lines start/end at x. The descending manifolds form a complex called a *Morse complex* of f, and the ascending manifolds define the Morse complex of $-f$. The set of intersections of ascending and descending manifolds creates the *Morse–Smale* complex of f. Each *cell* of the Morse–Smale complex is a union of integral lines that all share the same origin and the same destination. In other words, all the points inside a single cell have uniform gradient flow behavior. These cells yield a decomposition into monotonic, non-overlapping regions of the domain, as shown in Fig. 5.11b for a *two*-dimensional height function.

Jacobi Set for a Pair of Morse Functions. Given a generic pair of Morse functions, $f, g : \mathbb{M} \to \mathbb{R}$, their Jacobi set $\mathbb{J} = \mathbb{J}(f, g) = \mathbb{J}(g, f)$ is the set of points where their gradients are parallel or zero [23]. That is, for some $\lambda \in \mathbb{R}$,

$$\mathbb{J} = \{x \in \mathbb{M} \mid \nabla f(x) + \lambda \nabla g(x) = 0 \text{ or } \nabla g(x) + \lambda \nabla f(x) = 0\}. \quad (5.19)$$

The sign of λ for each x is called its *alignment*, as it defines whether the two gradients are aligned or anti-aligned. By definition, the Jacobi set contains the critical points of both f and g.

There exist several other descriptions of Jacobi sets [23, 26, 69]. One particularly useful description is in terms of the *comparison measure*, κ [26], which is a gradient-based metric to compare two functions. It plays a significant role in assigning an important value to subsets of a Jacobi set in terms of the underlying functions f and g by measuring the relative orientation of their gradients.

Reeb Graphs and Contour Trees. Let $f : \mathbb{X} \to \mathbb{R}^d$ be a generic, continuous mapping. Two points $x, y \in \mathbb{X}$ are *equivalent*, denoted by $x \sim y$, if $f(x) = f(y)$ and x and y belong to the same path-connected component of the pre-image of f, $f^{-1}(f(x)) = f^{-1}(f(y))$. The *Reeb space*, $\mathscr{R}(X, f) = \mathbb{X}/\sim$, is the quotient space contained by identifying equivalent points together with the quotient topology inherited from $\mathbb{X}$. A powerful analysis tool, the *Reeb graph*, is a special case when $d = 1$.

The Reeb graph of a real-valued function $f : \mathbb{X} \to \mathbb{R}$ describes the connectivity of its level sets. A *contour tree* is a special case of the Reeb graph if the domain

$\mathbb{X}$ is simply connected; see Fig. 5.11c. A *merge tree* is similar to the Reeb graphs and contour trees except that it describes the connectivity of sublevel sets rather than level sets. The Reeb graph stores information regarding the number of components at any function value as well as how these components split and merge as the function value changes. Such an abstraction offers a global summary of the topology of the level sets and connects naturally with visualization.

5.8   Color Spaces

> Although many great thinkers have held that an analytical or mathematical treatment of the subject is impossible or even undesirable, they have gradually deserted the field so that today and indeed throughout the past 50 years it has been generally recognized that a theory of color perception must be, both in form and content, a mathematical theory.
>
> *Howard L. Resnikoff, mathematician and business executive [81]*

Color is one of the central aspects of visualization and against common belief, a surprisingly mathematical one. Operations on color are an important aspect in many applications, e.g., color mapping, re-sampling of color images or movies, and image manipulations, such as stitching, morphing, or contrast adaption. These operations can be expressed through mathematical formulae if the colors themselves can be expressed as elements of mathematical space, in which certain concepts such as sums or distances have a meaning. However, as we will see, this is not easy.

The space of all colors is in principal infinite-dimensional because any function over the frequencies of the visible spectrum forms a color. Since, however, the human eye has only three receptors for color, the space of distinguishable colors for humans is only three-dimensional [44, 95]. Depending on the choice of the three basis dimensions, many different colorspaces were developed. In displays, the basic colors are usually red, green, and blue (RGB) and for printing, the standard is cyan, magenta, yellow, and key black (CMYK). The XYZ space by the Commission Internationale de L'Eclairage (CIE) is considered as the basis of all modern color spaces [45, 49]. It embeds all visible colors unambiguously into one space of three imaginary primaries [5, 34]. The chromaticity diagram in Fig. 5.12 is the result of projecting XYZ to the Maxwell triangle $x + y + z = 1$, which forms a representation of all visible hues and saturations.

A number of spaces, e.g., CIELAB, CIELUV, and DIN99, CIECAM [10, 49], were defined as transformations of XYZ to derive an ideal color space [55], where the Euclidean distance is proportional to the perceived color difference.

Human color perception has been known for a while to be non-Euclidean due to the principle called *hue superimportance* [54] (cf. Fig. 5.13). It refers to the fact that changes in hue are perceived more strongly than changes in saturation. The circumference of a circle of constant luminance and saturation would be estimated to measure about 4π for its radius, which cannot be embedded in a Euclidean plane.

Fig. 5.12 CIE XYZ
chromaticity diagram and a
path that represents a
colormap

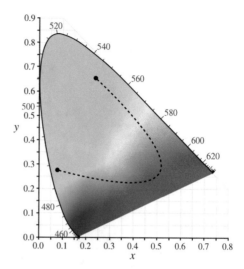

Please note that the length $l \in \mathbb{R}$ of a path $c : \mathbb{R} \to C$ is defined for arbitrary metric spaces C

$$l = \sup_{0=t_0,\ldots,t_n=1} \sum_{i=1}^{k} \Delta E(c(t_i), c(t_{i+1})). \tag{5.20}$$

Therefore, classic descriptions of color spaces, such as those of Von Helmholtz [95], Schrödinger [82], and Stiles [104], are based on Riemannian manifolds.

However, state-of-the-art research indicates that human color perception is also non-Riemannian, due to the further principle of *diminishing returns* [54]; see Fig. 5.13. In this context, diminishing returns refers to the phenomenon that when presented with two colors A and C and their perceived middle (average/mixture) B, an observer usually judges the sum of the perceived differences of each half greater than the difference of the two outer colors $\Delta(A, B) + \Delta(B, C) > \Delta(A, C)$. This effect is produced by a natural contrast enhancement filter employed into the human perceptual system to adapt to different viewing conditions. This property is dependent upon the distance between colors, especially for large distances.

As a result, modern color difference formulas (e.g., CIEDE1994, CIEDE2000) that were designed to match experimental data produce complicated spaces, which come with challenges. For example, they are not metric spaces. Being a metric is a very basic mathematical property that we would expect from a distance measure $d : C \times C \to \mathbb{R}$, i.e., that it suffices non-negativity $d(x, y) \geq 0$, identity of indiscernible $d(x, y) = 0 \Leftrightarrow x = y$, symmetry $d(x, y) = d(y, x)$, and the triangle inequality $d(x, z) \leq d(x, y) + d(y, z)$. The reasons for such a challenge are not in the experimental data but can be found in the mathematical models underlying the distance formulae [48, 64, 65]. An example of the violation of the triangle inequality is shown in Fig. 5.14.

Fig. 5.13 Illustration of hue superimportance with circumference of $\approx 4\pi r$ and diminishing returns $(\overline{AB} + \overline{BC} > \overline{AC})$

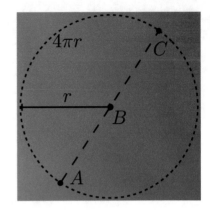

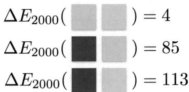

$$\Delta E_{2000}(\quad\quad) = 4$$

$$\Delta E_{2000}(\quad\quad) = 85$$

$$\Delta E_{2000}(\quad\quad) = 113$$

Fig. 5.14 Illustration of non-metric behavior of CIE ΔE_{2000}. Violation of the triangle inequality implies that the path over green RGB = (146,252,77) is shorter than the direct path from blue RGB = (0,0,255) to yellowish green RGB = (177,253,79), which is very counter intuitive

The difficulties, however, lie not only in the modeling of the color spaces but also in the visualization side. Mathematical operations on color become significantly harder in non-Euclidean spaces. As a basic example, consider linear interpolation where values are taken equidistantly on a straight line connecting two points. In non-Euclidean spaces, the concept of a straight line is, in general, undefined.

To overcome some of these difficulties, some authors generate spaces that are close to the original distance measure but are Euclidean or at least Riemannian [78, 94]. This, however, conflicts with the experimental results from the perceptual sciences. We believe that future color spaces will continue to better approximate human color perception and embrace its complicated non-Euclidean structure because our computational capacities will enable us to work with them despite those difficulties. We believe that the path forward lies in improving visualization algorithms so that they run on general non-Euclidean color spaces. A few results have been obtained recently for color interpolation [105] and colormap assessment [9].

Acknowledgements The authors would like to thank the organizers of the Dagstuhl Seminar 18041 in January 2018, entitled "Foundations of Data Visualization". Bei Wang is partially supported by NSF IIS-1910733, DBI-1661375, and IIS-1513616. Roxana Bujack is partially supported by the Laboratory Directed Research and Development (LDRD) program of the Los Alamos National Laboratory (LANL) under project number 20190143ER. Ingrid Hotz is supported through Swedish e-Science Research Center (SeRC) and the ELLIIT environment for strategic research in Sweden.

References

1. Amann, H.: Ordinary Differential Equations: An Introduction to Nonlinear Analysis. Studies in Mathematics, vol. 13. Walter de Gruyter, Berlin (2011). https://doi.org/10.1515/9783110853698
2. de Berg, M., van Kreveld, M., Overmars, M., Schwarzkopf, O.: Computational Geometry, Algorithms and Applications, 3rd edn. Springer, Berlin (2008). https://doi.org/10.1007/978-3-540-77974-2
3. Berger, M.: A Panoramic View of Riemannian Geometry. Springer, Berlin (2003). https://doi.org/10.1007/978-3-642-18245-7
4. Böttger, J., Schäfer, A., Lohmann, G., Villringer, A., Margulies, D.S.: Three-dimensional mean-shift edge bundling for the visualization of functional connectivity in the brain. IEEE Trans. Vis. Comput. Graph. **20**(3), 471–480 (2014). https://doi.org/10.1109/TVCG.2013.114
5. Broadbent, A.: Calculation from the original experimental data of the CIE 1931 RGB standard observer spectral chromaticity coordinates and color matching functions. Québec, Canada, Département de génie chimique, Université de Sherbrooke (2008)
6. Brun, A., Knutsson, H., Park, H.J., Shenton, M.E., Westin, C.F.: Clustering fiber traces using normalized cuts. In: Barillot, C., Haynor, D.R., Hellier, P. (eds.) Medical Image Computing and Computer-Assisted Intervention, pp. 368–375. Springer, Berlin (2004). https://doi.org/10.1007/b100265
7. Buhmann, M.D.: Radial Basis Functions: Theory and Implementations. Cambridge University Press, Cambridge (2003). https://doi.org/10.1017/CBO9780511543241
8. Bujack, R., Hotz, I., Scheuermann, G., Hitzer, E.: Moment invariants for 2D flow fields via normalization in detail. IEEE Trans. Vis. Comput. Graph. **21**(8), 916–929 (2015). https://doi.org/10.1109/TVCG.2014.2369036
9. Bujack, R., Turton, T.L., Samsel, F., Ware, C., Rogers, D.H., Ahrens, J.: The good, the bad, and the ugly: a theoretical framework for the assessment of continuous colormaps. IEEE Trans. Vis. Comput. Graph. **24**(1), 923–933 (2018). https://doi.org/10.1109/TVCG.2017.2743978
10. Büring, H.: Eigenschaften des farbenraumes nach din 6176 (din99-formel) und seine bedeutung für die industrielle anwendung. In: Proceedings of 8th Workshop Farbbildverarbeitung der German Color Group, pp. 11–17 (2002)
11. Butcher, J.C.: Numerical Methods for Ordinary Differential Equations. Wiley, New York (2016). https://doi.org/10.1002/9781119121534
12. Carr, H., Snoeyink, J., Axen, U.: Computing contour trees in all dimensions. Comput. Geom. **24**(2), 75–94 (2003). https://doi.org/10.1016/S0925-7721(02)00093-7
13. Celebi, M.E., Kingravi, H.A., Vela, P.A.: A comparative study of efficient initialization methods for the k-means clustering algorithm. Expert Syst. Appl. **40**(1), 200–210 (2013). https://doi.org/10.1016/j.eswa.2012.07.021
14. Chen, C., Wang, C., Ma, K., Wittenberg, A.T.: Static correlation visualization for large time-varying volume data. In: IEEE Pacific Visualization Symposium, pp. 27–34 (2011). https://doi.org/10.1109/PACIFICVIS.2011.5742369
15. Coddington, E.A.: An Introduction to Ordinary Differential Equations. Courier Corporation, Chelmsford (2012)
16. Coffin, J.G.: Vector Analysis: An Introduction to Vector-Methods and Their Various Applications to Physics and Mathematics. Wiley, New York (1911)
17. Cohen-Steiner, D., Edelsbrunner, H., Harer, J.: Stability of persistence diagrams. Discret. Comput. Geom. **37**(1), 103–120 (2007). https://doi.org/10.1007/s00454-006-1276-5
18. Correa, C.D., Lindstrom, P.: Towards robust topology of sparsely sampled data. Trans. Comput. Graph. Vis. **17**(12), 1852–1861 (2011). https://doi.org/10.1109/TVCG.2011.245
19. Desbrun, M., Kanso, E., Tong, Y.: Discrete differential forms for computational modeling. In: ACM SIGGRAPH 2006 Courses, pp. 39–54. ACM (2006). https://doi.org/10.1145/1198555.1198666
20. Desbrun, M., Polthier, K., Schröder, P., Stern, A.: Discrete differential geometry. In: ACM SIGGRAPH 2006 Courses, p. 1. ACM (2006)

21. Du, Q., Faber, V., Gunzburger, M.: Centroidal Voronoi tessellations: applications and algorithms. SIAM Rev. **41**(4), 637–676 (1999). https://doi.org/10.1137/S0036144599352836
22. Edelsbrunner, H.: Geometry and Topology for Mesh Generation. Cambridge Monographs on Applied and Computational Mathematics. Cambridge University Press, Cambridge (2001). https://doi.org/10.1017/CBO9780511530067
23. Edelsbrunner, H., Harer, J.: Jacobi sets of multiple Morse functions. In: Cucker, F., DeVore, R., Olver, P., Süli, E. (eds.) Foundations of Computational Mathematics, Minneapolis 2002, pp. 37–57. Cambridge University Press, Cambridge (2002)
24. Edelsbrunner, H., Harer, J.: Computational Topology: An Introduction. American Mathematical Society, Providence (2010). https://doi.org/10.1090/mbk/069
25. Edelsbrunner, H., Harer, J., Natarajan, V., Pascucci, V.: Morse-Smale complexes for piecewise linear 3-manifolds. In: Proceedings of the 19th ACM Symposium on Computational Geometry, pp. 361–370 (2003). https://doi.org/10.1145/777792.777846
26. Edelsbrunner, H., Harer, J., Natarajan, V., Pascucci, V.: Local and global comparison of continuous functions. In: IEEE Visualization, pp. 275–280 (2004). https://doi.org/10.1109/VISUAL.2004.68
27. Edelsbrunner, H., Harer, J., Patel, A.K.: Reeb spaces of piecewise linear mappings. In: Proceedings of the 24th Annual Symposium on Computational Geometry, pp. 242–250. ACM (2008). https://doi.org/10.1145/1377676.1377720
28. Edelsbrunner, H., Harer, J., Zomorodian, A.J.: Hierarchical Morse-Smale complexes for piecewise linear 2-manifolds. Discret. Comput. Geom. **30**, 87–107 (2003). https://doi.org/10.1007/s00454-003-2926-5
29. Edelsbrunner, H., Letscher, D., Zomorodian, A.J.: Topological persistence and simplification. Discret. Comput. Geom. **28**, 511–533 (2002). https://doi.org/10.1007/s00454-002-2885-2
30. Edelsbrunner, H., Morozov, D.: Persistent Homology: Theory and Practice. European Congress of Mathematics (2012). https://doi.org/10.4171/120-1/3
31. Edelsbrunner, H., Morozov, D.: Persistent homology. In: Goodman, J.E., O'Rourke, J., Tóth, C.D. (eds.) Handbook of Discrete and Computational Geometry. Discrete Mathematics and Its Applications, Chap. 24. CRC Press LLC, Boca Raton (2017)
32. Ester, M., Kriegel, H.P., Sander, J., Xu, X.: A density-based algorithm for discovering clusters a density-based algorithm for discovering clusters in large spatial databases with noise. In: Proceedings of the 2nd International Conference on Knowledge Discovery and Data Mining, pp. 226–231. AAAI Press (1996)
33. Estivill-Castro, V.: Why so many clustering algorithms: a position paper. ACM SIGKDD Explor. Newsl. **4**(1), 65–75 (2002). https://doi.org/10.1145/568574.568575
34. Fairman, H.S., Brill, M.H., Hemmendinger, H., et al.: How the CIE 1931 color-matching functions were derived from wright-guild data. Color Res. Appl. **22**(1), 11–23 (1997). https://doi.org/10.1002/(SICI)1520-6378(199702)22:1<11::AID-COL4>3.0.CO;2-7
35. Farin, G.: Curves and Surfaces for CAGD: A Practical Guide. The Morgan Kaufmann Series in Computer Graphics, 5th edn. Morgan Kaufmann Publishers, Burlington (2002)
36. Folland, G.B.: Introduction to Partial Differential Equations, 2nd edn. Princeton University Press, Princeton (1995)
37. Forman, R.: A user's guide to discrete Morse theory. Séminaire Lotharingien de Combinatoire **48**, (2002)
38. Gabriel, K.R., Sokal, R.R.: A new statistical approach to geographic variation analysis. Syst. Biol. **18**(3), 259–278 (1969). https://doi.org/10.2307/2412323
39. Georgii, H.O.: Stochastics: Introduction to Probability and Statistics. De Gruyter, Berlin (2008). https://doi.org/10.1515/9783110293609
40. Ghrist, R.: Three examples of applied and computational homology. Nieuw Archief voor Wiskunde (The Amsterdam Archive, Special issue on the occasion of the fifth European Congress of Mathematics) pp. 122–125 (2008)
41. Glassner, A.S.: Principles of Digital Image Synthesis. The Morgan Kaufmann Series in Computer Graphics and Geometric Modeling. Morgan Kaufmann Publishers Inc., Burlington (1995)

42. Goodman, R., Wallach, N.R.: Symmetry, Representations, and Invariants. Graduate Texts in Mathematics, vol 255. Springer, Berlin (2009)
43. Gosink, L., Anderson, J., Bethel, W., Joy, K.: Variable interactions in query-driven visualization. IEEE Trans. Vis. Comput. Graph. **13**(6), 1400–1407 (2007). https://doi.org/10.1109/TVCG.2007.70519
44. Grassmann, H.: Zur Theorie der Farbenmischung. Ann. Phys. **165**(5), 69–84 (1853). https://doi.org/10.1002/andp.18531650505
45. Guild, J.: The colorimetric properties of the spectrum. Philos. Trans. R. Soc. Lond. Ser. A **230**, 149–187 (1932). https://doi.org/10.1098/rsta.1932.0005
46. Hansen, C.D., Chen, M., Johnson, C.R., Kaufman, A.E., Hagen, H. (eds.): Scientific Visualization: Uncertainty, Multifield, Biomedical, and Scalable Visualization. Mathematics and Visualization. Springer, Berlin (2014)
47. Hatcher, A.: Algebraic Topology. Cambridge University Press, Cambridge (2002)
48. Huertas, R., Melgosa, M., Oleari, C.: Performance of a color-difference formula based on OSA-UCS space using small-medium color differences. JOSA A **23**(9), 2077–2084 (2006)
49. International Commission on Illumination: Colorimetry. CIE technical report. Commission Internationale de l'Eclairage (2004)
50. Ip, C.Y., Varshney, A., JaJa, J.: Hierarchical exploration of volumes using multilevel segmentation of the intensity-gradient histograms. IEEE Trans. Vis. Comput. Graph. **18**(12), 2355–2363 (2012). https://doi.org/10.1109/TVCG.2012.231
51. Jain, A.K., Murty, M.N., Flynn, P.J.: Data clustering: a review. ACM Comput. Surv. **31**(3), 264–323 (1999). https://doi.org/10.1145/331499.331504
52. James, I.M. (ed.): History of Topology. Elsevier B.V, Amsterdam (1999)
53. Jankowai, J., Hotz, I.: Feature level-sets: Generalizing iso-surfaces to multi-variate data. IEEE Trans. Vis. Comput. Graph. pp. 1–1 (2018). https://doi.org/10.1109/TVCG.2018.2867488
54. Judd, D.B.: Ideal color space: curvature of color space and its implications for industrial color tolerances. Palette **29**(21–28), 4–25 (1968)
55. Judd, D.B.: Ideal color space. Color. Eng. **8**(2), 37 (1970)
56. Jungnickel, D.: Graphs, Networks and Algorithms. Algorithms and Computation in Mathematics, 4th edn. Springer, Berlin (2012)
57. Kasten, J., Reininghaus, J., Hotz, I., Hege, H.C., Noack, B.R., Daviller, G., Morzynski, M.: Acceleration feature points of unsteady shear flows. Arch. Mech. **68**(1), 55–80 (2016)
58. Kindlmann, G., Scheidegger, C.: An algebraic process for visualization design. IEEE Trans. Vis. Comput. Graph. **20**(12) (2014). https://doi.org/10.1109/TVCG.2014.2346325
59. Kratz, A., Auer, C., Stommel, M., Hotz, I.: Visualization and analysis of second-order tensors: Moving beyond the symmetric positive-definite case. Comput. Graph. Forum - State Art Rep. **32**(1), 49–74 (2013). https://doi.org/10.1111/j.1467-8659.2012.03231.x
60. Kratz, A., Meyer, B., Hotz, I.: A visual approach to analysis of stress tensor fields. In: Hagen, H. (ed.) Scientific Visualization: Interactions, Features, Metaphors. Dagstuhl Follow-Ups, vol. 2, pp. 188–211. Schloss Dagstuhl–Leibniz-Zentrum fuer Informatik, Dagstuhl, Germany (2011). https://doi.org/10.4230/DFU.Vol2.SciViz.2011.188
61. Kühnel, W.: Differential Geometry: Curves - Surfaces - Manifolds. Student Mathematical Library. American Mathematical Society, Providence (2015). https://doi.org/10.1090/stml/077
62. Lamb, E.: What We Talk About When We Talk About Holes. Scientific American Blog Network (2014)
63. Liu, S., Maljovec, D., Wang, B., Bremer, P.T., Pascucci, V.: Visualizing high-dimensional data: advances in the past decade. IEEE Trans. Vis. Comput. Graph. **23**(3), 1249–1268 (2017). https://doi.org/10.1109/TVCG.2016.2640960
64. Luo, M.R., Cui, G., Rigg, B.: The development of the cie 2000 colour-difference formula: Ciede 2000. Color Res. Appl. **26**(5), 340–350 (2001). https://doi.org/10.1002/col.1049
65. Mahy, M., Eycken, L., Oosterlinck, A.: Evaluation of uniform color spaces developed after the adoption of CIELAB and CIELUV. Color Res. Appl. **19**(2), 105–121 (1994). https://doi.org/10.1111/j.1520-6378.1994.tb00070.x

66. Morton, K.W., Mayers, D.F.: Numerical Solution of Partial Differential Equations: An Introduction. Cambridge University Press, Cambridge (2005). https://doi.org/10.1017/CBO9780511812248
67. Munkres, J.R.: Elements of Algebraic Topology. CRC Press, Taylor & Francis Group, Boca Raton (1984). https://doi.org/10.1201/9780429493911
68. Munzner, T.: Visualization Analysis and Design. CRC Press, Taylor & Francis Group, Boca Raton (2014). https://doi.org/10.1201/b17511
69. Nagaraj, S., Natarajan, V.: Simplification of Jacobi sets. In: Pascucci, V., Tricoche, X., Hagen, H., Tierny, J. (eds.) Topological Data Analysis and Visualization: Theory, Algorithms and Applications, Mathematics and Visualization, pp. 91–102. Springer, Berlin (2011). https://doi.org/10.1007/978-3-642-15014-2_8
70. O'Connor, J.J., Robertson, E.F.: A History of Topology. MacTutor History of Mathematics (1996)
71. Peacock, T., Froyland, G., Haller, G.: Introduction to focus issue: objective detection of coherent structures. Chaos **25**, (2015). https://doi.org/10.1063/1.4928894
72. Peikert, R., Roth, M.: The "parallel vectors" operator - a vector field visualization primitive. Proc. IEEE Vis. **14**(16), 263–270 (1999). https://doi.org/10.1109/VISUAL.1999.809896
73. Polthier, K., Schmies, M.: Straightest geodesics on polyhedral surfaces. In: Hege, H.C., Polthier, K. (eds.) Mathematical Visualization, p. 391. Springer, Berlin (1998). https://doi.org/10.1007/978-3-662-03567-2_11
74. Post, F.H., Post, F.J., Walsum, T.V., Silver, D.: Iconic techniques for feature visualization. In: Proceedings of the 6th Conference on Visualization, p. 288. IEEE Computer Society, Washington, D.C. (1995)
75. Potter, K.: The visualization of uncertainty. Ph.D. thesis, University of Utah (2010)
76. Press, W., Teukolsky, S., Vetterling, W., Flannery, B.: Numerical Recipes in C: The Art of Scientific Computing, 3rd edn. Cambridge University Press, Cambridge (1992)
77. Raj, M.: Depth-based visualizations for ensemble data and graphs. Ph.D. thesis, University of Utah (2018)
78. Raj Pant, D., Farup, I.: Riemannian formulation and comparison of color difference formulas. Color Res. Appl. **37**(6), 429–440 (2012). https://doi.org/10.1002/col.20710
79. Reeb, G.: Sur les points singuliers d'une forme de pfaff completement intergrable ou d'une fonction numerique. Comptes Rendus Acad. Sci. Paris **222**, 847–849 (1946)
80. Renardy, M., Rogers, R.C.: An introduction to Partial Differential Equations, vol. 13. Springer Science & Business Media, Berlin (2006). https://doi.org/10.1007/b97427
81. Resnikoff, H.L.: Differential geometry and color perception. J. Math. Biol. **1**(2), 97–131 (1974). https://doi.org/10.1007/BF00275798
82. Schrödinger, E.: Grundlinien einer Theorie der Farbenmetrik im Tagessehen. Ann. Phys. **368**(22), 481–520 (1920). https://doi.org/10.1002/andp.19203682102
83. Schubert, E., Sander, J., Ester, M., Kriegel, H.P., Xu, X.: DBSCAN revisited, revisited: why and how you should (still) use DBSCAN. ACM Trans. Database Syst. **42**(3), 19:1–19:21 (2017). https://doi.org/10.1145/3068335
84. Schwichtenberg, J.: Physics from Symmetry. Undergraduate Lecture Notes in Physics, 2nd edn. Springer, Berlin (2017). https://doi.org/10.1007/978-3-319-19201-7
85. Shi, J., Malik, J.: Normalized cuts and image segmentation. IEEE Trans. Pattern Anal. Mach. Intell. **22**(8), 888–905 (2000). https://doi.org/10.1109/34.868688
86. Shirley, P.: Fundamentals of Computer Graphics. AK Peters Ltd, Natick (2005)
87. Snider, A.D., Davis, H.F.: Introduction to Vector Analysis, 7th edn. William C. Brown, Lowa (1987)
88. Sonka, M., Hlavac, V., Bohle, R.: Image Processing, Analysis and and Machine Vision, 3rd edn. Thomson, Stamford (2008)
89. Steward, J.: Multivariate Calculus, 7th edn. Brooks/Cole CENGAGE Learning (2019)
90. Strang, G.: Introduction to Linear Algebra, 5th edn. Wellesley-Cambridge Press, Cambridge (2016)

91. Suthar, N., jeet Rajput, I., kumar Gupta, V.: A technical survey on DBSCAN clustering algorithm. Int. J. Sci. Eng. Res. **4**(5) (2013)
92. Telea, A.C.: Data Visualization: Principles and Practice, 2nd edn. AK Peters Ltd, Natick (2015)
93. Tufte, E.R.: The Visual Display of Quantitative Information. Graphics Press, Cheshire (2001)
94. Urban, P., Rosen, M.R., Berns, R.S., Schleicher, D.: Embedding non-Euclidean color spaces into euclidean color spaces with minimal isometric disagreement. J. Opt. Soc. Am. A **24**(6), 1516–1528 (2007). https://doi.org/10.1364/JOSAA.24.001516
95. Von Helmholtz, H.: Handbuch der physiologischen Optik, vol. 9. Voss (1867)
96. Wang, J., Hazarika, S., Li, C., Shen, H.W.: Visualization and visual analysis of ensemble data: a survey. IEEE Trans. Vis. Comput. Graph. (2018). https://doi.org/10.1109/TVCG.2018.2853721
97. Wardetzky, M., Mathur, S., Kälberer, F., Grinspun, E.: Discrete laplace operators: no free lunch. In: Proceedings of the Eurographics Symposium on Geometry Processing, pp. 33–37 (2007)
98. Weyl, H.: Symmetry. Princeton University Press, Princeton (1952)
99. Whitaker, R.T., Mirzargar, M., Kirby, R.M.: Contour boxplots: a method for characterizing uncertainty in feature sets from simulation ensembles. IEEE Trans. Vis. Comput. Graph. **19**(12), 2713–2722 (2013). https://doi.org/10.1109/TVCG.2013.143
100. Wikipedia Contributors: Topology. Wikipedia, The Free Encyclopedia (2018)
101. Wong, P.C., Foote, H., Leung, R., Adams, D., Thomas, J.: Data signatures and visualization of scientific data sets. IEEE Comput. Graph. Appl. **20**(2), 12–15 (2000). https://doi.org/10.1109/38.824451
102. Woodring, J., Shen, H.: Multiscale time activity data exploration via temporal clustering visualization spreadsheet. IEEE Trans. Vis. Comput. Graph. **15**(1), 123–137 (2009). https://doi.org/10.1109/TVCG.2008.69
103. Wu, M.Q.Y., Faris, R., Ma, K.: Visual exploration of academic career paths. In: IEEE/ACM International Conference on Advances in Social Networks Analysis and Mining, pp. 779–786 (2013). https://doi.org/10.1145/2492517.2492638
104. Wyszecki, G., Stiles, W.S.: Color Science, vol. 8. Wiley, New York (1982)
105. Zeyen, M., Post, T., Hagen, H., Ahrens, J., Rogers, D., Bujack, R.: Color interpolation for non-Euclidean color spaces. In: IEEE Scientific Visualization Conference Short Papers. IEEE (2018)
106. Zhang, Y.J.: Geometric Modelng and Mesh Generation from Scanned Images. CRC Press, Taylor & Francis Group, Boca Raton (2016)

Chapter 6
Transformations, Mappings, and Data Summaries

Ross Whitaker and Ingrid Hotz

Abstract Fundamentally, data visualization is the process of placing dabs of ink or color on a 2D plane. However, the complexity of data is increasing so that we see large numbers of instances, dimensions, parameters, etc. Such data surpasses what can readily be shown on a 2D or 3D display. One solution to this challenge is the development of better or more complex interfaces, that include, for instance, linked views, large displays, dynamic visualizations, and sophisticated user interactions. The alternative and complementary approach is to develop sets of mathematical and statistical tools to transform, map, or summarize data and thereby reduce its complexity so that visualization and understanding of large and complex becomes more feasible. The role of visualization research, in this case, is to identify common use cases and develop methods and tools that can readily be adapted to particular applications. To address the challenges of complexity in the data, previous works have proposed reducing items and attributes and associated visualization conventions and practices. Here we take deeper (and complementary) look at the analytical frameworks and approaches for transforming data into forms that are appropriate for display devices, considered generally. The approach in this chapter is to begin by characterizing different types of data in a way that is well suited for this discussion. We will then focus on a few particular classes of data and different ways of summarizing and transforming data of those types. Finally, we will broaden the discussion to other types of data and how they map into the various methodologies.

R. Whitaker (✉)
University of Utah, Salt Lake City, USA
e-mail: Whitaker@CS.Utah.edu

I. Hotz
Linköping University, Norrköping, Sweden
e-mail: Ingrid.Hotz@LiU.se

© Springer Nature Switzerland AG 2020
M. Chen et al. (eds.), *Foundations of Data Visualization*,
https://doi.org/10.1007/978-3-030-34444-3_6

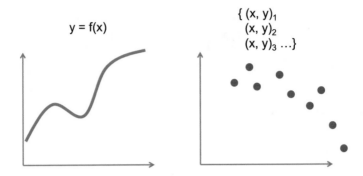

Fig. 6.1 In visualization, the placement of color or ink in 2D depicts instances that are generated from a process (scatterplot-right) or satisfy some constraint (graph of function-left)

6.1 Types of Data

Here we break different types of data down into a roughly hierarchical taxonomy and introduce some terminology and data properties that will facilitate the subsequent discussion. As we do so, we should bear in mind that this partitioning of data types and visualization goals is not unique and can be applied in different ways under different circumstances. As we shall see, even the same data set can be viewed as one type or another, depending on one's perspective or the goals of the analysis or visualization.

The first distinction to make is that of *instances* versus *mappings*. When visualizing *instances*, we are typically considering independent examples of data that share some common characteristics or sample space. Points in the x-y plane, shown as a scatterplot (each dot is an instance) in Fig. 6.1, are an example of a collection of instances. Alternatively, we are sometimes interested in visualizing a *mapping* that shows a relationship between two sets. A function $y = f(x)$, where $f : \Re \mapsto \Re$, is a special case of a mapping. Figure 6.1 shows the graph of a particular function $f(x)$. A graph of a function shares some properties with the scatterplot, because it shows, via ink on the page, all of the instances that satisfy the relation $y = f(x)$.

As we consider the distinction between instances and mappings, we should note that for a particular data set the difference may be in how we think about the data or the goals of the visualization. A discrete sampling of n points from of a function $y = f(x)$ could also be considered a set of instances, $(x_1, y_1), (x_2, y_2), \ldots, (x_n, y_n)$, but in general we are interested in different questions about these two cases. In the case of instances, there is typically an assumption (explicit or not) that these instances are generated from some kind of stochastic process or probability distribution, and one would like to understand, in geometric terms, the relative densities of data and the relationships between points. With functions, the structure of the independent axis (here the x-axis) is given, and we typically want to interrogate the geometric structure of $y = f(x)$, rather than densities and distances.

x	y	z
5.2mm	3.1mm	7.5mm
1.4mm	3.2mm	6.3mm
2.6mm	3.4mm	5.1mm
⋮		

(a)

Name	Age	City	Status
Kim	32	St. Louis	member
Samir	27	San Francisco	nonmember
Hari	45	Salt Lake City	senior member
		⋮	

(b)

Fig. 6.2 Data sets often consist of lists or arrays of *instances*, where the data for each instance may exist in a physical space with commensurate quantities (**a**), or may consist of heterogeneous quantities (**b**)

Another consideration for types of data is whether or not the data has an inherent *structure*. This distinction is most important in the case of instances. Some data sets consist of instances (or *points*) that have a consistent structure, e.g., each instance consists of values derived from a common set of *fields*. Each field might consist of a numerical value from a discrete or continuous space, a categorical value, or an ordinal value (ordered, but not quantitative). Often, structured data is defined in terms of a *data model*, where the model describes the structure of the fields, the possible values they can take, and their physical or semantic meaning. Figure 6.2 shows two examples of structured data sets. The first is a small set of records that contain points in three dimensions, and thus the attributes are all similar (e.g., same units, meters in this case) and where the space (three dimensions, $\mathfrak{R}^3$) has a special structure. The second example is a *heterogeneous* mixture of attributes—but where each record still contains the same attributes.

Alternatively, some data sets consist of *unstructured* data. In the case of unstructured data, instances each contain some set of data, but the data is not consistently organized into distinct fields with well-defined values, as it is in Fig. 6.2. Unstructured data is commonly text-heavy, but it often also contains other nontext data such as dates, numbers, and categorical attributes. A typical case of unstructured data comes in the form of free-form text, which one might see in online/electronic reviews, notes taken by a doctor/clinician in a medical exam, or other electronic communication, such text messages or email.

In the context of visualization, the type of data becomes important, because ultimately, visualization deals with the problem of how to assign colors to pixels in a 2D (or 3D) display. The choices of colors and where to put them are quantitative decisions; pixels are associated with 2D coordinates and colors are chosen from a multi-dimensional color palate. Similar decisions of placement, size, and color are important even when one is dealing with conventional visualization techniques such as graphs, glyphs, and various kinds of charts. *Virtually all data visualization strategies require one to represent instances or functions with relatively few quantities.*

In many or most cases, the data does not come in a form that maps directly onto these quantities, and typically a transformation from the original data into the desired visualization scheme is required. Even when the data lends itself to direct mapping into a 2D domain, as in the case of 2D, scalar fields (or images), one is often interested in some particular property of the data, rather than the entire data set, and this often entails some kind of transformation of the data to produce a set of relevant features.

6.2 Functions

Here we begin with a very brief overview of the transformations that are relevant for 2D, 3D, and high-dimensional functions. Please note that in the context of scientific visualization, such functions are also often called fields (Chap. 5). We do not discuss particular algorithms for fast or efficient rendering of such functions, but focus on the mathematics of the transformations. For functional data, we are considering mappings from $\Re^m$ to $\Re^k$, and we assume that k is relatively small. We also treat these objects, unless otherwise stated, as continuous mappings (e.g., the domain is continuous) and assume that discrete representations are suitably interpolated, as in Chap. 5, such that they are defined over continuous domains.

There are some trivial examples, such as $m = k = 1$, where one can simply graph the function to see its structure. Also, for $m = 2$ and $k = 1$, one can use the pixels on the 2D viewing plane to assign color values to points, thus treating the function as an *image*, and we can use the notation $f(x, y)$ to denote the values at each 2D point. While there are many interesting and important questions about displaying such scalar data using various *color maps*, it is a topic that is studied extensively in the literature [55]. It is also possible to graph such data as height-fields in 3D space and project the resulting surfaces on a 2D screen.

The topic of *transforming* 2D, scalar data for better visualization or interpretation is (or *was*) covered extensively in the field of image processing [13]. Here we only mention a few basic ideas. One of the main strategies is to transform the range with a function $g : \Re \mapsto \Re$, so that we obtain a new image, $f'(x, y) = g(f(x, y))$. Of course, $g(\cdot)$ could also be $g : \Re \mapsto \Re^3$ and thereby represent the operation of color mapping scalar values in a function.

Understanding such transformations entails studying the structure of $g(\cdot)$. Typical mappings will lighten or darken images. Another common operation is to increase or decrease the overall range of a function. For enhancing the contrast in images, often it is advisable to consider the histogram of values of the image (histogram of values in the range). There are a variety of methods for flattening histograms (e.g. histogram equalization), or targeting or matching certain histograms [13].

For visualization, a more challenging example is $k = 1$, and $m = 3$, where we have scalar values given in a 3D volume $f(x, y, z)$. This kind of data arises, for instance, in medical imaging in the case of MRI or CT or in physical simulations, e.g., of temperature fields. The challenge with volume data is that the dimensionality does not lend itself to direct display of the raw data. Graphs of such functions would require 4D displays and a direct display of values as colors would require a 3D display. Thus, mappings onto 2D grids or displays are important. One approach is to provide some *slicing* capability, often arranged along the grid axes by fixing one coordinate, for instance, to the kth slice. The resulting function $f'(u, v) = f(x = u, y = v, z = k)$ is defined over a 2D domain. More generally, arbitrary, 2D surfaces can be sampled from the 3D domain and then be displayed as (flat) images or rendered as texture-mapped surfaces, illustrated in Fig. 6.4a.

More commonly, 3D functions are rendered after some kind of *projection* onto a 2D viewing plane. Typically, the projection is a line integral following a ray from each point in the view plane into the 3D volumes, as illustrated in Fig. 6.3. The simplest case is:

$$f'(u, v) = \int f(u, v, \alpha) d\alpha, \qquad (6.23)$$

which is a projection along the z axis to form a 2D function, which is then mapped onto pixel/display intensities. The specific bounds for the integration are a visualization decision, where the bounds should include some finite viewing frustrum. Other views can be obtained by applying a coordinate transformation, $\phi : \Re^3 \mapsto \Re^3$ (which should probably smooth and invertible),

$$f'(u, v) = \int f(\phi(u, v, \alpha)) d\alpha. \qquad (6.24)$$

The transformation ϕ could include rotations and translations, but also could encode a perspective projection, or even nonlinear *curves* through the volume, effectively warping the 3D data. In the discussion that follows, we will leave off this coordinate transformation for simplicity (and without a loss of generality).

Another simple projection of 3D functions that is useful is, for instance, the maximum intensity projection, which takes the maximum value of $f(\cdot)$ along the rays associated with each pixel (along the z direction in this case),

$$f'(u, v) = \sup_\alpha f(u, v, \alpha). \qquad (6.25)$$

Fig. 6.3 3D functions or volumes are often transformed into 2D functions by accumulating data along rays that intersect the volume and a viewing plane (in blue)

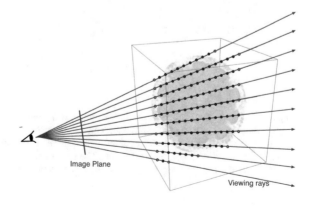

Image Plane

Viewing rays

The field of *volume rendering* [10] addresses various levels of complexity for these kinds of projections. A typical formulation of the volume rendering equation is (Fig. 6.4):

$$f'(u, v) = \int G\left(u, v, \alpha, f(\cdot), Df(\cdot), \ldots, O(\cdot)\right) d\alpha, \qquad (6.26)$$

where the "·" notation indicates volume coordinates (u, v, α). The *occlusion function* $O(u, v, \alpha)$ quantifies how much a point contributes to the rendering. It also is a line integral (from the viewing plane to the 3D point) depending on the opacity of the volume that lies between the point and the viewing plane. The range of $G(\cdot)$ is mostly a 3D color space. The positional information, u, v, α, can also be used for view-dependent lighting effects. The first derivatives of f, denoted Df, indicates the local gradient vector, which provides normals, for lighting/shading, or edge enhancement in volumes, which are characterized by high gradients. Many other parameters have been considered in the function $G(\cdot)$, indicated by the ellipses $(\ldots)$, for instance higher order derivatives of $f(\cdot)$ [25]. The mapping of values of $f(\cdot)$ and its derivatives into colors and opacities is called a *transfer function*. The transfer function defines which parts of the data will be visible in the final rendering and essentially contributes to a good visualization result [32].

The integral in Eq. 6.26 describes many of the most basic options for high-quality volume rendering. Research beyond this basic formulation has focused on fast methods for volume rendering, e.g., on specialized hardware [49] and more *realistic* models and volume illumination. Early work focuses on methods for efficiently approximating light transport by restricting the type and number of light sources, e.g., the seminal method by Kniss et al. [26]. Deep shadow maps by Hadwiger et al. [16] enable complex lighting models in interactive direct volume rendering

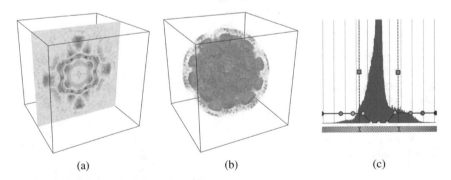

(a)   (b)   (c)

Fig. 6.4 Volume visualization of an electron microscopy data set of a feline calicivirus. **a** A slice through the data sets shows the entire data range in the respective slice. **b** Volume rendering using ray casting highlights selected scalar values in the data set. **c** Transfer function used for the volume rendering, overlaid with the data-histogram. The scalar values are mapped to the x-axis, and the y-axis shows the transparency assigned to the scalar values. Visualization: Martin Falk, https://inviwo.org

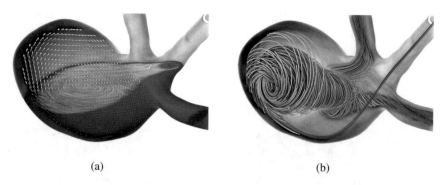

(a)   (b)

Fig. 6.5 Visualization of the blood flow in an aneurysm. **a** Vector-like glyphs represent the flow on a vertical slice through the aneurysm. A texture shows the flow an a horizontal slice through the aneurysm. **b** A selected set of streamlines illustrates the overall flow behavior. Visualization: Wito Engelke

(DVR). Early approaches aiming at full global illumination include the work by Hernell et al. [18]. More recently, volumetric illumination with multiple scattering based on photon mapping and Monte Carlo ray tracing has been introduced [20]. For a fuller account of the development of illumination in DVR, see the survey by Jönsson et al. [21].

For domains of higher-dimension, e.g., $m = 4$, the situation becomes even more challenging. If the dimensions are space and time, as is often the case, then there is a natural mapping into a dynamic visualization (e.g. a cine of a 3D rendering). For other situations, visualizations often depend on application-dependent choices of 2–3 coordinates to render, with some interaction or dynamics to convey the behavior across other coordinates.

For functions with higher-dimensional range $k > 1$, there are several approaches, with some depending on the specific application. An example from imaging is color images for which extensions for volume rendering exist [11]. Another special case is that of *vector fields*, where commonly $m = k \in \{2, 3\}$, and the domain and range are the same space. Direct visualization using vector-like glyphs is often feasible. However, such representations easily suffer from clutter or miss important details of the data. An alternative strategy is to map the vector field onto a scalar quantity, such as the magnitude (or length) of the vector or its orientation. Most commonly used methods are integration-based. They represent a set of lines (e.g., streamlines) following the vector field through the domain or generate textures conveying the directional properties of the field [37]. Figure 6.5 shows some examples of basic vector field visualizations. More advanced vector field visualization methods have been motivated through the task of *flow analysis* with very domain-specific demands. The analysis includes questions related to material transport and characteristic flow structures, as vortices, which are often expressed by derived scalar fields which will be discussed in the next section that discusses the *features*.

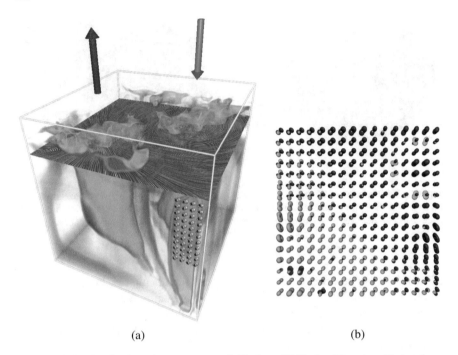

(a)   (b)

Fig. 6.6 Basic visualization of a stress tensor field of a solid block with one pushing and one pulling force. **a** Hybrid visualization: volume rendering of a derived scalar field, here an anisotropy measure, a slice with a texture highlighting the principal stress directions and glyphs in a selected region. **b** A slice showing glyphs (Reynolds glyphs) displaying the entire tensor information in selected locations. Visualization: Jochen Jankoway, https://inviwo.org

Another example is that of *tensor fields*, which often arise from physical processes, such as diffusion [2] or mechanical deformations and stresses [29]. The most direct visualization of tensors is displaying glyphs (e.g., [23, 28]) in selected positions. Glyphs represent the entire tensor information but are limited to low resolution. Continuous visualization methods entail the extraction of scalar values from the tensors, such as tensor magnitudes, eigenvalues, anisotropy, or orientations of eigenvectors. Tensor lines following the main eigenvector direction or textures are used [19] to emphasize the directional character of the tensors. Most commonly used visualizations are hybrid methods combining glyphs with textures and volume rendering of scalar fields [27]. Figure 6.6 shows an example of some basic tensor field visualizations.

More advanced methods consider physically derived fields of tensors or vectors, which often resemble derivatives in their mathematical structure and as such are invariants (e.g., to coordinate transformations), of these objects are particularly interesting, as described in the next section.

There is some work on more general, higher-dimensional transfer functions. These would typically be defined with user input and require effective controls and inter-

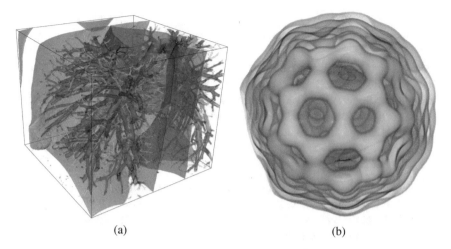

(a)   (b)

Fig. 6.7 Isosurface rendering. **a** CT imaging of a human lung, isosurfaces for two different densities emphasizing the vessel structure in the lung; **b** nested isosurface for a Fullerene molecule. Visualization: Martin Falk, https://inviwo.org

faces, e.g., [35]. The other option is to perform dimensionality reduction on this data, treating the collection of pixel positions in the range as *instance data*, and using some of the techniques in Sect. 6.4 to find lower-dimensional proxies for the data in the range. One can also deal with such data by visualizing a lower-dimensional *feature* extracted from the function, rather than the function directly; this is the topic of the next section.

6.3   Extracting Features from Functions

Often, functions are best understood in terms of specific structural attributes, rather than a description or depiction of the complete function. These special, or meaningful, attributes of a function often consist of subsets of the domain and are referred to as *features*. They can come in the form of points, curves, surfaces, or regions in the image domain. They sometimes include attributes associated with the original function data.

Perhaps the most common or prevalent derived feature associated with the visualization of functional data is *isocontours* (*isosurfaces* for 3D domains), also called *level sets*. In 2D, these contours can help show qualitative features such as high and low points (e.g., their locations and shapes), as well as ridges and valleys. In 3D, these features form surfaces, which allow the use of 3D rendering tools associated with graphics conventions and protocols to facilitate their display. Figure 6.7 shows two examples of level set visualization.

Mathematically, the specification of level sets of functions is stated as a subset of the domain that satisfies a constraint. In 3D, for the kth level set (or isosurface), we have

$$\mathscr{S} = \{(x, y, z) \in \mathscr{D} | f(x, y, z) = k\}, \tag{6.27}$$

which means that the isosurface is the set of points in the domain of f such that the function evaluates to k at those points. Often, when discussing level sets of functions, we consider for simplicity only the zero sets of a function, with the understanding that the kth level set of $f(\cdot)$ is the zero level set of $f'(\cdot) = f(\cdot) - k$. Often, we only consider functions that are considered *generic*, which means that the functions have nonzero derivatives almost everywhere and that the level sets follow certain structures. Level sets in any dimension have several important properties:

- Level sets are closed, except at the boundaries of the domain.
- If f is smooth and generic, level sets are smooth almost everywhere.
- Level sets of different values of k cannot cross and they are nested (enclose each other) according to the values of k, Fig. 6.7b.

Figure 6.9 depicts the general structure of the level sets and the particular examples of singularities for 2D domains.

This focuses on mathematical transformations, rather than specific numerical algorithms, but here we mention that extraction and representation of level sets from functions is itself an important consideration. The most common way to represent level sets is to construct a mesh of simplicies, which are edges in 2D and triangles in 3D. These discrete geometric objects are typically computed from a continuous representation of $f(\cdot)$. There are several strategies. The most common approach is to cover the domain of $f(\cdot)$ with a regular background grid (for instance, squares or cubes) and to identify the cells where the level intersects the boundaries of those cells. From those intersections, the algorithm typically infers some connectivity to insert simplices within the cell that appropriately intersect the cell boundaries. For instance, in 2D, the 2D grid lines intersect the 2D level sets at points, and line segments are used to connect those points within the cell according to a case table (as in Fig. 6.8). For 3D domains, the conversion of cubic or hexahedral intersections with isosurfaces into small patches of a triangular mesh forms the underlying machinery of the marching cubes algorithm [33].

Some other methods for identifying and representing level sets are: placing mesh vertices in cells/cubes that are adjacent to level sets and deforming the resulting mesh onto the level set [56]; placing systems of particles or points near level sets and attracting them to the level set [38]; and finding level-set points and growing surface representations outward from such points [30].

When one varies the value of k, the result is a family of level sets of f, parameterized by k. One can study or visualize the behavior of these sets as f continuously varies. The sets can split, merge, disappear, or appear with different values of k. These behaviors are well defined and these events, combined with the nesting structure of the levels sets, have led to a family of visualization algorithms that represent functions as the family of nested level sets and the *special* events that occur when these sets exhibit isolated, not smooth, behaviors such as merging or splitting, as illustrated in Fig. 6.9.

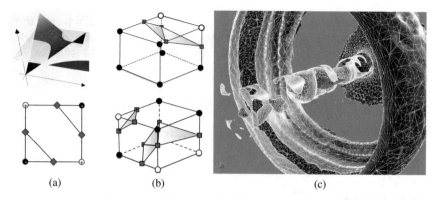

(a)   (b)   (c)

Fig. 6.8 Marching cubes algorithm for contour and isosurface computation. **a** Height field of 2D scalar field over one cell and its discrete approximation using lines. **b** Examples of two case for the approximation of isosurface using triangular simplicies. **c** Example of a mesh resulting from a marching cubes computation

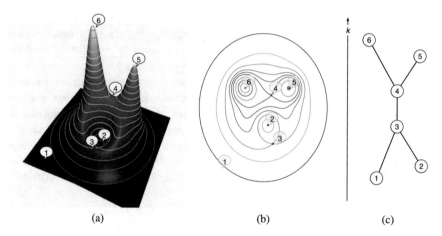

(a)   (b)   (c)

Fig. 6.9 Level sets or isocontours of a 2D analytical data set. **a** Displayed as a height field over the domain; **b** nested contours are shown in the domain; the red dots show the points where the gradient is zero. In maxima and minima, the contours degenerate to points. In saddle points, contours merge or split. **c** The contour tree tracks the changes of the contours when changing the isovalue k

Beyond level sets of $f(\cdot)$, it is also helpful to consider the derivatives of $f(\cdot)$. Here we use the notation D to represent the set of partial derivatives in vector/tensor format, so that in 3D we have:

$$Df = \begin{pmatrix} c\frac{\partial f}{\partial x} \\ \frac{\partial f}{\partial y} \\ \frac{\partial f}{\partial z} \end{pmatrix} \text{ and } D^2f = \begin{pmatrix} \frac{\partial^2 f}{\partial^2 x} & \frac{\partial^2 f}{\partial x \partial y} & \frac{\partial^2 f}{\partial x \partial z} \\ \frac{\partial^2 f}{\partial x \partial y} & \frac{\partial^2 f}{\partial^2 y} & \frac{\partial^2 f}{\partial y \partial z} \\ \frac{\partial^2 f}{\partial x \partial z} & \frac{\partial^2 f}{\partial y \partial z} & \frac{\partial^2 f}{\partial^2 z} \end{pmatrix} \tag{6.28}$$

The magnitude or norm of Df, denoted $||Df||$, is a conventional measure of the contrast a point in the image domain. It is sometimes used to filter level sets of f, so that we can restrict the set to include only those locations in a region that have sufficiently high contrast:

$$\mathscr{S}_{\mathscr{T}} = \{(x, y, z) \in \mathscr{D}| f(x, y, z) = k \text{ and } ||Df|| \geq T\}. \tag{6.29}$$

The derivatives of f also help to define the extrema or singularities of f, which are the set of points:

$$\mathscr{E} = \{(x, y, z)|Df(x, y, z) = 0\}, \tag{6.30}$$

where the comparison with zero indicates that all partial derivatives are zero. This operation can be used to produce a set of singularities that can be viewed as an intersection of level sets. That is, the zero-crossing of each partial derivative produces a level set (on a derivative), or isosurface, and the intersections of those isosurfaces (there are 2 in 2D and 3 in 3D) consist of points that represent the singularities.

These extremal points come in several different forms, depending on the dimension of the domain. In 2D, there are minima, maxima, and saddle points. One can categorize these by examining the eigenvalues of the matrix $D^2 f = DD^T f$ (at every point), which is also called the *Hessian* of f, see also Chap. 5.

When considering or computing features on functions using derivatives, one must keep in mind how these features transform under some basic operations. For instance, the choice of axes for independent coordinates is often arbitrary (e.g., spatial coordinates), and therefore one would expect the features not to depend on that choice. For this reason, we often consider *differential invariants*, that is, features that commute with the rigid transformations (rotation, translation). If we denote the transformation as $T : \mathfrak{R}^3 \mapsto \mathfrak{R}^3$, the invariant feature operator G would behave as follows (in 3D) :

$$g(x, y, z) = G \circ f(x, y, z) \leftrightarrow g(T(x, y, z)) = G \circ f(T(x, y, z)). \tag{6.31}$$

One can easily confirm, for instance, that the length of the gradient vector, $||Df||$, does not change with a change in coordinates. Likewise, singular points are also invariant to rotations/translations of the domain.

Several other aspects of differential operators and invariants are important for visualization. First, many features are developed to characterize *local contrast or variation in function values*. This is true, for instance, of the gradient magnitude, $||Df||$, which is typically considered a place of high contrast in f, also called *edge*. Second, zero crossings of differential operators are often used to find points or features in the domain that are extremal in some property of f.

For instance, the famous *Canny* edge [3] is defined, mostly simply, as the zero crossings of the directional derivatives of $||Df||^2$, in the particular direction of Df. This gives the following condition for these extremal points (in 3D):

$$\mathscr{C} = \{(x, y, z)|g(x, y, z) = 0\} \text{ where } g = (Df)^T D^2 f(Df), \tag{6.32}$$

and thus we see that *edges* are zero level sets of a differential invariant. For robust edges, one typically imposes addition criteria, such as a minimal value (threshold) of $||Df||$ and that the directional derivative of g be negative (for a maximum), although this is rarely necessary when using a threshold.

There are many such invariants and features that can be derived by either thresholding them or finding zero crossings of their derivatives. Other examples include:

- Edges of various types by considering zero sets of second derivatives, as in the Canny edge above, and alternatives such as $D^T Df = 0$, as proposed by Marr and Hildredth [36].
- Extremal points of level sets (local max/min of curvature) by considering level sets of second derivatives of the gradient (third derivatives of f) along the direction(s) perpendicular to the level set. Special care must be taken in 3D, where there is a tangent plane to the level set.
- Ridges on f by considering extremal points of the eigenvalues of $D^2 f$ in various directions. There are various choices here, described extensively by Eberly [7].

Kindlmann [24] gives a compelling overview of this strategy along with various practical considerations. In particular, derivatives are prone to high-frequency artifacts (amplify the magnitudes of small features) and can increase the effects of noise and errors associated with approximations from a discrete grid. The solution to this is usually some combination of smoothing and approximating functions with higher order smoothness guarantees.

In considering these differential invariants, transformations and features on vector fields are also important. For this discussion, we consider vector fields of the form $\mathbf{v} : \Re^m \mapsto \Re^m$, and where the domain and the range are the same space (e.g., the vector is expressed in the same coordinates as the domain). Often, in 2D this would entail $\mathbf{v}(x, y) = v_x(x, y), v_y(x, y)$, and where the subscripts represent components of $\mathbf{v}$ associated with those coordinate directions. This representation is important because it means that the domain and range of $\mathbf{v}$ transform with the same operations (e.g., rotations affect both the domain and range).

The magnitude of such a vector field, $||\mathbf{v}||$, is an invariant. So too are singularities, where $\mathbf{v} = 0$ (once again, the crossing of level set curves/surfaces). Often, such vector fields are the output of a physical simulation, and they represent a physical quantity such as a fluid flow or a mechanical deformation. In these cases, the second derivatives are also important, and they are characterized by the Jacobian matrix:

$$
J = D\mathbf{v}^T = \begin{bmatrix} \frac{\partial \mathbf{v}_x}{\partial x} & \frac{\partial \mathbf{v}_y}{\partial x} \\ \frac{\partial \mathbf{v}_x}{\partial y} & \frac{\partial \mathbf{v}_y}{\partial y} \end{bmatrix} \tag{6.33}
$$

The Jacobian of a vector field bears a resemblance to the Hessian of a scalar function—indeed, the Hessian is a special case. It is the Jacobian of the *gradient field* of a function. While the Hessian is symmetric, Jacobians in general need not be.

A typical strategy in computing features from the Jacobian is to compute invariants of this matrix. For instance, the eigenvalues of the Jacobian, which might be complex-valued, are invariant (to coordinate transformations) and of interest because the real parts describe how the field is pushed in or away from a point, where the imaginary part describes the rotation of the field (around a point). There has been a considerable amount of visualization work that has sought to identify singularities in flow fields (where the flow is zero) and to characterize the rotations and compressions/expansions around those points.

Generally, there is a mathematical system for computing invariants of the Jacobian. Two invariants of particular interest are the trace of J, which is $\text{Tr}(J) = J_{11} + J_{22}$ and is also the sum of the eigenvalues. The norm, which is $\text{Tr}(J J^T) = \sum_{k,l} J_{kl}^2$, is the sum of the squared magnitudes of the eigenvalues. The determinant of J, also the product of eigenvalues, is also relevant to understanding the structure of the field.

In the context of displacement fields, we are often interested in the total amount of deformation pointwise, which is captured in the symmetrized Jacobian

$$\varepsilon = \frac{1}{2}\left(J + J^T\right), \tag{6.34}$$

and the norms of ε produce scalars that summarize this deformation.

In the context flows, the vorticity of the vector field gives the rate rotation at each point—i.e, how would in infinitesimal circle/sphere rotation if its surface followed the flow. In two-dimensions, the vorticity is

$$\omega_z = \frac{\partial v_x}{\partial y} - \frac{\partial v_y}{\partial x}, \tag{6.35}$$

and it has the convention of being a vector perpendicular to the plane (either inward- or outward-facing, depending on the sign). In 3D, the vorticity is written as the *curl* of the velocity

$$\boldsymbol{\omega} = \nabla \times \mathbf{v}, \tag{6.36}$$

and the vector is along the axis of rotation (with direction defined according to the right-hand rule). In both the 2D and 3D cases, vorticity computation results in another scalar or vector field, respectively. Scalar invariants such as magnitude or the acceleration magnitude combined with extremal analysis or level sets produce subsets that allow for the visualization of vortex structures in flows [22]. It is worth noting that in many fluid applications the velocity fields are dynamic, and are functions of space and time, e.g., $\mathbf{v}(x, y, t)$. In such cases, the analysis of vorticity and other flow properties over time becomes important but is beyond the scope of this discussion [15].

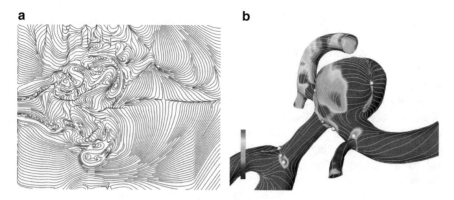

Fig. 6.10 Topology guided uniform streamline placement. **a** 2D jet flow. **b** Surface blood flow of an aneurysm. The background color represents wall shear stress. Visualization: Olufemi Rosanwo, Amira

6.3.1   Integral Curves of Functional Data

In addition to transformations that rely on derivatives of functions, many transformations done for visualization rely on *integrals* of vector fields. Here we consider $\mathbf{v} : \Re^3 \mapsto \Re^3$, as above.

In analyzing or visualizing such flow fields, the integral lines of $\mathbf{v}$. Thus, one can define a streamline $\mathbf{u}$, parameterized with s as:

$$\frac{\partial \mathbf{u}}{\partial s} \times \mathbf{v}\,(\mathbf{u}(s)) = 0, \tag{6.37}$$

which says that the tangents to the curve $u(s)$ are parallel to the vector field. In practice, these curves are usually computed via integration:

$$\mathbf{u}(s) = \int_0^s \mathbf{v}\,(\mathbf{u}(\alpha))\,d\alpha, \tag{6.38}$$

where $\mathbf{u}(0) = \mathbf{u}_0$ is the starting point of the streamline. A typical streamline visualization of a vector field consists of a rendering (in 2D or 3D) of a collection of polylines or tubes that give the overall structure of the field, see Fig. 6.5b. The placement of the initial points for these lines requires some care, so that streamlines are too sparse or too cluttered, and this is an area of significant attention in the visualization community [46], see Fig. 6.10.

When considering dynamic vector fields (vector fields that are functions of time), several more options for integrating curves arise. One option is to let s in the integral above be t, the dynamic parameter of $\mathbf{v}(x, y, t)$, and let the vector field change with t. This is called a *pathline*, and it is the equivalent of letting a particle loose in the flow and rendering the path it travels. A second option is to simulate the path of a

continuous stream of particles placed into the flow at a point. This dynamic curve is
called a *streakline*.

A special kind of vector field is a *gradient field*, which arises when **v** is the gradient of f, i.e., $\mathbf{g} = Df$, where we use **g** to denote this special case. Under these
circumstances, the field is *curl free* by construction (and the vorticity of such a field
would always be zero). Because of this property, the integral curves of such a gradient field always connect singularities, where $Df = 0$ These singularities consist
of different types: minima, maxima, and saddle points. The status of a singularity is
determined by evaluating the eigenvalues of the Hessian. If we consider the eigenvalues in descending order, $k_1, \ldots, k_m$ for an m-dimensional domain, we have the
following:

$$
\begin{aligned}
k_1 \ldots, k_m &> 0 \text{ minimum} \\
k_1 \ldots, k_j > 0, \ k_{j+1}, \ldots, k_m &< 0 \text{ saddle} \\
k_1 \ldots, k_m &< 0 \text{ maximum}
\end{aligned}
\tag{6.39}
$$

Notice that we do not normally consider cases where the eigenvalues are zero, because
these are not considered *generic* or *regular* points, and they show up with very low
probability (in theory). In practice, special care must be taken to avoid the numerical
problems associated with data sets that do not meet these criteria.

Virtually every point in the domain of a (generic, regular) function has a gradient.
The integral curve of the gradient field from that point terminates at a maximum. A
relatively few points will terminate at a saddle. From saddle points, one can trace
curves (in the directions of the eigenvectors of the Hessian) toward sets of maxima
(e.g., pairs in 2D). This same analysis extends to toward minima/saddle points if one
integrates the negative of the gradient field, $-\mathbf{g}$. Note that a very similar concept can
be applied to general vector fields. Here, limit sets play the role of critical points. As
for the gradient field, there are locally defined limit sets. These are sources, sinks,
and saddle points. In addition, there are, however, also nonlocal limit sets. In 2D,
these are periodic orbits; in 3D, more complex configurations are possible [17].

Using these ideas, we can partition the domain of the image into regions that each
share the same minimum. The set of points in the domain whose integral curves of
$-\mathbf{g}$ lead to the same minimum is often called a *watershed*, because if the function
were treated as a topographical surface and water were to flow toward minima (due
to gravity), the water falling (e.g., in a rainstorm) on that region of the domain would
all flow toward the same location. This kind of *watershed segmentation* has shown
up extensively in the image processing literature (and software) for partitioning
images around edge-like features, such as the gradient magnitude, as in Fig. 6.11.
In performing this kind of analysis, one must recognize that the number of minima
and/or maxima in a field of data (function) can be arbitrarily large, especially in
regions of the image where the gradients are small (nearly flat regions). To address
this, we typically filter this partitioning of the domain, and combine regions based on
the *depth* of the watershed, as in Fig. 6.11. Each watershed has along its boundaries a
sequence of maxima and saddle points. The difference between the function value at
the minimum and the saddle point of least value is the watershed depth. More recent
work has referred to this depth as the *persistence* of a watershed region, and which

shows stability under certain conditions [8]. Watersheds that are not deep (shallow) are often combined with adjacent watersheds to form larger, deeper regions. This can be done interactively by users, with an appropriate interface [5].

Several other aspects of this kind of *topological analysis* are important for visualization. First, if one considers the ascending and descending integral curves (integrating negative and positive gradient fields), they (almost always) terminate at maxima and minima points, respectively. The sets of points that share maxima and minima (terminations of descending and ascending gradient flows) also form a partitioning of the domain, which is sometimes called the *Morse–Smale complex* and the individual elements (region and min/max pair) are *crystals*. The boundaries of these regions

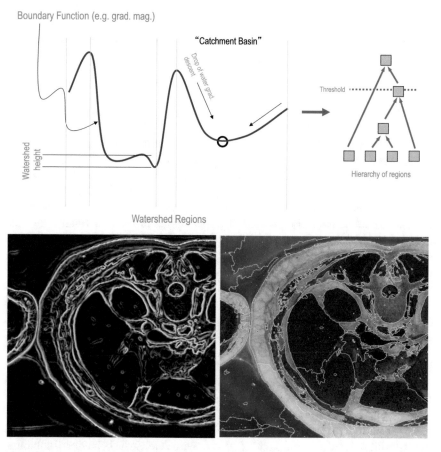

Fig. 6.11 Top: a watershed decomposition of a function tracks regions for which the integrals of the gradient fields terminate in a common minimum (or maximum). Bottom-left: the gradient magnitude of an anatomical image indicates boundaries of regions. Bottom-right: a partitioning of the function (overlaid with white lines on the original color image) shows watershed regions

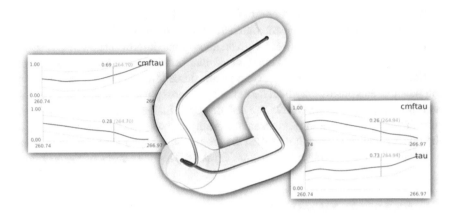

Fig. 6.12 Figures from Gerber et al. [12] show a rendering of a Morse–Smale (M–S) decomposition of the high-dimensional parameter space associated with climate simulations, including selected parameter values from the two M–S crystals

consist of ascending/descending integral curves (surfaces, or families of curves, in 3D) that pass through saddle points.

This strategy, of reducing a function to it singularities (minima, maxima, and saddles) and connecting those singularities by either the Morse–Smale crystals or the curves that connect saddles along the boundaries, has been proposed as a way of visualizing the structure of complex or high-dimensional functions [31]. Much like level sets and streamlines, this kind of analysis produces a discrete set of geometric objects that are more easily rendered than the original function. Virtually any visualization method that relies on this kind of topological analysis must include some manner, as described above, of removing/combining shallow, small, or otherwise insignificant regions.

An example of this kind of topological analysis is the work of Gerber et al. [12], where they visualize high-dimensional scalar functions by rendering the function as a graph, with extremal points as vertices, connected by edges, rendered as curves/tubes, that represent the structure of the Morse–Smale crystals that connect those extrema. See Fig. 6.12. The method relies on embedding discrete sets of singularities into lower-dimensional spaces (2D or 3D) as described later in this chapter.

Also important to these topological analysis methods are the methods that combine the analysis of singularities with levels sets (or *contours*). If one considers the level sets of a function at some value k, then the family of level-set curves or surfaces forms patterns that adhere to certain rules. For k increasing and considering a contour to be a curve with an interior defined with $f(\cdot) < k$, we can track the behavior of contours:

A. Topologically separate contours form/begin at minima (of value k), as points and then isolated, closed contours (curves or surfaces).
B. Contours join/merge at saddle points, and the new structures can achieve alternative/complex topologies (e.g., holes) as they merge.
C. Isolated holes in contours contract to points (and annihilate) at maxima.

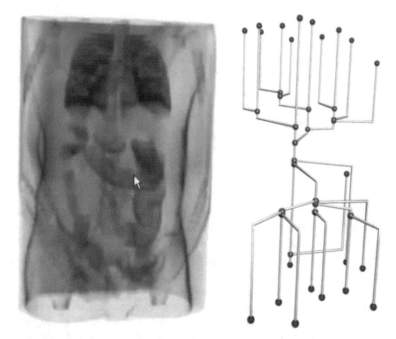

Fig. 6.13 The structure of a 3D function is characterized by a volume rendering and a rendering of the associated Reeb graph. Image courtesy of Vijay Natarajan and Harish Doraiswamy

This kind of analysis [9] forms a graph (sometimes referred to as the *Reeb graph*) that facilitates the visualization of a function in terms of its associated Reeb graph [34], as in Fig. 6.13.

6.4 Visualizing Instances by Dimensionality Reduction

A typical visualization problem is as follows. A data set consists of a number of instances of structured data. In the following discussion, we also refer to an instance as a *data point*. One would like to visualize these points to understand the following:

- Do the points group together or form *clusters*? If so, how distinct are these clusters, how many are there, etc.?
- Are there trends or relationships among points and variables that could give qualitative or quantitative insights into the collection of data?
- Does the data conform to expectations of samples from known probability distributions, such as normal distributions?
- Does the set contain instances that are unusual or very different from the other instances? How different and how many are there?

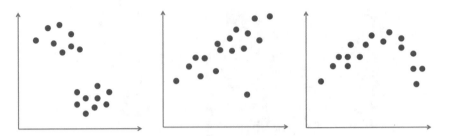

Fig. 6.14 Scatterplots show clustering, correlation, and nonlinear structure

If the data points are samples in a 2D (or even 3D) space, one can typically rely on direct visualization via a *scatterplot*, where individual points are represented via the positions of symbols or glyphs (e.g., dots, squares) on a 2D graph (or a 3D cloud within a 3D or interactive display). Figure 6.14 shows examples of 2D scatterplots that demonstrate some of the properties above.

Of course, as we consider the analysis of data points, we must be aware of the opportunity and/or need for quantitative analyses. For instance, often when looking for relationships among variables, one considers the correlations among variables or the best-fitting linear model (e.g., fitting a line in the 2D case). Anscombe [1] describes a quartet of examples where best-fitting lines for 2D data points can be misleading, as a motivation for direct data visualization. This danger, of being potentially misled (or at least underinformed) by a simple model, is a very general threat to people using and analyzing data; it goes beyond linear models. For instance, people will often consider the mean and (co)variance of a distribution, which often misses important aspects of a data set (such as outliers, skew), and the *whisker plot* (or *box plot*) is a common visualization tool for 1D points, using rank statistics, that helps evaluate properties beyond mean and variance. In general, virtually any parameterization or low-dimensional model of a set of instances risks missing some important aspects of the data. Yet, for high-dimensional data, direct visualization is often impossible. Thus, a complementary approach that combines visualization and analysis is often required.

One of the most common methods for visualizing point sets (instances) of more than 2 dimensions (and assuming a metric space) is to *project* the data onto a 2D subspace. The most widely used method for this is *principal component analysis* (PCA), which is equivalent to finding the k-dimensional, linear subspace that minimizes the projection distance onto subspace. The procedure, mathematically, is as follows: A point set X is represented as a matrix

$$X = \begin{bmatrix} x_{11} & x_{12} & \cdots & x_{1n} \\ x_{21} & x_{22} & \cdots & x_{2n} \\ \vdots & & \vdots & \\ x_{m1} & x_{m2} & \cdots & x_{mn} \end{bmatrix}, \tag{6.40}$$

where x_{ij} is the ith coordinate of the jth data point. The data is first *centered*, so that the mean is zero. Thus, we have $\hat{X} = X - \bar{x}$, where $\bar{x} \in \Re^m$ is the mean across the data. That is

$$\hat{x}_{ij} = x_{ij} - \frac{1}{n} \sum_j x_{ij}. \tag{6.41}$$

From the centered data, next compute the inner product, or correlation matrix

$$C = \hat{X}\hat{X}^T. \tag{6.42}$$

The k-dimensional *basis* for the projection consists of the first k eigenvectors (ordered by decreasing eigenvalue) of C, which we denote, $E = e_1, \ldots, e_k$. The lower-dimensional coordinates for centered data points are the *loadings* of the data onto this new basis:

$$Y = E^T \hat{X}. \tag{6.43}$$

These coordinates can then be used for visualization, e.g., when $k = 2$.

The new coordinates, Y are in terms of the basis vectors, V, which form a k-dimensional, hyperplane in $\Re^m$. The hyperplane coordinates for the data are computed as

$$X_p = EE^T \hat{X} + \bar{x}, \tag{6.44}$$

where X_p is the projection of the data onto the best-fitting, k-dimensional hyperplane.

This kind of transformation, of finding a lower-dimensional space (and a smaller set of coordinates) to represent a set of points etc. is sometimes called an *embedding* of the data, because it assumes that the original nD data is positioned on a kD manifold (in this case, a hyperplane) that is embedded in the higher-dimensional space. As we consider this process, it is important to keep several things in mind. First is the accuracy of the representation. For PCA, the *projection error* of the points onto the kD hyperplane is given by the root of the sum of squares of the eigenvalues associated with the $n - k$ smaller eigenvectors. To visualize the effects of projection, we often use a *scree plot*, which shows the percentage of the total variance captured in the first k eigenvalues, as shown in Fig. 6.15. Also note that PCA is the optimal choice of kD hyperplanes to model data since it minimizes this projection error, or residual. Also worth noting is that in some cases the number of samples is smaller than the dimensionality of the ambient space (i.e., $m < n$). In this case, the better computational strategy is to work on the dual of the original problem, which operates on the linear subspace defined by the data points. For this, we conduct the eigenanalysis on the matrix $C' == \hat{X}\hat{X}^T$, which has the same eigenstructure, E', as C, defined in (Eq. 6.42). The basis is obtained by multiplication with the data itself:

$$E = X\Lambda^{-\frac{1}{2}} E', \tag{6.45}$$

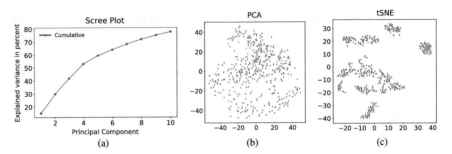

Fig. 6.15 a A scree plot of MNIST handwritten digit data depicting percentage of variance explained by PCA modes. **b** A 2D embedding/layout using PCA loadings (units are arbitrary). **c** A 2D embedding using t-SNE

where *Lamda* is the diagonal matrix of eigenvalues. In cases where m and n are both very large, the construction of the associated (large) covariance matrix can be prohibitive. In these cases, the largest eigenvectors/values can be found through iterative methods, such as *power methods*, that do not require explicit construction or decomposition of the matrix (e.g., the power method).

Notice that the dual formulation of the PCA problem relies on the analysis of the $n \times n$, inner product matrix (also called a *Gramm matrix*). This opens up the possibility of embedding data points that may not be given in a conventional, metric space, but for which there exists only an inner product (or similarity) operator. This situation arises, for instance, when using the *kernel method* for analysis of data.

An alternative method for formulating the embedding of data points into low-dimensional spaces is via the matrix of distances between all pairs of data points, which we denote as D, with elements d_{ij}. The goal is to find parameters $Y \in \mathfrak{R}^k$ so that the distances between points in $\mathfrak{R}^k$ match, as closely as possible, those given in D. This problem is sometimes called *multi-dimensional scaling* (MDS), and MDS is often used to refer to the family of methods that try to find such coordinates, Y. The *classic* approach to MDS (indeed, called *cMDS*) is an algorithm that centers the matrix D and then computes the eigenvectors, as in PCA. The algorithm is:

A. Construct the squared distance matrix, $D^{(2)}$, where $d_{ij}^{(2)} = d_{ij}^2$.
B. Center and negate the squared distance matrix:

$$B = -\frac{1}{2} J D^{(2)} J \quad \text{where} \quad J = I - \frac{1}{n} 11^T \tag{6.46}$$

and 1 is a vector with values of 1 and length n.
C. The coordinates are $Y = E_k \Lambda_k^{frac12}$.

Note that cMDS minimizes the normalized strain of the embedding in the case where the original distances are from a Euclidean space (e.g. $X \in \mathfrak{R}^m$). However, this same algorithm is used in many other settings for visualizing data, e.g., $k = 2, 3$, with useful results.

There are many other approaches to MDS, but one that it also very widely used is to directly minimize the stress associated with the embedding. The stress is typically

$$\text{Stress}_D(Y) = \left(\sum_{ij} (d_{ij} - ||y_i - y_j||)^2 \right)^{\frac{1}{2}}, \qquad (6.47)$$

where y_i, the ith column of Y, are the new, embedded coordinates for the ith data point. This function is bounded by an approximation that is quadratic in the unknowns, and minimization is solved efficiently using the iterative *SMACOF* algorithm.

There is a deep relationship between (squared) distance matrices and the matrix of inner products, also called the *Gramian matrix*, which is used in the dual formulation of PCA. Under certain conditions (e.g., distances/products in Euclidean space) one can be derived from the other. For embedding points for the purposes of visualization, these two types of matrices are used in a similar manner; their eigenvectors are used to construct coordinates in a new, lower-dimensional space. Thus, for many visualization applications, practitioners will use either distance or inner product matrices, depending on what is available and appropriate for the original data.

The ability to compute low-dimensional coordinates using only distances (or similarities, inner products) gives rise to some important technologies for visualizing collections of points, even if these data points do not have well-defined coordinates in some metric space. Here we give several examples of how this is useful.

When modeling high-dimensional data, it is sometimes useful to treat the data as existing on a lower-dimensional, curved (or nonlinear) surface, or manifold. The so-called *manifold learning* problem has received a great deal of attention in machine learning and statistics, but has become less important in recent years because of advances in machine learning technologies that can learn directly on complex, high-dimensional data sets. However, for visualization, manifold learning is still a useful dimensionality reduction method. A relatively easy to use for discovering manifold coordinates is the method of *isomap* [51]. The isomap algorithm is designed to construct an approximate distance matrix that captures distances *within the manifold*, which is embedded in a high-dimensional, mD, space. The isomap algorithm works as follows:

A. Compute the distance matrix D for the original data.
B. Determine the K nearest neighbors (kNN) for each point (K is a free parameter), and construct the KNN graph with edge lengths being distance.
C. Compute the distance between each pair of points on the KNN graph (e.g., using breadth-first search), and construct a new distance matrix D'.
D. Determine new coordinates from MDS (any method) on D'.

This algorithm has been shown, in some cases, to learn the manifold structure from very curved or convoluted manifolds in high-dimensions. The main challenge with the isomap algorithm is the selection of the number of nearest neighbors K, because this determines which jumps on the NN graph will be considered within the manifold.

If K is too small, the graph becomes disconnected (and the eigenstructure of D' shows this), and if K is too large (in the limit), the method produces results that resemble MDS on the original distances.

An alternative to preserving distances between points is to pose the embedding problem as preserving *data density*, which is the strategy of the *t-sne* embedding method [54]. Local probability densities are computed with nonparametric density estimation, and the target coordinates Y are constructed so that every point has a similar nearby density of points. The method is widely used and generally effective, but has the effect of preserving or enhancing clusters in the data (which have higher data density).

Another case in which MDS-like methods are useful is when the data points have intrinsically no coordinates, but where distances or similarities are readily available. This comes often in the context of graph layout. The edges of a graph often have weights that are associated with either dissimilarity (approximate distance) or similarity (inner products). This happens, for instance, when vertices have associated signals, such as voting patterns for politicians, weather patterns for cities or stations, or, in biology, interactions between genes, molecules, or organisms. In these cases, it is sometimes helpful to embed the graph vertices in 2D as part of the visualization. As above, the affinities/similarities or distances are part of the computation of 2D coordinates for the vertices. Stress minimization, as above, is often the method of choice, in part because it can be combined with other criteria. The problem of graph layout is widely studied, and effective solutions often address concerns in addition to distance, such as edge crossings, and edge/vertex density.

6.5   Data Summaries

In many cases, we are presented with a collection of *instances*, where each individual instance may be a data point or some more complex object, such as function, unstructured document/record, or a graph. Organizing these data as points or icons in a 2D display, as in the previous section, can often be helpful, it is sometimes effective to summarize the overall structure and relationships of these instances. This section describes some methods of summarizing collections of instances. Here we assume that the data is homogeneous and structured, and we leave the discussion of more complex data types for the next section. Of course, the choices we make in summarizing data depend on the kinds of questions we are trying to answer and the applications we have in mind. Here we give some examples of the most widely used strategies.

A typical problem in visualization of instances or points sets is to understand the relationships among samples and if the data naturally form groups or clusters (and the nature of those clusters). The notion of *clustering* is a longstanding problem in pattern recognition and data analysis, and there are a wide variety of approaches. Typically, data is said to consist of clusters if there are two or more subsets of the data for which the point-to-point distances are smaller than the distances to nearby groups.

A typical formulation of the clustering problem is as follows. Inter- and intragroup distances are quantified (e.g., as sums over distances between point pairs), giving rise to an objective (or energy) function that can be optimized. Because the energy involves assignments to groups, it is combinatoric and nonconvex, and thus it is often solved iteratively. A widely used and generally effective clustering algorithm for points in a metric space (distances can be computed) is *K-means*, which is the following algorithm, for input data $X = \{x_1, \ldots, x_K\} \in \mathcal{D}$, where $\mathcal{D}$ is the domain of the data points:

A. The user decides on the number of clusters K, and the cluster centers $C_1, \ldots, C_K$ $\in \mathcal{D}$ are initialized (usually at random).
B. Each data point is assigned to the nearest cluster center.
C. The cluster centers are updated and assigned to the average of the data points to which they are assigned.
D. If the update is sufficiently small, terminate, otherwise, go to step B.

The output is the positions of the cluster centers and their assigned data points.

This kind of clustering or grouping of data points can impact visualization in several ways. In some cases, it is useful to visualize the clusters themselves. Thus, the clusters would be embedded (e.g., using MDS, as above) in 2D, and each cluster would be represented with a mark or glyph, which might also encode information about the cluster, such as its number of elements, extent/variability, or center element. Another use for clusters is to modify the glyphs in a visualization of a 2D embedding/projection, which can help identify differences in data along dimensions that are not well captured in the embedding. Finally, some embedding algorithms are designed to preserve information about clusters in the data, where the separation of clusters (detected as a preprocessing step) is a criterion that is built into the objective function that is optimized for the embedding. The general strategy of clustering data is useful in other contexts, where collapsing or summarizing groups of instances aids in interpretation. For instance, clustering of edges in large graphs has been used for *edge bundling*, to reduce complexity in graph visualization [6].

The problem of clustering data points has also been examined from a very different point of view—using hierarchies of graphs. We consider the ε-graph of a data set as consisting of a vertex for each data point and edges connecting every vertex pair x_i, x_j if and only if $d_{ij} \leq \varepsilon$. Then we can consider the connected components of that graph to be individual clusters. This is sometimes called *single-linkage clustering* and it is known to be unstable when one makes small perturbations to the data points. However, if one considers the hierarchy of clusters as a function of ε, as in the *dendritic tree*, this can provide information about the texture/structure of the point set. Also, clusters that are stable or persistent through a wider range of ε values might be considered more important (e.g., more robust to perturbations in data), sharing the same mathematical underpinnings as the watershed depth, described in the previous section.

This kind of basic, distance-based, cluster analysis is the simplest example of a very rich set of methods in computational topology called *persistent homology* [4, 8]. Here we give only a high-level view of the methodology. First, one extends the notion of the ε-graph, as above, to include a filtration (nested sequence) of *simplices*, which are not only vertices and edges, but also triangles, tetrahedra, etc. There are several approaches for constructing those filtrations of simplices. One can then compute, in a very precisely defined manner, topological summaries, including not only the number of connected components, but also tunnels (holes in one-dimension, loops in higher-dimensions), and cavities (2D holes, hollow regions enclosed by a surface). We can also track changes in these summaries (or the corresponding feature, such as a hole) as ε increases. These summaries are sometimes visualized as a collection of stacked horizontal line segments (or bars), where the ends of each bar correspond to the appearance (or birth) and disappearance (or death) of the associated feature, as a function of ε.

6.5.1 Statistical Summaries

In visualizing sets of instances of data, it is sometimes difficult to make sense of the raw data, especially if the individual data points have an inherently complex structure or if there are especially many of them. In many cases, one would like to get a high-level or *big picture* view of the data. This kind of visual analysis is often for quality control or to inform some other type of quantitative analysis. For this reason, it is often useful to construct statistical summaries of data and to visualize those summaries either instead of or in addition to the raw data.

A typical summarization strategy is to compute the mean an variance of a data set. For instance, if we consider functions $f_1(\mathbf{x})$, $f_2(\mathbf{x})$, ..., $f_n(\mathbf{x})$, with 1-2-3D domains, there is a sample mean $\bar{f}(\mathbf{x}) = (1/n) \sum_i f_i(\mathbf{x})$ and an associated covariance. Here we avoid a discussion of the technical issues associated with variance in functions spaces, and instead assume that f_i has a finite-dimensional representation (e.g., values evaluated on a regular grid, as with an image or volume), and each instance is represented as a vector of length n. The covariance structure can be very high-dimensional ($n \times n$ matrix), and difficult to visualize, so simplifications or approximations are common. The most common simplification is to compute the variance of f pointwise over the ensemble for every $\mathbf{x}$ in the domain. This is equivalent to considering only the diagonal of the covariance matrix, and it ignores the correlations between points. Figure 6.16 shows this mean and pointwise covariance for a data set from a fluid simulation. Alternatively, one can visualize the eigenvectors of the full covariance matrix (typically, one would use the dual method described in section about dimension reduction) and visualize the eigenvectors of the covariance.

With functions, one is often interested in features, and how those features behave within a set or ensembles of functions. As in Sect. 5.4, many interesting features can be represented as zero-crossing of the function itself or fields of data derived from that function and its derivatives. Several researchers have proposed to extend the computation of level sets to the probabilistic setting. The problem can be stated

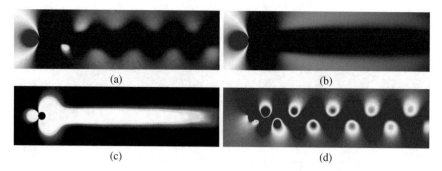

Fig. 6.16 Analysis of a set of pressure data from a fluid simulation, with flow left to right across a circular obstacle. **a** One example of a pressure field from this simulation (purple-low, green-high). **b** The mean pressure field from an ensemble of 300 samples. **c** The pointwise variance (heat map). **d** The first eigenvector of the covariance matrix

as computing the probability of a level set passing through or between a set of grid points (or pixels), given a stochastic model of nearby function values. This is the strategy behind probabilistic marching cubes and several variants [42].

The strategy of computing the mean and (co)variance to summarize data has limitations, in that it reduces the ensemble of (possibly) complex data sets to a relatively small number of values, and these summaries are sensitive to data outliers, and, as in the case of point-wise variances, ignore global relationships (e.g., correlations) in the high-dimensional data. Indeed, these are often the very things that we are attempting to detect or understand as we visualize such ensembles. Thus, nonparametric is *descriptive* approaches to summarizing data are also important.

The descriptive approach to summarizing data is well motivated by one of the most widely used of all visualization tools—the box or whisker plot, proposed by Tukey [53] and shown in Fig. 6.17. The whisker plot typically shows a summary of rank statistics of a set of 1D data points, with bars or icons to indicate the median (and often the mean as well), various percentile ranks (e.g., 25 and 75%), and outliers. These rank statistics are computed from an *ordering* of the data along a single axis. The extension of this kind of visualization to more complex data requires two developments. The first development is the generalization of rank statistics to multidimensional and nonmetric data. For this, several researchers have proposed the use of *data depth*, which is a tool from descriptive statistics that constructs a *center outward* ordering of a collection of data points. In such a scheme, the median of a data sent would typically be the *deepest* among the given ensemble. There are several methods for computing the depth of a data point within an ensemble, but a useful strategy is the method of *band depth*, where the depth of a data point is computed as the probability that it lies between a small, random selection of the data. The notion of *between* must be defined for each data type, depending on the application, and the probability of lying within a band formed by a random set of samples is computed with a Monte Carlo approach, which is a sample average is computing by choosing small subsets from the given data.

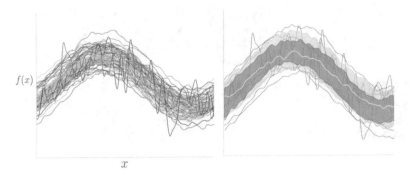

Fig. 6.17 Left: an ensemble of functions have some common structure and show significant variability. Right: a functional box plot, as proposed by [50]

The data depth approach is developed for functions, where the band is formed by the min/max values at each point in the domain for a small set of j functions (chosen at random) [41]. Given a set of functions, one can compute the median and the min/max extent of the functions within a certain rank of the data (e.g., 50%). Sun and Genton [50] use function band depth to construct function box plots, which are the natural extension of whisker plots to functions, as in Fig. 6.17.

For points in $\Re^n$, band depth is the probability that a given point lies in the simplex or convex hull of $K > n$ randomly chosen points. For large n, the availability of sufficiently many K-sized subset is often prohibitive, and alternatives, such as half-space depth [52] and spatial depth [48], become desirable. For points in $\Re^2$, a depiction of the 50% band and an inflated version of that, with outliers marked, is called *bagplot*.

Several researchers have proposed extensions of data depth and associated visualizations, extensions to box plots, to more complex data types. The method applies to 2D, scalar functions [50], as well as curves in 2D and 3D. Whitaker et al. [57] have extended data depth to sets and show box plots for level sets (contours) in 2D and 3D [44]. Figure 6.18 shows some examples of 2D and 3D box plots for different objects. Raj and Whitaker have extended data depth to vertices and paths on graphs [45].

In dimensionality reduction, the projection of data into lower-dimensions often obscures the relative depth of a data point, and thus outliers can be misrepresented as being central to the dataset. Raj and Whitaker [43] have proposed dimensionality reduction techniques that preserve data depth, in addition to distance (or density), as in Fig. 6.19.

6.6 Transformations on Unstructured and Discrete Data

Virtually all of the methods in the previous sections of this chapter rely on quantitative relationships between samples or points in the range or domain of a function.

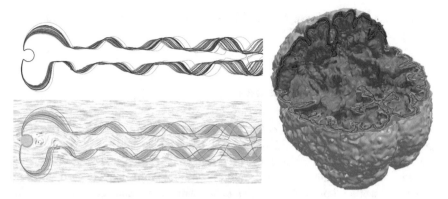

Fig. 6.18 Left: contour box plots of isocontours of pressure in an ensemble of fluid simulations. Right: 3D contour box plots from an ensemble of registered brain images

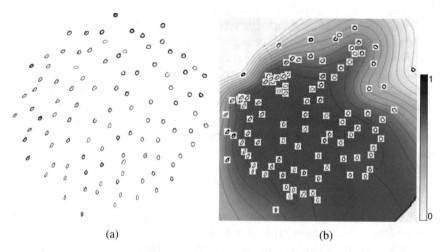

(a)   (b)

Fig. 6.19 2D layouts of the MNIST "0" digit data. **a** Layout using an MSD embedding. **b** Layout with depth-aware embedding that organizes by depth with contours/colors that show relative depth of samples

However, many data sets come in forms that are not well suited to the quantification of distances, similarities, or coordinates. A very common example is a *corpus* of text documents, as one may have from a collection of news articles or emails. Each instance, in this example, is a document consisting of words, spaces, and punctuation (we will ignore emojis for this discussion :-)). One might like to visualize the corpus of documents and understand how they relate to each other. Typically one might like to know if there trends over time, if they form clusters, if there are outliers, etc. These kinds of questions might benefit from a scatterplot visualization, a clustering, or topological analysis. However, these methods will require distances or coordinates for each sample (each document in this case). This chapter discusses some of the methods by which unstructured data, such as a text document, are encoded for subsequent analysis or visualization.

6.6.1 Organizing Data in Bags

A text document consists of a collection of words in a particular order. The actual words used and the order in which they appear give the document its meaning. However, practitioners of natural language processing have noticed that some information about a document can be discerned from the types of words that are used, and their frequency, while ignoring the ordering of the words, the sentences, grammar, etc., in the document. This leads to a way giving coordinates to a document—we simply count the number of times each word occurs in a document and the resulting histogram becomes a quantitative descriptor of the document, which induces a distance and/or inner product computation with other documents. Of course, there are a great many words in any particular language, and typically the word count strategy ignores very common words that are present in large numbers in all documents, such as articles, conjunctions, and prepositions. Through histograms of meaningful words, documents can be clustered or embedded in kD spaces for analysis and visualization. Because of the large number of words (bins in the histogram), this analysis is often combined with PCA to produce a smaller set of descriptors, which make subsequent computation more tractable. Figure 6.20 shows an example of an embedding of a corpus of news articles by word counts. Notice that this kind of analysis can generally give information about the general *topic* of a document, but it loses the meaning of the document, because to discern meaning one must typically examine the semantics of individual sentences.

This general strategy for documents is referred to as a *bag of words* because it treats each document as a container (bag) for words that ignores the ordering of words, as well as the construction of phrases and sentences. This *bag* strategy has been used in a variety of contexts to deal with large sets of unstructured instances. For instance, in images, local features (measured through some time of detector and descriptor, such as corners or textures) are counted, and their location ignored, to determine the environment of an image or to quantify the similarity/difference between images in a large collection. Graphs are often unstructured and difficult to compare, but one can construct a *fingerprint* of a graph by quantifying different types of local neighborhood structures around vertices. Researchers have compared histograms of valences of vertices, numbers of cliques of different sizes, or, if the vertices have labels, categories of vertices based on their neighborhood structure [40].

If we consider a text document as a string of tokens (words), then the bag approach has been modified for many variations. For instance, besides documents, one might also need to quantify or give coordinates to words (this also helps in documents). In deciding if two words are similar, we can quantify how often they occur with or are near other words. In this way, cooccurrence (defined at some scale—phrase, sentence, document) becomes a signature for comparing words. This strategy has

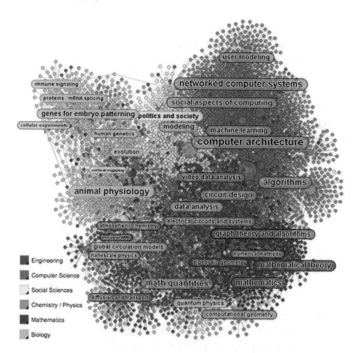

Fig. 6.20 A visualization of a corpus of articles from [14] are organized in 2D based on associations with (probabilities) topics, which are derived from vectors of word counts, i.e., *bag of words* analysis

been incorporated into various neural net approaches, as described in subsequent sections.

6.6.2   Edit Distance

Another common way of quantifying distances/similarities between unstructured or complex data types is to consider the cost of converting one instance to another. This is often done by describing a set of atomic *editing* operations on the data object and assigning a cost to each type of edit. For instance, in comparing the lexicographic words (ignoring their semantics), we could assign a cost to changing a letter in a word, as well as adding or deleting letters. Thus, to convert the word "Sunday" to "Saturday", we notice that first characters are the same, as are the last three. To convert the middle "un" to "atur", we might replace "n" with "r", and then insert a "t" and an "a". The precise edit sequence would depend on the cost of each operation, but the *edit distance* is typically the cost of the *least expensive edit* that converts one object to another. This edit distance is used extensively in genetics to compare genetic sequences (which are, essentially, strings). If the cost structure is properly constructed, this edit distance is computed efficiently using Dykstra's algorithm.

Another example of edit distance is in the analysis of graphs. Unaligned graphs are graphs where the vertices are not uniquely identified from one graph to another. Graphs with different types of nodes can be compared by removing, introducing, or changing the labels on nodes, and by allowing similar edits on edges. This kind of edit distance can then be used to cluster ensembles of graphs or embed them in lower-dimensional spaces, or visualize their evolution in time. Graph edit distance is computationally challenging; it is NP-hard in general. However, special cases of graphs (e.g., acyclic) are compared more tractably, and approximate solutions are often quite effective.

6.6.3 Kernel Methods

A very useful tool in data analysis is to construct a similarity measure between pairs of instances for a particular data type and then rely on the *kernel method* or *kernel trick* to conduct analysis in the space induced by this similarity measure. Mercer's theorem states that for any *kernel*, $k(x_1, x_2)$, operating on pairs of data points/instances that it is guaranteed to produce a positive-definite inner product matrix, there is a corresponding Euclidean space for which this kernel is the Euclidean inner product (dot product). This technique allows one to define inner products to create high-dimensional spaces for which there might not be an explicit representation.

If the data points in the analysis have an associated metric space, then monotonically decreasing functions of point-to-point distance from a Mercer kernel. Indeed, a widely used kernel is the Gaussian function of distance:

$$k(x_1, x_2) = \exp\left(-\frac{|x_1 - x_2|^2}{2\sigma^2}\right), \qquad (6.48)$$

where σ is a free parameter that must be tuned to a specific application. The Euclidean space associated with this kernel is not finite-dimensional, and has no explicit set of coordinates. All operations in the kernel space are represented in terms of inner products with the given data ensemble. However, this lifting of the data into the kernel space provides opportunities for

A variety of methods have been adapted for kernel spaces, including PCA (called *kernel PCA* [47]), clustering (also *spectral clustering*), regression, and classification (e.g., *support vector machines*). Kernels or inner products can be defined using bag-of-words strategies (kernel is typically a function of the product of the histograms), or some other distance measures on structured or unstructured data. The lifting of data into the kernel space has the advantages of (i) analysis operations without an explicit distance measure and (ii) moving the data into spaces where simpler models (e.g., linear) for regression, clustering, and classification are often more effective. Likewise, this kind of separation of data in the kernel space can aid in visualizing trends or clusters.

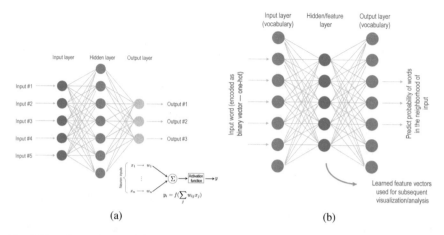

Fig. 6.21 **a** A neural network is a sequence of layers consisting of individual elements (neurons) that linearly combine outputs of the previous layer and perform a nonlinear activation (e.g., smooth threshold). **b** The skip-gram architecture for assigning words to vectors develops a feature vector that is effective at predicting the *context* of a given word—i.e., the probability of nearby words

6.6.4 Neural Networks

Recently, technology in the training and application of artificial neural networks has provided new opportunities for transforming data and embedding data sets into spaces that are well suited for analysis and visualization. There are a great many introductions or tutorials on neural networks, and here we assume that the reader is familiar with the basic technology, which we review briefly.

A neural network (NN) is a set of processing units, each of which performs a linear combination (weighted sum) of inputs and produces an output, which is a nonlinear function of that weighted sum:

$$y = \phi \left(\sum_j w_j x_j \right), \tag{6.49}$$

where x_j is a vector of inputs and w_j are the weights associated with the inputs to this particular neuron. The nonlinear *activation function*, $\phi : \Re \mapsto \Re$, is typically some kind of soft threshold. These individual elements are arranged in layers (where the elements of a layer share the same inputs, x_j), and the network transforms data by passing it through a sequence of layers, as in Fig. 6.21.

Most NNs are trained in a *supervised* manner, and the weights are modified incrementally so that the outputs of the network approximate the training data for any particular input. The training of NNs has an associated, and very extensive, set of methods, research, and theory.

The conventional wisdom is that the layers of a neural network perform a sequence of transformations on the input data, making the data progressively better suited to the task of the final layer, which must produce the desired output from a linear combination of input data. Thus, the network in its intermediate layers is successively transforming the data into spaces that are well suited to the task, and thus the intermediate layers represent transformations of the data that can aid in analysis and visualization. Besides the architecture and training of the network, the important issues for embedding are how to encode the inputs and outputs and how to define the supervised problem. Here we give an example that demonstrates the principles.

A classic problem in the analysis of text and documents is the vectorization of words, which is the assignment of each word to a coordinate in a space where distances reflect the similarities between the meaning and usage of words. Recent work in neural networks has addressed this problem by constructing and training NNs that learn associations between words in sentences. There are several versions of this architecture and here we give the *skip-gram* version of the method. The strategy, depicted in Fig. 6.21, is to train the network to predict nearby words in a sentence from a single word input. Words are typically coded as binary or *hot* vectors ("1" indicates the presence of the word), and thus the size of the network is proportional to the size of the vocabulary. The output is a vector (again the size of the vocabulary), where the signal indicates the probability that a given word appears within a window of nearby words (5–10 nearby words, typically). The hidden layer is constructed to be 100–1000 units and the output of the hidden layer (a vector) is used as the embedding for subsequent processing.

This example demonstrates the general strategy for using NNs to construct transformations. First, one must encode the input in a general manner that has the appropriate symmetries. Categorical data, for instance, is often best encoded with binary vectors, which also allows for sets or bags of examples. Second, the network must be constructed with a hidden layer appropriate for output into some other part of the visualization or analysis pipeline. This is sometimes called a *latent representation*. Third, the NN must be trained with a task that is appropriate for the transformations. In the word2vec method, the task is the prediction of context words (nearby words in sentences), which captures the meaning and usage of words (indeed, the ability of one word to replace another in a context), which has been shown to capture similarities between words.

References

1. Anscombe, F.J.: Graphs in statistical analysis. Am. Stat. **27**(1), 17–21 (1973). http://www.jstor.org/stable/2682899
2. Bihan, D.L., Mangin, J., Poupon, C., Clark, C.A., Pappata, S., Molko, N., Chabriat, H.: Diffusion tensor imaging: concepts and applications. J. Magn. Reson. Imaging **13**, 534–546 (2001)
3. Canny, J.: A computational approach to edge detection. IEEE Trans. Pattern Anal. Mach. Intell. **8**(6), 679–698 (1986)
4. Carlsson, G.: Topology and data. AMS Bull. **46**(2), 255–308 (2009)

5. Cates, J.E., Whitaker, R.T., Jones, G.M.: Case study: an evaluation of user-assisted hierarchical watershed segmentation. Med. Image Anal. **9**(6), 566–578 (2005)
6. Cui, W., Zhou, H., Qu, H., Wong, P.C., Li, X.: Geometry-based edge clustering for graph visualization. IEEE Trans. Vis. Comput. Graph. **14**(6), 1277–1284 (2008). https://doi.org/10. 1109/TVCG.2008.135
7. Eberly, D.: Ridges in Image and Data Analysis. Kluwer Academic Publishers, Boston (1996)
8. Edelsbrunner, H., Letscher, D., Zomorodian, A.: Topological persistence and simplification. Discret. Comput. Geom. **28**, 511–533 (2002)
9. Edelsbrunner, H., Harer, J., Natarajan, V., Pascucci, V.: Morse-Smale complexes for piecewise linear 3-manifolds. In: SCG'03: Proceedings of the 19th Annual Symposium on Computational Geometry, pp. 361–370. ACM, New York, NY, USA (2003)
10. Engel, K., Hadwiger, M., Kniss, J.M., Rezk-Salama, C., Weiskopf, D.: Real-Time Volume Graphics. A. K. Peters Ltd., Natick (2006)
11. Falk, M., Hotz, I., Ljung, P., Treanor, D., Ynnerman, A., Lundström, C.: Transfer function design toolbox for full-color volume datasets. In: Proceedings of the IEEE PacificVis'17 (2017)
12. Gerber, S., Bremer, P.T., Pascucci, V., Whitaker, R.: Visual exploration of high dimensional scalar functions. IEEE Trans. Vis. Comput. Graph. **16**(6), 1271–1280 (2010)
13. Gonzales, R.C., Woods, R.E.: Digital Image Processing, 2nd edn. Prentice Hall, Englewood Cliffs (2013)
14. Gretarsson, B., O'Donovan, J., Bostandjiev, S., Höllerer, T., Asuncion, A.U., Newman, D., Smyth, P.: Topicnets: visual analysis of large text corpora with topic modeling. ACM TIST **3**, 23:1–23:26 (2012)
15. Günther, T., Theisel, H.: The state of the art in vortex extraction. Comput. Graph. Forum **37**(6), 1–24 (2018)
16. Hadwiger, M., Kratz, A., Sigg, C., Bühler, K.: GPU-accelerated deep shadow maps for direct volume rendering. In: Proceedings of ACM SIGGRAPH/EUROGRAPHICS Symposium on Graphics Hardware, GH'06, pp. 49–52. ACM, New York, NY, USA (2006)
17. Helman, J.L., Hesselink, L.: Visualizing vector field topology in fluid flows. IEEE Comput. Graph. Appl. **11**, 36–46 (1991)
18. Hernell, F., Ljung, P., Ynnerman, A.: Interactive global light propagation in direct volume rendering using local piecewise integration. In: Eurographics/IEEE VGTC Symposium on Volume and Point-Based Graphics, pp. 105–112. Eurographics Association (2008)
19. Hotz, I., Feng, L., Hagen, H., Hamann, B., Jeremic, B., Joy, K.I.: Physically based methods for tensor field visualization. In: VIS'04: Proceedings of IEEE Visualization 2004, pp. 123–130. IEEE Computer Society Press (2004)
20. Jönsson, D., Kronander, J., Ropinski, T., Ynnerman, A.: Historygrams: enabling interactive global illumination in direct volume rendering using photon mapping. IEEE Trans. Vis. Comput. Graph. **18**(12), 2364–2371 (2012)
21. Jönsson, D., Sundén, E., Ynnerman, A., Ropinski, T.: A survey of volumetric illumination techniques for interactive volume rendering. Comput. Graph. Forum **33**(1), 27–51 (2014)
22. Kasten, J., Reininghaus, J., Hotz, I., Hege, H.C., Noack, B.R., Daviller, G., Morzynski, M.: Acceleration feature points of unsteady shear flows. Arch. Mech. **68**(1), 55–80 (2016)
23. Kindlmann, G.: Superquadric tensor glyphs. In: Proceedings of the Joint Eurographics - IEEE TCVG Symposium on Visualization, pp. 147–154 (2004)
24. Kindlmann, G., San Jose Estepar, R., Smith, S.M., Westin, C.F.: Sampling and visualizing creases with scale-space particles. IEEE Trans. Vis. Comput. Graph. **15**(6), 1415–1424 (2009)
25. Kniss, J., Kindlmann, G., Hansen, C.: Interactive volume rendering using multi-dimensional transfer functions and direct manipulation widgets. In: VIS'01: Proceedings of the Conference on Visualization'01, pp. 255–262. IEEE Computer Society (2001)
26. Kniss, J., Kindlmann, G., Hansen, C.: Multidimensional transfer functions for interactive volume rendering. IEEE Trans. Vis. Comput. Graph. **8**(3), 270–285 (2002)
27. Kratz, A., Auer, C., Stommel, M., Hotz, I.: Visualization and analysis of second-order tensors: moving beyond the symmetric positive-definite case. Comput. Graph. Forum - State Art Rep. **32**(1), 49–74 (2013)

28. Kratz, A., Auer, C., Hotz, I.: Tensor invariants and glyph design. In: Westin, C., Burgeth, B., Vilanova, A. (eds.) Visualization and Processing of Tensors and Higher Order Descriptors for Multi-valued Data (Dagstuhl'11), Mathematics and Visualization, pp. 17–33. Springer (2014)
29. Kratz, A., Schöneich, M., Zobel, V., Burgeth, B., Scheuermann, G., Hotz, I., Stommel, M.: Tensor visualization driven mechanical component design. In: Proceedings of Pacific Vis Conference (2014)
30. van Kreveld, M., van Oostrum, R., Bajaj, C., Pascucci, V., Schikore, D.: Contour trees and small seed sets for isosurface traversal. In: SCG'97 Proceedings of the 13th Annual Symposium on Computational Geometry (1997)
31. Liu, S., Maljovec, D., Wang, B., Bremer, P.T., Pascucci, V.: Visualizing high-dimensional data advances in the past decade. In: Borgo, R., Ganovelli, F., Viola, I. (eds.) Eurographics Conference on Visualization (EuroVis) - STARs, pp. 127–147. The Eurographics Association (2015)
32. Ljung, P., Krüger, J., Gröller, E., Hadwiger, M., Hansen, C.D., Ynnerman, A.: State of the art in transfer functions for direct volume rendering. In: Maclejewsk, R., Ropinski, T., Vilanova, A. (eds.) Computer Graphics Forum - STAR, vol. 35, p. 23 (2016)
33. Lorensen, W.E., Cline, H.E.: Marching cubes: a high resolution 3D surface construction algorithm. In: ACM SIGGRAPH Computer Graphics and Interactive Techniques, pp. 163–169 (1987)
34. Maadasamy, S., Doraiswamy, H., Natarajan, V.: A hybrid parallel algorithm for computing and tracking level set topology. In: Proceedings of the 19th International Conference on High Performance Computing, pp. 1–10 (2012). https://doi.org/10.1109/HiPC.2012.6507496
35. Maciejewski, R., Chen, W., Woo, I., Ebert, D.: Structuring feature space - a non-parametric method for volumetric transfer function generation. IEEE Trans. Vis. Comput. Graph. 15(6), 1473–1480 (2009)
36. Marr, D., Hildreth, E.: Theory of edge detection. Proc. R. Soc. Lond. Ser. B Biol. Sci. 207(1167), 187–217 (1980)
37. McLoughlin, T., Laramee, R.S., Peikert, R., Post, F.H., Chen, M.: Over two decades of integration-based, geometric flow visualization. Comput. Graph. Forum - State Art Rep. 29(6), 1807–1829 (2010)
38. Meyer, M., Kirby, R.M., Whitaker, R.: Topology, accuracy, and quality of isosurface meshes using dynamic particles. IEEE Trans. Vis. Comput. Graph. 12 (2007)
39. Munzner, T.: Visualization Analysis and Design. CRC Press, Taylor & Francis Group, Boca Raton (2014)
40. Neumann, M., Garnett, R., Bauckhage, C., Kersting, K.: Propagation kernels: efficient graph kernels from propagated information. Mach. Learn. 102(2), 209–245 (2016). https://doi.org/10.1007/s10994-015-5517-9
41. Pintado, S., Jörnsten, R.: Functional analysis via extensions of the band depth. IMS Lect. Notes-Monogr. Ser. 54 (2007). https://doi.org/10.1214/074921707000000085
42. Pöthkow, K., Weber, B., Hege, H.C.: Probabilistic marching cubes. Comput. Graph. Forum 30(3) (2011)
43. Raj, M., Whitaker, R.T.: Visualizing multidimensional data with order statistics. Comput. Graph. Forum 37, 277–287 (2018). https://doi.org/10.1111/cgf.13419
44. Raj, M., Mirzargar, M., Kirby, R.M., Whitaker, R.T., Preston, J.S.: Evaluating alignment of shapes by ensemble visualization. IEEE Comput. Graph. Appl. 36 (2015). https://doi.org/10.1109/MCG.2015.70
45. Raj, M., Mirzargar, M., Ricci, R., Kirby, R., Whitaker, R.T.: Path boxplots: a method for characterizing uncertainty in path ensembles on a graph. J. Comput. Graph. Stat. 26 (2016). https://doi.org/10.1080/10618600.2016.1209115
46. Rosanwo, O., Petz, C., Hotz, I., Prohaska, S., Hege, H.C.: Dual streamline seeding. In: IEEE Pacific Visualization Symposium'09, pp. 9–16 (2009)
47. Scholkopf, B., Smola, A.J.: Learning with Kernels: Support Vector Machines, Regularization, Optimization, and Beyond. MIT Press, Cambridge (2001)

48. Serfling, R.: A depth function and a scale curve based on spatial quantiles. In: Dodge, Y. (ed.) Statistical Data Analysis Based on the L1-Norm and Related Methods, pp. 25–38. Birkhäuser, Basel (2002)
49. Stegmaier, S., Strengert, M., Klein, T., Ertl, T.: A simple and flexible volume rendering framework for graphics-hardware–based raycasting. In: Proceedings of the International Workshop on Volume Graphics'05, pp. 187–195 (2005)
50. Sun, Y., Genton, M.G.: Functional boxplots. J. Comput. Graph. Stat. **20**(2), 316–334 (2011). https://doi.org/10.1198/jcgs.2011.09224
51. Tenenbaum, J.B., de Silva, V., Langford, J.C.: A global geometric framework for nonlinear dimensionality reduction. Science **290**(5500), 2319–2323 (2000). https://doi.org/10.1126/science.290.5500.2319
52. Tukey, J.W.: Mathematics and the picturing of data. In: James, R. (ed.) Proceedings of the International Congress of Mathematicians, pp. 523–531 (1975)
53. Tukey, J.W.: Exploratory Data Analysis. Addison-Wesley, Reading (1977)
54. van der Maaten, L., Hinton, G.: Visualizing high-dimensional data using t-SNE. J. Mach. Learn. Res. **9**, 2579–2605 (2008)
55. Ware, C.: Information Visualization, 3rd edn. Morgan Kaufmann, San Francisco (2013)
56. Whitaker, R.T.: Reducing aliasing artifacts in iso-surfaces of binary volumes. In: IEEE Symposium on Volume Visualization (2000)
57. Whitaker, R.T., Mirzargar, M., Kirby, R.: Contour boxplots: a method for characterizing uncertainty in feature sets from simulation ensembles. IEEE Trans. Vis. Comput. Graph. **19**, 2713–2722 (2013). https://doi.org/10.1109/TVCG.2013.143

Part II
Empirical Studies in Visualization

Empirical studies play a crucial role in developing the foundations of data visualization, and can take many forms, including controlled and semi-controlled experiments; structured or free-text surveys and questionnaires; interviews, focus group discussions, and think aloud sessions; laboratory and field observation and so on. While many empirical studies have been designed to evaluate hypotheses and facilitate the validation or falsification of proposed theories, many others have been conducted to explore unknown causalities about human performances and kindle new fundamental discoveries through collecting new data, posing new questions and hypotheses, and simply being inspired by phenomena in practical settings. In the field of visualization, empirical studies have also been designed and conducted to evaluate and compare the utilities of different visual representations, algorithmic and interactive techniques, and tools and systems.

This part of the book consists of six chapters focusing various aspects of empirical research in visualization:

- In Chap. 7, Abdul-Rahman, et al. present the first survey on the variables used in controlled and semi-controlled empirical studies in visualization, revealing a diverse range of variables, and a huge scope of research questions, most of which have only been sparsely covered by the existing empirical studies.
- In Chap. 8, moderated by Matković, Wischgoll and Laidlaw present engaging discourses on whether it is necessary to include empirical evaluations in application papers that report interdisciplinary collaboration between visualization researchers and domain experts. Through a series of dialogic correspondences, they explore this contentious topic in a rational and evidence-based manner.
- In Chap. 9, Preim and Joshi make a compelling argument for long-term case studies as a form of empirical studies for evaluating visualization researches and applications. While the chapter features practical examples and logical rationales, it also challenges the status quo for requiring evaluation conclusions resulting from short-term empirical study methods.
- In Chap. 10, Weiskopf makes an informed observation of the difficulties in using traditional empirical study methods, and outlines the need for, and feasibility of, using visualization techniques to aid data-rich empirical studies. Using the figurative term *visualization for visualization* (Vis4Vis), he inspires visualization

researchers to utilize our own tools in making fundamental discoveries in our own field.

- In Chap. 11, Chen and Edwards present a collection of schools of thought in visualization and juxtapose them with those in computer science and psychology. By reflecting the different options and beliefs in our current understanding about visualization, the chapter provides a thought-provoking motivation for conducting more empirical studies in order to build better foundations of data visualization.
- In Chap. 12, Ziemkiewicz et al. describe five major challenges in empirical visualization research and outline possible approaches for addressing these challenges, providing a coherent and stimulating summary of the intensive discussions on empirical studies during Dagstuhl Seminar 18041 *Foundations of Data Visualization* in January 2018.

Chapter 7
A Survey of Variables Used in Empirical Studies for Visualization

Alfie Abdul-Rahman, Min Chen and David H. Laidlaw

Abstract This chapter provides an overview of the variables that have been considered in the controlled and semi-controlled experiments for studying phenomena in visualization. As all controlled and semi-controlled experiments have explicitly defined independent variables, dependent variables, extraneous variables, and operational variables, a survey of these variables allows us to gain a broad prospect of a major aspect of the design space for empirical studies in visualization.

7.1 An Overview of Empirical Studies in Visualization

Empirical studies are an integral part of the research activities in visualization, in a recent survey by Kijmongkolchai et al. [22], some 80 papers on empirical studies, which were published in visualization journals and conferences, were categorized. This is the largest collection to date of papers reviewing controlled empirical studies in visualization, though there are no doubt many more in the literature to be discovered. Many of these empirical studies have provided verifiable means for evaluating different visual designs and visualization techniques, and many others focused on controlled experiments designed to gain some understanding or measurement about specific phenomena in visualization, such as color perception, the effect of emotion, or the use of knowledge.

A. Abdul-Rahman (✉)
King's College London, London, UK
e-mail: alfie.abdulrahman@kcl.ac.uk

M. Chen
University of Oxford, Oxford, UK
e-mail: min.chen@oerc.ox.ac.uk

D. H. Laidlaw
Brown University, Providence, UK
e-mail: dhl@cs.brown.edu

© Springer Nature Switzerland AG 2020
M. Chen et al. (eds.), *Foundations of Data Visualization*,
https://doi.org/10.1007/978-3-030-34444-3_7

All controlled empirical studies are designed to study the impacts of the variations of a number of conditions. Mathematically, the individual aspects of the conditions that are being changed during an experiment are defined as *independent variables*, while the effects to be measured are defined as *dependent variables*. Meanwhile, because the variation of an effect could potentially be caused by many variables, each experiment usually has to minimize the impact of some potential variables in order to maintain the total number of conditions being studied at such a level that all conditions can be sampled adequately. The methods for controlling a variable other than the predefined independent and dependent variables typically include (a) setting it to a constant (e.g., using the same room) and (b) making its instances reasonably random (e.g., ordering different conditions randomly). In some empirical studies, such as Web-based crowdsourcing studies, there are well-defined independent and dependent variables, but the impact of some potential variables cannot be fully controlled (e.g., the computer or the room used for the study). They are commonly referred to as *semi-controlled studies*.

There are many forms of empirical studies that do not predefine a set of independent and dependent variables, including free-text questionnaires, observation diaries, focus group discussions, think aloud sessions, interviews, and so on. One of the goals of such a study is to identify, in an open-minded manner, some independent and dependent variables that may offer potentially the most meaningful explanation about a causal relation in visualization.

In this chapter, we survey the independent and dependent variables that have been studied in controlled and semi-controlled empirical studies in the visualization literature, while examining how extraneous variables were controlled in three case studies. In the remainder of this chapter, we will first give more precise definitions of the main categories of variables. This is followed by a collection of examples for each category. We then detail how variables are defined in three case studies. We offer our summary observation and concluding remarks at the end of the chapter.

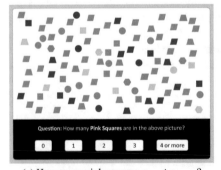
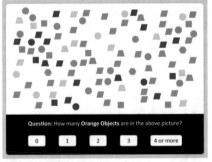

(a) How many pink squares $c_{\text{pink}} \wedge s_{\text{square}}$?   (b) How many orange objects $c_{\text{orange}} \wedge (s_i \in \mathbb{S})$?

Fig. 7.1 Two example stimuli that may be used for a traditional experiment on visual search

7.2   Independent, Dependent, and Other Variables

A *variable V* in an empirical study is a conceptual entity that may change during an experiment, and such an entity can be a piece of stimulus information, a characteristic attribute, an experimental condition, a measurement, or other entity that may vary. For example, in a basic visual search experiment, participants may be shown stimuli similar to the two shown in Fig. 7.1. There are many variables that may change during an experiment, such as:

a. the color of an object in a stimulus;
b. the shape of an object in a stimulus;
c. the size of an object in a stimulus;
d. the position of an object in a stimulus;
e. the number of objects in a stimulus;
f. the aspect ratio of the display area of a stimulus;
g. the questions that may appear in conjunction with a stimulus;
h. the ways in which a question may be answered (e.g., multiple choice buttons, pull down menu, free text, etc.);
i. the number of options available for a multiple choice answer;
j. the ordering of the options available for a multiple choice answer;
k. the type of computing devices used for the experiment;
l. the venue where the experiment is conducted;
m. the time of day when the experiment is conducted;
n. the gender of a participant;
o. the age of a participant;
p. the visual capabilities of a participant;
q. the education background of a participant;
r. the knowledge of a participant that may be used to complete a trial in the experiment;
s. the time taken by a participant to complete a trial in the experiment;
t. the correctness of a participant's response to a stimulus in a trial;
u. the average time taken by all or a specific group of participants to complete the same type of trials in the experiment;
v. the average accuracy of the responses given by all participants or a specific group of participants to the same types of stimuli in the experiment;
w. ...

While, we are almost running out of the letters in the English alphabet, it is not difficult to add to the above list. Some variables can be further decomposed into simpler variables. For example, the sight variable may be decomposed to elementary variables of shortsightedness, color blindness, etc., and the education background may be decomposed to elementary variables of levels, subjects, language, etc.

An empirical study is usually designed to evaluate one or a few hypotheses. Each trial in the experiment is a process that instantiates a causal relation from a set of variables to another set of variables. As a tradition of empirical studies, such a causal relation is usually expressed negatively as a *null hypothesis*. Let a null hypothesis be defined as follows:

Null Hypothesis: Varying variables of $X_1, X_2, \ldots, X_m$ will not have impact on variables of $Y_1, Y_2, \ldots, Y_n$ $(m > 0, n > 0)$.

In general, it is very difficult for an empirical study to evaluate a hypothesis that depends on many variables. Consider, for example, the two stimuli shown in Fig. 7.1. There are 100 objects in each stimulus, and each object may have one of the five colors (i.e., c_{green}, c_{grey}, c_{orange}, c_{pink}, and c_{purple}) and in one of the five shapes (i.e., s_{circle}, $s_{hexagon}$, $s_{parallelogram}$, s_{square}, and $s_{trapezoid}$). Hence, each object may appear in one of the 25 color-shape combinations. When considering the 100 objects collectively as a group, the group of objects may appear in 25^{100} color-shape combinations. In other words, if the 100 objects were placed on a fixed regular grid (e.g., a 10×10 grid), there would be 25^{100} possible stimuli. When one takes other variations into consideration, such as the number of the objects, the size of the objects, and the positions of the objects, and so on, the number of possible stimuli will increase rapidly. Since an empirical study can only have a limited number of trials, only a limited number of stimuli can be selected from a vast number of all possible stimuli.

The word "controlled" thus plays a vital role in designing every collected empirical study. Firstly, one has to select a small number of variables of $X_1, X_2, \ldots, X_m$ in a hypothesis by controlling m such that it is a relatively small number (typically $m < 5$). These selected variables are commonly referred to as *independent variables* in the literature of empirical studies. Those variables, which may or may not have an impact on the participants' performance but are not included in the set of, are referred to as *extraneous variables* [7, 21], *nuisance variables* [21], or *potential confounding variables* [13]. An extraneous variable becomes reprehensible as an actual confounding variable, when it is known to have a *confounding effect* on the participants' performance but has not been adequately controlled.

Secondly, one has to control the number of variations or optional values that each independent variable can have. For example, although there are many different colors and shapes that could be used in designing the stimuli in Fig. 7.1, one has to exercise some control to restrict the number of colors and the number of shapes. To manifest the limited sampling of an independent variable X, one may consider it as an *alphabet* $\mathbb{X}$, which is a term for variable in information theory. The limited number of variations in a variable X is thus the number of letters in the corresponding alphabet $\mathbb{X}$. Therefore, for the experiment illustrated in Fig. 7.1, the alphabet for the sampled colors $\mathbb{C}$ has five letters, i.e., $\mathbb{C} = \{c_{green}, c_{grey}, c_{orange}, c_{pink}, c_{purple}\}$. The alphabet for the sampled shapes $\mathbb{S}$ also has five letters, i.e., $\mathbb{S} = \{s_{circle}, s_{hexagon}, s_{parallelogram}, s_{square}, s_{trapezoid}\}$, assuming that the stimuli used in all trials feature the same alphabets $\mathbb{C}$ and $\mathbb{S}$.

Thirdly, one has to control the impact of the extraneous variables, typically by setting each of them to a constant. For example, in the case of the experiment illustrated in Fig. 7.1, the number of objects is fixed to 100 for all stimuli, and the filled areas of all objects are fixed to the same size. However, not all variables can be fixed to some constants. In many empirical studies, some variables may be sampled randomly, or may appear to be sampled randomly (commonly referred to as pseudo-randomly). For example, the 100 objects in Fig. 7.1 may appear to be placed in the display area randomly. In fact, they are positioned pseudo-randomly to manifest a reasonably uniform distribution of the objects while avoiding any overlapping, because varying the spatial distribution of the objects would introduce another independent variable, while varying the amount of occluded part area of each object would undermine the aforementioned control of the object size.

Finally, many variables can neither be fixed to constants nor be sampled randomly or pseudo-randomly. For example, it would be very difficult to fix the ages and education backgrounds of the participants to some constants, or to recruit participants in a way reflecting a uniform distribution. In such cases, the common wisdom is to record the variations of such variables in an empirical study and discuss their potential impact in the report of the experiment. In some situations, one may determine whether or not such a variable has an impact on the hypothesized causal relation. In most other cases, one may have to leave such conclusions to some future empirical studies.

The set of variables of $Y_1, Y_2, \ldots, Y_n$ in the general formation of a null hypothesis are referred to as *dependent variables*. There are two main classes of dependent variables. The variables that are to be measured in individual trials are *measured dependent variables*. The most elementary dependent variable is a binary variable. Most experiments for studying just-noticeable difference (JND) ask participants to choose whether the attribute of one stimulus is above or below that of another stimulus. Many experiments for testing rapid reaction or decision capacities also use binary variable, such as "yes" or "no", "on" or "off", "action" or "no action", and so on.

The slightly more complicated dependent variable is a set of multiple choices, typically implemented as multiple command buttons, radio buttons, or selectable visual objects in a stimulus. The examples in Fig. 7.1 show five command buttons. Hence, when the multiple choices are considered as letters of an alphabet, we have an alphabet for the answer $\mathbb{A} = \{a_0, a_1, a_2, a_3, a_{4+}\}$.

Some empirical studies have much more complicated alphabets as measured dependent variables. For example, selecting a location on a map from n optional locations or entering a real number with high precision involves a very large alphabet. Later in Sect. 7.4.3, we will see an empirical study that captures 14 time series as measured dependent variables.

From one or more measured variables, one may define a *derived dependent variable*. For example, one may define the correct answer of Fig. 7.1a is a_0 and that of Fig. 7.1b is a_3. With such defined ground truth information, one may define the *correctness* of each trial as a *derived dependent variable*. By aggregating the correctness values of a group of trials, one can define accuracy (in percentage) as a derived dependent variable for the group. Similarly, the response time of a participant in each individual trial is a measured dependent variable, while the mean response time for a group of trials is a derived dependent variable. The way in which the trials are grouped together depends on the hypothesis concerned, the definition of independent variables, and the control of extraneous variables.

In order to compute a derived dependent variable from a measured variable, one has to use some additional definitions (e.g., the ground truth) and additional functions (e.g., statistical or algorithmic functions). The variation of such a definition or a function would have an impact on the derived dependent variable concerned. Hence, these definitions and functions are also variables, which are referred to as *operational variables* or *operational definitions* [21].

7.3   Examples of Variables Used in Empirical Studies

In this section, we first provide three lists of typical variables resulting from our surveys of the papers collected by Kijmongkolchai et al. in [22]. In particular, we conducted the close reading of 32 papers that report controlled and semi-controlled empirical studies in visualization and identified all variables in these papers. In Sect. 7.4, we will detail our analysis of the variables in three examples of empirical studies, which represent quite different study designs.

7.3.1   Independent Variables

There are numerous independent variables that have been studied in different empirical studies. It is not feasible to list all these variables exhaustively. Following a careful reading of 32 papers on visualization-related empirical studies, we identified some 50 variables, and categorize them into five classes.

7.3.1.1   Varying Values in a Single Visual Channel or Varying Types of Visual Channels

The first class is *elementary visual channels* (or *elementary visual variables*), which have been often featured in studies that investigate the attentiveness, distinguishable values, and metaphoric association of different visual channels, as well as the differentiation and interaction between them. The following list gives a number of examples used in several empirical studies. Each item listed, X, can be read as "varying X in the stimuli."

We note that many empirical studies feature stimuli with different visual channels. When a study was not designed to evaluate any hypothesis suggesting that varying such visual channels might have an impact, we do not consider them independent variables of the study. For example, Szafir [32] conducted an empirical study to investigate whether varying the size of graphical primitives impacts color perception. There were extraneous variables associated with the graphical primitives, which were not part of the hypotheses. A polyline primitive, for instance, features many data points, which are extraneous variables that determine the shape of the polyline. The study focused on the thickness of polylines as an independent variable, while controlling other extraneous variables such as the overall height and width, the number of data points, and so on.

- color differences (their levels) [32];
- colors (of glyphs) [12];
- shapes (of glyphs) [12];
- sizes (of glyphs) [12];
- sizes (of graphical primitives) [32];
- types of visual channels (for indicating grouping) [2];
- types of visual channels (for values of missing data) [30];
- vector magnitude [36].

7.3.1.2   Varying Visual Objects Featuring Multiple Visual Channels or the Characteristic Attributes of the Combined Variations

This class of independent variables features variations of multiple visual channels of some visual objects in stimuli. The goal of such a study is typically to investigate the interaction or the combined effects of more than one visual channel. In some cases, the experimenters may focus on a single independent variable that characterizes the combined variations of multiple visual channels, such as the ordering of colors in a colormap [29]. Because the variation of the ordering in this case is more complicated than the variation of a single color, we consider such an independent variable falls into this class.

- bi-variate channels (the shape-color combinations) [12];
- bi-variate channels (the shape-size combinations) [12];
- bi-variate channels (the size-color combinations) [12];
- continuous colormaps (their key colors) [4];
- continuous colormaps (their ordering of key colors) [29];
- discrete colormaps (palette sizes) [14];
- discrete colormaps (palette scoring functions) [14];
- discrete colormaps (user-generated vs. software recommended vs. random) [14];
- discrete colormaps (with semantic association or not) [29];
- multivariate channels (the combinations of 25 channels used for indicating grouping); [2];
- multivariate channels (for map textures) [24].

7.3.1.3   Varying Visual Patterns Made of Multiple Visual Objects or the Characteristic Attributes of the Visual Patterns

This class features variations of what one common referred to as "patterns". A pattern is considered to be made of multiple visual objects. Typical examples include a cluster in a scatter plot or dot plots, an ego or focal node in a network visualization, a volatile section in a time series plot, etc. In general, the variation of patterns involves the simultaneous variations of several visual objects and is thus considered to be more complicated than the variation of a few visual channels of the same visual object as discussed in Sect. 7.3.1.2.

Because the possible number of such variations is usually excessively large and their distribution in a context is often not well established, it is difficult to create a set of stimuli that constitute an unbiased sampling of the space of such variations. It is thus common to control the sampling by introducing some characteristic attributes (e.g., levels of complexity or sparseness, types of ordering or configuration, and so on), and making such attributes as the independent variables.

- data characteristics (level of deviation from a trend-line) [9];
- data characteristics (densities) [18];
- data characteristics (gap, flow-type outlier, spike) [10];
- data characteristics (levels of noise) [28];
- data characteristics (trend types) [9];
- feature patterns (in dot plots) [27];
- feature patterns (ordering of visual objects) [37];
- feature patterns (simple vs. complex) [22];
- highlighting methods (color, leader line) [16];
- levels of appearance fidelity (of virtual human avatars) [34];
- levels of negative emotions (time-steps) [34];
- pixel patterns (block resolutions) [4];
- pixel patterns (block sizes) [15];
- pixel patterns (pixel sizes) [15];
- pixel patterns (subset configurations) [15];
- pixel patterns (levels of variety) [17];
- pixel patterns (types of variety: color or motion) [17];
- pixel patterns (types of variety: local vs. global) [17];
- word-tag patterns (area of words) [11];
- word-tag patterns (colors of word tags) [11];
- word-tag patterns (densities of word tags) [11];
- word-tag patterns (lengths of word tags) [11];
- word-tag patterns (area of words and types of word spacing) [11].

7.3.1.4   Varying Plot Types or Plot-Level Visual Designs

The independent variables in this class define variations at the plot level and are typically used to compare different visual representations or significant variations of visual designs of a type of plots.

- multi-plots (multi-view compositions: map with scatter plot versus map with parallel coordinates plots) [16];

- plot attributes (aspect ratios) [18];
- plot attributes (chart height and virtual resolution) [19];
- plot attributes (chart height and gridline spacing) [18];
- plot types (nine types of plots) [18];
- plot types (braided graph, horizon graph, line graph, small multiples) [20];
- plot types (density plot, gap-detection histogram, dot plot) [10];
- plot types (graph, scatter plot, storyline, treemap) [33];
- plot types (filled line chart, mirrored chart, 2-band horizon chart) [19];
- plot types (line graph, colorfield) [8];
- plot types (scatter plot, line graph, area) [9];
- visual designs (2D flow visualization) [23];
- visual designs (bar charts and difference overlays) [31];
- visual designs (for map-based flow visualization) [35];
- visual designs (with or without embellishment) [3]).

7.3.1.5 Varying Variables Not in the Depicted Data

The effectiveness of visualization does not only depend on the depicted data values, the selection of visual channels or the design of visual representations, but also on many other factors such as user, task, application, and so on. This class thus includes all independent variables that are used to study the impact of such factors on visualization processes.

- display types (mono, stereo) [36];
- display types (MacOS, others) [18];
- teaching methods (bottom-up, top-down) [33];
- application contexts [22];
- color compensation configurations [26];
- statistical measures (min/max, mean, stdev) [22];
- learning approaches (passive, active) [33];
- visualization tasks (many studies, e.g., [3, 4, 18, 23, 31]).

7.3.2 Dependent Variables and Derived Variables

The variables that are to be measured in individual trials are *measured dependent variables*. In most cases, the collected values of some dependent variables are processed to yield some numerical quantities or categorical values, using, e.g., statistics or algorithms, we consider the corresponding variables as *derived dependent variables*.

There are a number of measured dependent variables that commonly defined in many empirical studies, including:

- response time (RT) of a trial;
- a selection out of k choices ($k \geq 2$);
- a value entered using a 1D scroll bar;
- a position in a 2D map entered using a pointer device (often with many optional locations);
- a location in a 3D real or virtual environment entered using a 3D input device;
- a sequence of action records (e.g., user interactions, and navigation actions in a virtual environment);
- an eye-tracking record;
- one or more time series records of EEG (electroencephalography);
- one or more imagery records of fMRI (functional magnetic resonance imaging or functional MRI).

During an empirical study, some measured dependent variables may be used to compute some derived variables dynamically. Perhaps the most common variable derived dynamically is correctness indicated by a measured value in order for the experiment system to give a feedback to the participant. For example, a system for facilitating trials with multiple choice questions may maintain the ground truth answer for each trial and use it to determine the correctness of an answer. A system for eye-tracking may maintain a set of areas of interest, and use these to determine if a participant's gaze has been fixated on any of the areas of interest.

Because these derived dependent variables are obtained using some predefined operational variables such as ground truth values, threshold values, quantization bands, etc., they are not only dependent on the input stimuli and the human actions during trials, but also on these operational variables. Hence, it is helpful to consider them as derived dependent variables in order to be mindful about the variations of the underlying operational variables and functions that could affect the findings of the study.

In almost all empirical studies, the analysis of the results involves derived dependent variables defined through statistical aggregation and analysis. The most commonly used derived dependent variables are:

- accuracy and error rate (percentage values calculated based on a collection of correctness values);
- precision and recall (for information retrieval tasks);
- just-noticeable difference (JND);
- average response time (mean RT, often abbreviated as RT);
- basic statistical measures for a collection of measured or derived values (e.g., mean, max, mean median, mode, range, correlation coefficient, mutual information, etc.);
- measures resulting from processes of statistical analysis, such as t-test, χ^2-test, ANOVA (analysis of variance), and so on.

In experiments designed with some specific apparatus, there are usually some specialized dependent variables. For example, in eye-tracking experiments, one may define (a) time from the start of a trial to the first fixation at an area of interest and (b) the number of fixations during a trial as derived dependent variables computed based on gaze records [16]. A number of studies measured specific types of participants' judgment, such as alpha contrast optimization [18], discriminability rate [32], perceived complexity [28], perceived data quality [30], and so on. Using an electro-dermal activity (EDA) sensor, one may obtain an EDA dataset as measured dependent variables, and may compute differential emotions scale (DES) as a derived dependent variable [34]. In a recent empirical study, the traditional accuracy and mean RT variables were transformed to information-theoretic measures of benefit and cost as a new pair of derived dependent variables [22].

In general, determining a collection of variables that may affect a design provides a means for defining a design space. In visualization, some notable publications (e.g., [5]) proposed and discussed design spaces of visual representations. One may wonder if there might be a design space for controlled empirical studies in visualization. Our enumeration of experimental variables here may begin to inform the description of such a space. However, given the level of complexity that arises from just this simple initial step, formulating a structured description of such a design space seems to be out of reach at the moment. We hope that a design space for empirical studies in visualization will emerge in the future.

7.4 Case Studies

In this section, we present three case studies to show how one may extract the information of independent, dependent, and constrained variables. In psychology, many papers reporting empirical studies define independent and dependent variables explicitly. In those papers that do not offer explicit definitions of the study variables, it is usually not too difficult to extract such information indirectly. In general, the stimuli used in visualization-related experiments are more complicated, and it is not always easy to extract the definitions about such variables. For the three papers discussed in this section, the authors of this chapter first read the papers and wrote down the independent and dependent variables individually. They then compared the notes and agreed on a common set of variables (Fig. 7.2).

7.4.1 A Study on Using Visual Embellishments in Visualization

Borgo et al. presented a study on the impact of visual embellishment on participants' ability (a) to remember the numerical data depicted, (b) to perform visual search of visual objects, and (c) to grasp the concept conveyed by the text shown in visualization images [3]. Because the tasks in such a study had to be reasonably simple in order to control the potential confounding effects and the length of each trial, they anticipated that the impact might not be easily detectable if the participants were paying attention

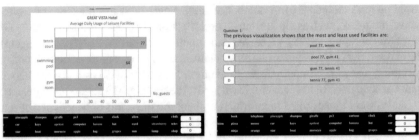

(a) A stimulus screen and its follow-on question screen

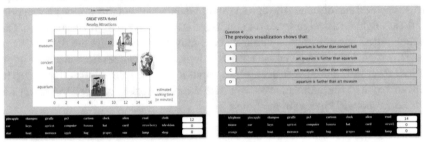

(b) A related stimulus screen and its follow-on question screen

Fig. 7.2 Two related trials in the empirical study. The top 80% of each screen was used for the primary task and the bottom 20% was used for the secondary task. A stimulus without any visual embellishment is shown in **a**, and a stimulus with a similar visual representation and a similar amount of information as well as with some visual embellishment is shown in **b**. The two trials were distributed pseudo-randomly among others to minimize any learning effect

to their tasks. They thus designed a dual-task experiment, where a secondary task was used to restrict the amount of the cognitive capability available to the primary task in each trial, allowing the trials with embellishment and those without more differentiable.

For the primary tasks, all stimuli were designed in pairs, one with visual embellishment and one without. Hence, this binary variable was the most important independent variable being studied. The experimenters had four hypotheses and they divided the stimuli into four sections. Since these four sections were conducted within the same empirical study, the four topics, i.e., working memory, long-term memory, visual search, and concept grasping were the four values of a variable about tasks. Thus, there were two independent variables for the primary tasks.

To avoid learning effects, different stimuli had to feature different data values. These are extraneous variables that should be controlled. The experimenters carefully selected these values to ensure a similar level of complexity within each pair of stimuli, while having different levels of complexity across different pairs for each section of the experiment. Both measures provided means of controlling the potential confounding effects due to the variations of the data values.

Similarly, there were variations of the designs for different visual embellishment across different stimuli. Such variations were unavoidable since each trial featured a different dataset and it was necessary to change the semantics featured the datasets to avoid learning effect. The experimenters controlled the potential confounding effects due to such an extraneous variable by using the same approach for dealing with the variations of data values.

For the primary task, each trial presented participants with a question and four optional answers (one correct answer and three distractors). Hence, the measured dependent variables were the selection out of four options and the time taken to make this selection. By predefining the ground truth value of each trial (i.e., an operational variable), the experimenters obtained a derived dependent variable for the correctness of the selected answer. Similar to numerous empirical studies, from the correctness and response time of each trial, the two commonly used dependent variables were derived, i.e., the mean accuracy and the mean response time.

The stimuli for the secondary task ran continually in parallel with the stimuli for the primary task throughout the experiment. In a rectangular area at the bottom of the screen, a sequence of words moved horizontally from left to right, with new words appearing from the left continually. Participants were required to point and click at any fruit word that appeared in that area. When each word displayed was considered as a visual object, from the perspective of the secondary task, each word had only two states, a fruit word or a non-fruit word. Thus, the independent variable was a binary variable. When a participant selected a word, the dependent variable was correctness. The software for the experiment showed three counters on the screen, keeping the count of how many fruit words had been correctly selected, how many had been missed, and how many words had been wrongly selected. These counters were derived dependent variables.

In addition to the aforementioned effort for controlling the potential confounding effects due to the variations of data values and visual embellishment, the experimenters also discussed an effort for controlling other extraneous variables, such as knowledge bias, ordering bias, and attention bias.

7.4.2   A Study on Visual Semiotics and Uncertainty Visualization

MacEachren et al. presented two controlled empirical studies on aspects of uncertainty visualization [25]. The first experiment was designed to obtain measurements about the participants' judgment as to the suitability of visual representations for a given category of uncertainty. They defined their first independent variable for ten categories of uncertainty, which were referred to as ten series in their paper [25]. The alphabet $\mathbb{X}_{series}$ thus consists of 10 letters: (x_1) general, (x_2) spatial accuracy, (x_3) spatial prevision, (x_4) spatial trustworthiness, (x_5) temporal accuracy, (x_6) temporal precision, (x_7) temporal trustworthiness, (x_8) attribute accuracy, (x_9) attribute precision, and (x_{10}) attribute trustworthiness.

The letters x_2-x_{10} were defined over two elementary alphabets. One alphabet defined three categories of data to be displayed (i.e., space, time, and attribute), and the other defined three types of uncertainty associated with the data (i.e., accuracy, precision, and trustworthiness). The letters x_2-x_{10} were the nine combinations of the letters of these two elementary alphabets.

The second independent variable $\mathbb{X}_{level}$ defined the two levels of abstraction of the symbol sets: namely abstract or iconic. Each symbol set consisted of k glyphs that represented different levels of uncertainty. In this experiment, k was considered as an extraneous variable, which was fixed to $k = 3$.

The experimenter designed 76 symbol sets for 76 trials. They were used primarily as repeated measures of the two levels of abstraction. For series x_1, 22 symbol sets were used, and the symbol sets were designed based on different visual channels (e.g., color, size, shape, etc.). For each of series x_2-x_{10}, six symbol sets (three abstract and three iconic) were used. The variation of symbol sets was a variable difficult to control, because it was not easy to define the design space of the symbol sets. The experimenters made a good effort to design various symbol sets considered to be the most representative and sensible designs heuristically. They recorded and reported the impact of individual symbol set on the participants' judgment, exhibiting the best practice for handling such an extraneous variable.

The measured dependent variable was the subjective judgment in each trial by a participant. The corresponding alphabet $\mathbb{Y}_{judgment}$ consists of seven levels of intuitiveness of a symbol set, i.e., $\mathbb{Y}_{judgment} = \{1, 2, 3, 4, 5, 6, 7\}$. It was implemented using a set of clickable numbers for the seven multiple choices. From this measured dependent variable, the experimenters computed a set of derived dependent variables, including five statistical measures (i.e., min, max, mean, median, and mode) and two measures of the Mann–Whitney test (i.e., W and p-value).

In addition, the experimenters obtained a measured dependent variable of the time taken to complete each trial. They reported three derived dependent variables resulting from the independent two-group t-test with Welsh df-modification (i.e., t, df, and p-value).

The second experiment was designed to obtain the measurements about the effectiveness of the symbol sets through a typical task in map visualization. Participants were asked to assess and compare the aggregated uncertainty in two map regions based on the glyph representation of uncertainty in each location. The experiment featured one independent variable that defines 20 symbol sets selected based on the results of the first experiment. In other words, it was an alphabet with 20 letters. The goal of the experiment was to determine the relative merits among these 20 symbol sets.

In the context of this experiment, a map being visualized can be considered as a background image, and the uncertainty glyphs can be placed on a $w \times h$ grid superimposed on top of the background map. The variations of the map image and the grid resolution would manifest variables with very large sampling spaces. The experimenters considered them as extraneous variables and controlled both of them by using constants. The background image was simply removed from all stimuli, while the grid resolution was fixed to 3×3.

The stimulus in each trial depicted two regions, each with 3×3 uncertainty glyphs. All 18 glyphs in each stimulus were selected from the same symbol sets. The task of each participant was to aggregate the nine uncertainty values in each of the two regions and select the region that was less certain. Since each symbol set had three glyphs representing three uncertainty values, there were a total of 3^{18} possible variations of the stimuli. The experimenters controlled this extraneous variable using pseudo-randomness by predefining 12 configurations that represented a relatively uniform sampling of the stimuli space. Although varying the 12 configurations could be considered as an independent variable, they were featured in the experiment design as an extraneous variable for supporting repeated measures for each symbol set. Together, the experiment had a total of 240 trials (20 symbol sets and 12 configurations).

The measured dependent variables were the correctness of a participant's selection and the response time in each trial. The derived dependent variables reported in [25] included the accuracy of 20 symbol sets, and the accuracy value for each symbol set was an aggregation of the 360 correctness values (the 30 participants and 12 configurations). In addition, the experimenters applied the Pearson's χ^2 test with Yates' continuity correction to the correctness values, yielding three derived dependent variables χ^2, df, and p-value; and applied the independent two-group t-test with Welsh df-modification to response time, yielding three derived dependent variables t, df, and p-value.

7.4.3   An EEG Study on Visualization Effectiveness

Anderson et al. presented an empirical study on participants' cognitive load during visualizing different visual designs of box plots [1]. In each trial, a participant was shown two types of box plots with different data and was asked to choose the distribution with a larger inter-quartile range.

The main independent variable defined the variations among six visual designs of box plots. The corresponding alphabet had six letters. Most box plots typically depicted five statistical measures computed over a data sample, including (i) minimum, (ii) median, (iii) maximum, (iv) the 25th percentile, and (v) the 75th percentile. Most visual designs allowed the viewers to estimate the min-max range and the inter-quartile range (between the 25th and 75th percentile). Some box plots also depicted the distribution of data values in the sample using a visual representation based on histogram or a density map. The experimenters selected three visual designs with a density map and three without. This additional independent variable allowed the evaluation of a hypothesis related to the absence/presence of the distribution information.

The summary statistical measures depicted by a box plot were computed from n values in a data sample. The variations of the data sample determined the variations of its statistical measures, and hence the corresponding box plot. The data space for n values was exponentially related to n. The experimenters had to control such variations. In this experiment, the extraneous variable of data samples was controlled firstly by fixing the number of data values to 100 and the distribution of the sample to uniform, and secondly by using randomly generated data values with controlled ranges for the mean and standard division of the sample. A total of 500 samples were generated, hence there was a pool of 500 box plots. In the study, each participant performed tasks in 100 trials, each of which showed two box plots selected from the pool.

Given two samples, the experimenters estimated the task difficulty, in the range of [0, 1], of comparing the two corresponding box plots. This variable was not explicitly featured in the stimuli but was used in results analysis as a possible cause that might impact the cognitive load. One could consider this as an independent variable.

For each trial, the experiment captured a number of measured dependent variables, including (a) the response time and (b) the electroencephalography (EEG) signals in the form of 14 time series. The experimenters used a numerical function for transforming the 14 time series to a derived dependent variable to as the estimated cognitive load per trial. From the estimated values for cognitive load for all trials, two further derived dependent variables were computed, namely constant- and Gaussian-weighed averages. They also applied two-tailed t-tests to compare the cognitive load values estimated for every pair of visual designs, and obtained the 15 p-values for the corresponding derived dependent variables.

7.5 Conclusions

In this chapter, we have conducted a survey on independent and dependent variables used in controlled or semi-controlled empirical studies on the subject of visualization. In particular, we analyzed the variables considered in 32 publications on such studies. We categorized independent variables into five categories. We noticed that there is no shortage of studies on independent variables in each category. We consider this a particularly encouraging sign, because this shows that visualization researchers are asking many research questions about visualization at different levels of visual designs and from many different perspectives. Meanwhile, it also suggests that there are many more research questions yet to be asked or answered, and the scope of visualization-related empirical studies is huge.

Meanwhile, when an independent variable is examined in one study, it can be an extraneous variable to be controlled in another study. The variety of independent variables that have already been examined in the previous studies indicate the challenge in alleviating confounding effects since controlling many extraneous variables is not a trivial undertaking in most visualization-related empirical studies.

The large number of variables and potential experimental designs also brings up the point that designing experiments is a creative process. As with any process that involves design, there are many choices to be made in many trade-offs that need to be balanced in making those choices. There is no one best design, just as there is no one best painting, building, or software application. Learning to design good experiments is a matter of study and practice, and there are numerous books and other resources that teach how to do it. We have touched on a few of the design decisions and trade-offs that we identified in the visualization literature, but this survey is only a sparse sampling of the rich space of experimental design.

It is hence necessary for the experiment designers to be aware of the potential impact of different extraneous variables is important, while it is helpful for the reviewers to appreciate the challenge of alleviating confounding effects. Occasionally, some of us in the community may wish the stimuli in some empirical studies to be more complex or more realistic without appreciating that more complex or more realistic stimuli would likely introduce more confounding effects that could undermine the statistical significance of the experiment results. In other occasions, some of us in the community may wish that the stimuli in some empirical studies could feature fewer independent variables or extraneous variables could be controlled more stringently without being aware of the experimenters' intention to examine the impact of variables at a higher level (e.g., multi-object patterns or plot-level visual designs).

It may thus be desirable for the visualization researchers who conduct empirical studies to be more coherently organized, instead of being distributed sparsely in Info-Vis, SciVis, VAST, and other areas of visualization. This will allow these researchers to share their expertise (e.g., in the review processes) more easily and to formulate research agenda in a more ambitious and structured manner. If one considers different schools of thought in visualization (see Chap. 11 [6]) as high-level hypotheses, there are indeed many ambitious research questions that may be answered using empirical studies. By providing some opportunities to bring all these researchers together, we may soon see the emergence of a new area of visualization psychology.

References

1. Anderson, E.W., Potter, K.C., Matzen, L.E., Shepherd, J.F., Preston, G.A., Silva, C.T.: A user study of visualization effectiveness using EEG and cognitive load. Comput. Graph. Forum **30**(3), 791–800 (2011)
2. Bae, J., Watson, B.: Reinforcing visual grouping cues to communicate complex informational structure. IEEE Trans. Vis. Comput. Graph. **20**(12), 1973–1982 (2014)
3. Borgo, R., Abdul-Rahman, A., Mohamed, F., Grant, P.W., Reppa, I., Floridi, L., Chen, M.: An empirical study on using visual embellishments in visualization. IEEE Trans. Vis. Comput. Graph. **18**(12), 2759–2768 (2012)
4. Borgo, R., Proctor, K., Chen, M., Jänicke, H., Murray, T., Thornton, I.M.: Evaluating the impact of task demands and block resolution on the effectiveness of pixel-based visualization. IEEE Trans. Visualization & Computer Graphics **16**(6), 963–972 (2010)
5. Card, S.K., Mackinlay, J.: The structure of the information visualization design space. In: Proceedings of IEEE Symposium on Information Visualization, pp. 92–99 (1997)
6. Chen, M., Edwards, D.J.: Isms in visualization. In: Chen, M., Hauser, H., Rheingans, P., Scheuermann, G. (eds.) Foundations of Data Visualization. Springer (2019)
7. Cohen, P.R.: Methods for Artificial Intelligence. MIT Press, Cambridge (1995)

8. Correll, M., Albers, D., Franconeri, S., Gleicher, M.: Comparing averages in time series data. In: Proceedings of ACM SIGCHI Conference on Human Factors in Computing Systems (CHI), pp. 1095–1104 (2012)
9. Correll, M., Heer, J.: Regression by eye: Estimating trends in bivariate visualizations. In: Proceedings of ACM SIGCHI Conference on Human Factors in Computing Systems (CHI), pp. 1387–1396 (2017)
10. Correll, M., Li, M., Kindlmann, G., Scheidegger, C.: Looks good to me: visualizations as sanity checks. IEEE Trans. Vis. Comput. Graph. **25**(1), 830–839 (2019)
11. Correll, M.A., Alexander, E.C., Gleicher, M.: Quantity estimation in visualizations of tagged text. In: Proceedings of ACM SIGCHI Conference on Human Factors in Computing Systems (CHI), pp. 2697–2706 (2013)
12. Demiralp, C., Bernstein, M.S., Heer, J.: Learning perceptual kernels for visualization design. IEEE Trans. Vis. Comput. Graph. **20**(12), 1933–1942 (2014)
13. Eysenck, M.W.: Psychology: A Student's Handbook. Psychology Press (2000)
14. Gramazio, C.C., Laidlaw, D.H., Schloss, K.B.: Colorgorical: creating discriminable and preferable color palettes for information visualization. IEEE Trans. Vis. Comput. Graph. **23**(1), 521–530 (2017)
15. Gramazio, C.C., Schloss, K.B., Laidlaw, D.H.: The relation between visualization size, grouping, and user performance. IEEE Trans. Vis. Comput. Graph. **20**(12), 1953–1962 (2014)
16. Griffin, A.L., Robinson, A.C.: Comparing color and leader line highlighting strategies in coordinated view geovisualizations. IEEE Trans. Vis. Comput. Graph. **21**(3), 339–349 (2015)
17. Haroz, S., Whitney, D.: How capacity limits of attention influence information visualization effectiveness. IEEE Trans. Vis. Comput. Graph. **18**(12), 2402–2410 (2012)
18. Heer, J., Bostock, M.: Crowdsourcing graphical perception: Using mechanical turk to assess visualization design. In: Proceedings of ACM SIGCHI Conference on Human Factors in Computing Systems (CHI), pp. 203–212 (2010)
19. Heer, J., Kong, N., Agrawala, M.: Sizing the horizon: the effects of chart size and layering on the graphical perception of time series visualizations. In: Proceedings of ACM SIGCHI Conference on Human Factors in Computing Systems (CHI), pp. 1303–1312 (2009)
20. Javed, W., McDonnel, B., Elmqvist, N.: Graphical perception of multiple time series. IEEE Trans. Vis. Comput. Graph. **16**(6), 927–934 (2010)
21. Kantowitz, B., Roediger III, H., Elmes, D.: Experimental Psychology. Wadsworth Publishing, Belmont (2014)
22. Kijmongkolchai, N., Abdul-Rahman, A., Chen, M.: Empirically measuring soft knowledge in visualization. Comput. Graph. Forum **36**(3), 73–85 (2017). https://doi.org/10.1111/cgf.13169
23. Laidlaw, D.H., Davidson, J.S., Miller, T.S., da Silva, M., Kirby, R.M., Warren, W.H., Tarr, M.: Quantitative comparative evaluation of 2D vector field visualization methods. In: Proceedings of IEEE Visualization, pp. 143–150 (2001)
24. Livingston, M., Decker, J.: Evaluation of trend localization with multi-variate visualizations. IEEE Trans. Vis. Comput. Graph. **17**(12), 2053–2062 (2011)
25. MacEachren, A.M., Roth, R.E., O'Brien, J., Li, B., Swingley, D., Gahegan, M.: Visual semiotics & uncertainty visualization: an empirical study. IEEE Trans. Vis. Comput. Graph. **18**(12), 2496–2505 (2012)
26. Mittelstädt, S., Keim, D.A.: Efficient contrast effect compensation with personalized perception models. Comput. Graph. Forum **34**(3), 211–220 (2015)
27. Pandey, A.V., Krause, J., Felix, C., Boy, J., Bertini, E.: Towards understanding human similarity perception in the analysis of large sets of scatter plots. In: Proceedings of ACM SIGCHI Conference on Human Factors in Computing Systems (CHI), pp. 3659–3669 (2016)
28. Ryan, G., Mosca, A., Chang, R., Wu, E.: At a glance: pixel approximate entropy as a measure of line chart complexity. IEEE Trans. Vis. Comput. Graph. **25**(1), 872–881 (2019)
29. Schloss, K.B., Gramazio, C.C., Silverman, A.T., Parker, M.L., Wang, A.S.: Mapping color to meaning in colormap data visualizations. IEEE Trans. Vis. Comput. Graph. **25**(1), 810–819 (2019)

30. Song, H., Szafir, D.A.: Where's my data? evaluating visualizations with missing data. IEEE Trans. Vis. Comput. Graph. **25**(1), 914–924 (2019)
31. Srinivasan, A., Brehmer, M., Lee, B., Drucker, S.M.: What's the difference?: evaluating variations of multi-series bar charts for visual comparison tasks. In: Proceedings of ACM SIGCHI Conference on Human Factors in Computing Systems (CHI), pp. 304:1–304:12 (2018)
32. Szafir, D.A.: Modeling color difference for visualization design. IEEE Trans. Vis. Comput. Graph. **24**(1), 392–401 (2018)
33. Tanahashi, Y., Leaf, N., Ma, K.L.: A study on designing effective introductory materials for information visualization. Comput. Graph. Forum **35**(7), 117–126 (2016)
34. Volante, M., Babu, S.V., Chaturvedi, H., Newsome, N., Ebrahimi, E., Roy, T., Daily, S.B., Fasolino, T.: Effects of virtual human appearance fidelity on emotion contagion in affective inter-personal simulations. IEEE Trans. Vis. Comput. Graph. **22**(4), 1326–1335 (2016)
35. Yang, Y., Dwyer, T., Goodwin, S., Marriott, K.: Many-to-many geographically-embedded flow visualisation: an evaluation. IEEE Trans. Vis. Comput. Graph. **23**(1), 411–420 (2017)
36. Zhao, H., Bryant, G.W., Griffin, W., Terrill, J.E., Chen, J.: Validation of splitvectors encoding for quantitative visualization of large-magnitude-range vector fields. IEEE Trans. Vis. Comput. Graph. **23**(6), 1691–1705 (2017)
37. Zheng, L., Wu, Y., Ma, K.L.: Perceptually-based depth-ordering enhancement for direct volume rendering. IEEE Trans. Vis. Comput. Graph. **19**(3), 446–459 (2013)

Chapter 8
Empirical Evaluations with Domain Experts

Krešimir Matković, Thomas Wischgoll and David H. Laidlaw

Abstract Over the past thirty years, the visualization community has developed theories and models to explain visualization as a technology that augments human cognition by enabling the efficient, accurate, and timely discovery of meaningful information in data. Along the way, practitioners have also debated theories and practices for visualization evaluation: How do we generate durable, reliable evidence that a visualization is effective? Interestingly, there is still no consensus in the visualization research community how to evaluate visualization methods. The goal of this chapter is to rise awareness of still open issues in the visualization evaluation and to discuss appropriate evaluations suitable for different visualization approaches. This includes user studies and best practices to conduct them but also other approaches for suitable evaluation of visualization. The chapter is structured as a moderated dialog of two visualization experts.

8.1 Introduction

Over the past thirty years, the visualization community has developed theories and models to explain visualization as a technology that augments human cognition by enabling the efficient, accurate, and timely discovery of hidden information in data. Along the way, practitioners have also debated theories and practices for evaluation of visualizations: How do we generate durable, reliable evidence that visualization

K. Matković (✉)
VRVis Zentrum für Virtual Reality und Visualisierung Forschungs-GmbH,
Vienna, Austria
e-mail: matkovic@vrvis.at

T. Wischgoll
Wright State University, Dayton, OH, USA

D. H. Laidlaw
Brown University, Providence, RI, USA

© Springer Nature Switzerland AG 2020
M. Chen et al. (eds.), *Foundations of Data Visualization*,
https://doi.org/10.1007/978-3-030-34444-3_8

is effective—that visualization facilitates obtaining insight into the data in ways that are demonstrably beneficial to the user, and that it perfectly complements automatic methods in cases where problems and queries are ill-defined or hard to specify?

User studies seem like the first choice for evaluation of visualization as it is really a human-centric technique. Somewhat paradoxically, user studies are both taken for granted and controversial among visualization practitioners. On the one hand, it is difficult to get an application paper accepted for conference or journal publication without a "user evaluation" section at the end of a manuscript. At the same time, user studies are often haphazardly executed and presented, leading researchers (and reviewers) to question whether user studies are helping move the field forward in any meaningful way. A proper evaluation typically is an important aspect of a visualization publication. Since visualization involves a visual interface combined with other user interface elements, it appears natural to deploy some form of user study to evaluate the system. However, this is only one type of evaluation. There can be several reasons why a user study is not appropriate for a specific visualization approach. In addition, the term "user study" is not always used correctly in the visualization community. Besides its strict definition, many visualization researchers consider almost any user experiment to be a user study. It is our goal to provide guidance toward appropriate evaluations suitable for visualization approaches. As pointed out earlier, user studies are only one way of evaluating a visualization approach.

There are guidelines on how to do proper user studies in various fields. As its name clearly implies, a user study is impossible without a user. The user, one of the fundamental players in the visualization, has to be taken with great caution when evaluating visualization research. The visualization solutions are designed for a specific user group. The target audience could be casual users on the Web on one end of a spectrum or highly trained domain experts solving very specific problems on the other end. Can we evaluate visualization solutions for those two with the same methods? It hardly seems possible.

Finding the users for a user study on case visualization is, in general, simpler. Crowdsourcing mechanisms make it possible to recruit a large number of users in a relatively short time. Motivating a large number of experts who deal with a specific problem is close to impossible. First, there are not many experts that deal with very specific problem, and secondly, their time is usually too precious and they cannot afford to participate in a lengthy study. If the study has to be repeated for any reason, it is also impractical as it uses up even more of the expert's valuable time. Once, we had to design an interactive exploratory visualization system geared toward systems designers for fuel injection into a diesel engine [18]. We evaluated it with two experts with whom we collaborated on the project. It was impossible to perform a large user study on such a specific topic. However, does this still count as an evaluated solution?

Nevertheless, we should strive to evaluate our solutions. Evaluation should focus on lessons learned and take-home messages for visualization researchers. This might be the most difficult part of the evaluation. We should evaluate it for the given task and users, but we should also strive to generalize it or at least to reflect on important findings from the visualization research perspective. Without such a reflection, the research is of less interest for the visualization community (it still can be of a great interest for other scientific domains and worth publishing in their journals). At the same time, there is also a certain threat in trying to generalize everything. If a solution cannot be generalized, it still can represent valuable research for the visualization community.

There are multiple orthogonal ways to evaluate a visualization approach, user studies being one of them. Isenberg et al. [9] provide a review of evaluation means for visualization. Some other chapters of this book also deal with evaluation [1, 28]. The following sections address different questions about user studies and evaluation in general. As the evaluation and user studies are often a hot topic of formal and informal discussions at many visualization conferences, we decided to structure this chapter as a conversation. We are fully aware that none of the extreme approaches ("user studies are a must" and "we do need user studies at all") is appropriate. We hope that the following dialog can help in clarifying the evaluation needs of visualization research.

8.2   A Conversation About Empirical Evaluation of Visualization Approaches

Krešimir Matković

We start this section with position statements on evaluation in visualization research from two visualization scientists Thomas Wischgoll and David Laidlaw with decades of combined visualization research experience. Thomas Wischgoll is an expert in flow visualization [12, 27], medical visualization [4], virtual reality [26], and areas of information visualization [5]; whereas David Laidlaw is an expert in multi-valued volume visualization [11], applications of visualization to science [7], virtual reality for visualization [24], visualization design [10], and visualization evaluation [3, 6, 13, 14]. Both are experts in working with domain specialists to successfully apply visualization algorithms to various disciplines. We continue the section as a moderated dialog. So, let us start with your viewpoints on evaluation in visualization research? Do we always need it, is it an unnecessary add-on which is required by reviewers, or do you think, we should stubbornly omit it whenever possible?

Thomas Wischgoll

A proper evaluation typically is an important aspect of a visualization publication. Since visualization involves a visual interface combined with other user interface elements, it appears natural to deploy some form of user study to evaluate the system. However, this is only one type of evaluation. There can be several reasons for why a user study is not appropriate for a specific visualization approach.

The purpose of visualization is by definition to involve humans as their target audience and to provide better insight into the data that is to be visualized. There are numerous aspects of providing better insight, however. A visualization approach could be better in terms of providing insight more quickly or in a more comprehensible way. The visualization could also be more user-friendly or intuitive, albeit those are both fairly subjective criteria. Hence, there are several different aspects one could focus on for a user evaluation. For example, Lam et al. [15] list seven scenarios for empirical evaluations: evaluating visual data analysis and reasoning, evaluating user performance, evaluating user experience, evaluating environments and work practices, evaluating communication through visualization, evaluating visualization algorithms, and evaluating collaborative data analysis. This underscores the complexity of user evaluations as well as the different aspects a user evaluation can be used to test for with respect to a visualization approach. There is quite a number of publications in the literature to provide guidelines for evaluations. Munzner [20] suggests a nested, four-tiered model to assist in the validation of design studies. Meyer et al. [19] refined this model by adding blocks for additional flexibility to describe activities within the tiers of the original model. Focusing on design studies, Sedlmair et al. [21] propose a nine-stage model for the entire life cycle of the visualization system and provide guidance and common pitfalls. Lam et al. [16] analyzed papers from the information visualization area between 2009 and 2015 to develop a framework for breaking down goals of a project to individual tasks, whereas Chen and Ebert [2] provide an ontological framework to assist in the design and evaluation of visual analytics systems.

At the same time, it is imperative to provide a proper evaluation of some sort. There are research areas outside of visualization that suffer from issues of lack of reproducibility. One reason for this is improper use of statistical methods during the evaluation or the selection of the participants which leads to misleading results. However, once a paper that applied such improper statistical methods is published, it is considered factual and researchers may not question the results despite the fact that there is a high probability that the results are invalid. This stresses the importance of properly executed user studies or any kind of evaluation for that matter. It does not serve the visualization community to publish papers with user studies or other types of evaluations that are flawed in a significant way, thereby making the findings questionable. We therefore need to find a way to make the evaluation of visualization approaches easier for researchers that ensures meaningful results. Some additional guidance may be needed, and the visualization researchers, if they want to go for a user evaluation, need to make sure they know what they want to test for and how to execute the user study properly in accordance with those goals.

David Laidlaw

I would modify Thomas's first statement, "a proper evaluation typically is an important aspect of a visualization publication," to say that a visualization publication needs to clearly state how it extends human knowledge, successfully arguing for both the novelty and the significance of that new knowledge. Empirical evaluations can be a part of this argumentation in two ways.

First, an empirical evaluation can serve as a measure of the significance of a new visualization artifact, by which I mean an algorithm, interactive technique, or software system. There are numerous examples of this kind of user evaluation in the literature as well as a number of papers that describe the process and organize examples of it. These example evaluations or studies range from small numbers of expert users sharing their opinions about a visualization artifact to quantitative performance comparisons among several artifacts that are similar enough to compare. The scope and type of evaluation and evaluators is a research design consideration, and there is no single best choice. In particularly young areas, there may not be a clearly related artifact to compare with a newly created one. In such a case, the opinions of a few domain science experts as evaluators may be sufficient to establish that a particular system or technique holds enough promise to share in a publication. But an evaluation is always stronger when there is a comparison of some kind to what has already been published. This kind of anchoring of a research result to the rest of the literature is something that our visualization field could do more consistently. There is always a most closely related artifact that has been described in a publication. In most cases, if an approach serves a need that is already being addressed, however inefficiently, it can be compared to the current inefficient approach. Such a comparison may be sufficiently self-evident that it does not even require an experiment. It does require explicit statement.

Second, an empirical study may establish new knowledge about how humans and computers interact. This kind of research is often hypothesis-driven, and the user study employed serves to test the hypothesis. A hypothesis might state that certain visual cues are more easily perceived by users. Or it might state that one approach for a particular task is more efficient than another approach. What then emerges is new knowledge about whether the hypothesis is supported or falsified by experimental testing. In the best cases, the emergent knowledge can be generalized and helps to guide future research as well as the design and use of visualization artifacts.

I do not think that an empirical evaluation is a requirement for a visualization publication. However, there are a number of publications I have seen where the novelty or significance could have been established much better with an empirical evaluation. Too often those of us who are engineers create a software system or a visualization technique that is new and presume that that is enough to warrant a publication. After all, it was a lot of work! But software documentation, even of a novel piece of software, is typically insufficient to extend human knowledge in a significant way. It requires an explanation of how it is significant. Is it faster? Does it scale better? Does it more efficiently use screen real estate? Questions like these can be answered without a user study, but they do require some testing, analysis, and argumentation. And, as with a user study, in order to be demonstrably faster or more efficient, there has to be something specific to be compared to. There are other ways for something novel to be significant. Do users like it more? Does it speed their work? Does it make them more accurate? Is the experience of using it more pleasurable? If these are the ways something is significant, then some kind of empirical evaluation is likely to be essential. If the claim of significance includes "more," then the empirical evaluation likely needs to include a concrete comparison.

Some of what I have said probably sounds abstract and some perhaps even grandiose. But I believe that extending human knowledge is truly the bar for a research publication. With that context, perhaps we can converge on some conclusions.

Thomas Wischgoll

I actually agree to a great deal with what David lists here. I see the evaluation as the aspect of the publication that illustrates in what way the chosen approach is an improvement over existing work. I use the term *evaluation* fairly loosely here as it could be any means of showing the benefits of the presented work. A properly executed user study can be an effective way of accomplishing that. But there are certainly others as well as both David and I tried to hint at in our previous statements. Weber et al. [25] list 12 different ways in which an application paper can contribute to the area of visualization each of which could be shown with different empirical measures. Personally, I like quantitative and objective measures, such as the execution time of an algorithm in a very well defined test environment. On the one hand, such measures are easier to determine. But at the same time, they cannot be refuted easily either. However, given the fact that visualization algorithms are geared toward making human beings more effective at specific tasks, a user study may be the only way to prove a certain measure.

This leads me to one of the issues with user studies as it is of utmost importance for a user study to be designed, set up, and executed properly to bring value to a publication. This involves the number of people to include, how to recruit participants, and the analysis of the data collected through those participants. One issue can be with domain experts as there may only be a fairly low number of experts suitable for such a user study depending on who the visualization approach is designed for and these domain experts may not be accessible to everyone which makes reproducibility difficult. These domain experts may have a very specific mindset already based on the day-to-day work and that bias may be different compared to another group of domain experts.

Since user studies by definition involve human beings, they are susceptible to different types of biases. Hence, the selection of the study participants is critical to the success of the study. A good user study typically documents very well the selection criteria and processes that were used as well as the entire procedure used to conduct the user study itself. This then aids in improving the reproducibility and should enable the reader to at the very least better understanding of what was tested.

Ultimately, I would like to reiterate the importance of a properly executed user study. If the user study was poorly executed, it does not only provide little to no value to the field. But to make matters worse, it actually provides misleading data. For example, it could suggest that an approach works better than it actually does and the user study then provides a false sense of confidence in the approach. There are other research areas outside of visualization that suffer from exactly that problem, and in the end, the trustworthiness in that research area suffers from it.

David Laidlaw
I think that we are in agreement that any experiment should be properly designed. That is easy to say, but quite difficult to do. The design of an experiment can be as intricate and complex as the design of a software system or other artifact. Thomas mentioned several design elements. One is the number of participants. There is no universally proper number of participants—in some cases, a single participant is sufficient for an experiment to be well designed.

I think that our field could benefit from two changes. The first change is to better educate ourselves on how to design experiments. It is hard to know how to do everything, especially in a field like ours where we need to communicate across multiple disciplines. But if we are going to use experiments as a core element of our research, we need to know how to design them. We do not need to be the best experimental designers in the world, but we do need to know the basics.

The second change is to better appreciate the good parts of an imperfectly designed experiment. As with any creative artifact, a viewer (or reviewer) can always find ways in which it can be changed or improved. With an algorithm or even a software system, small changes are typically easily implemented. With an experimental design, small changes to the experimental procedures mean doing all of the experiments again. Reviewers often do not seem to weigh the cost of suggested changes against the marginal benefits. I think that most researchers who submit experimental work would like their work to be judged on its merits, not critiqued for re-execution.

Krešimir Matković

I have noticed that Thomas used the term "user studies" while David used the term "empirical evaluation." Both of you have reached an agreement that "user studies" are not essential for a research paper involving domain experts. I am wondering what is the place of other empirical evaluation methods, such as surveys, discussion groups, think aloud, user testimony, observation diaries, and so on. Should an application paper be published in top journals in the field without any empirical evaluation?

Thomas Wischgoll

This obviously depends on the application paper. In some cases, a user study or other form of empirical evaluation can be very helpful in terms of evaluating the proposed application technique. The list of empirical evaluations is certainly quite extensive which increases the likelihood of one of them being an appropriate evaluation technique. Which one to choose depends on various factors, such as the target audience and what solution the technique is trying to solve. If the targeted audience is fairly small, a user testimony may be more feasible than discussion groups, for example.

However, I do believe that it should be possible to get an application paper published in a top journal without an user-based evaluation. If the authors can make the case for their method to be more effective for a particular application in some way that would be a valid evaluation that shows the usefulness of the approach. For example, the approach could utilize some optimization that makes it perform faster leading to a more effective use of the user's time. I do, however, recognize the fact that lines get blurry fairly fast when we talk about the user's time. If the time saving comes from a more efficient user interface in some way, a more thorough empirical evaluation would be warranted. On the other hand, if the increase in efficiency solely stems from algorithmic improvements, other forms of evaluations can be sufficient.

David Laidlaw

I used "empirical evaluation" because it is in the title of this part of the book. I consider the choice of which type of empirical evaluation to be a central part of the experimental design. I stand by my assertion that empirical evaluation can be a part of a visualization paper, but I do not think that it must be. If the novelty and significance of the work are compelling without an empirical evaluation, then a paper does not need the evaluation. Some examples of visualization papers with no empirical evaluation and over 500 citations are: Force-Directed Edge Bundling for Graph Visualization, by Holten and Van Wijk [8], Marching Cubes, by Lorensen and Cline [17], and The Application Visualization System: A Computational Environment for Scientific Visualization by Upson et al., including me [23].

Krešimir Matković

Somehow along the lines of both of you, I remember the capstone talk of Jarke van Wijk at IEEE VIS 2013. He said, "Develop new methods/interface/software that are so awesome, cool, impressive, compelling, fascinating, and exciting that reviewers, colleagues, users are totally convinced just by looking at your work and some examples." Should we advise our younger colleagues (and the whole community) to focus on research which does not need evaluation? Can we say that a need for evaluation indicates a lack of awesomeness, coolness, impressiveness in our research? Further, Smith and Pell in their famous paper in the medical domain [22] argue that randomized control trials are sometimes simply not needed and, still, considered a must in medical research. They illustrate their point on a fictitious case of controlled trials on parachute usage in prevention of death and major trauma related to gravitational challenge. Are you aware of the cases when a visualization paper has been rejected due to a missing study in spite of obvious benefits of the proposed method?

Thomas Wischgoll

If it is a ground-breaking new technique that is proposed and can stand on its own, I do think that it can be a valuable contribution even without a formal evaluation. If you look at the history, even for VIS, there are a number of publications that fall into that category, most of them probably earlier in history than later.

But this is where it gets tricky: The authors would still have to provide some indication as to why this method is ground-breaking in some aspect. If the benefits can be easily described for the reader to follow, then one may get by without a more formal evaluation. But if not then some form of evaluation in the form of a user study or some other metrics would be warranted.

Looking at the historical context of some of the more successful VIS papers, it used to be relatively common to describe a novel method based on a sample use case or application. The application was then used as some form of evaluation to showcase the utility of the method. The approach then would be picked up by other researchers and extended to different applications who then may include additional evaluation. Over time, this can build a very thorough use case analysis of a visualization algorithm and thus provide great additions to the state-of-the-art in its entirety.

Part two of your question refers to some of the points I was trying to make earlier. But I do believe that it is worthwhile stressing some of those aspects more thoroughly. I think most of us know of recent cases of papers being rejected due to some reviewers considering the evaluation insufficient. In the medical field, randomized control studies seem to be the gold standard for some types of research. However, there are a lot of issues starting with the question as to how one would pick a truly randomized group of people that at the same time reflects the average composition of the population. Sometimes, the size of the group is used as a measure to guarantee this. Other times, statistical methods are applied to reduce the fact that the random-ized group was not as reflective of the population as desired. For example, the effect of smoking is sometimes eliminated statistically for that reason. In that case, it would be important to disclose the exact methods used to perform that elimination step in my view. The current debate about reproducibility particularly in the medical domain supports this need. What this example shows is that there are a lot of reasons for why a randomized user study may not show what the authors say they do if the user study was not carefully planned and all the steps taken described clearly in the publication. This is why I would always prefer an evaluation based on some quantitative metrics. However, this is not feasible in many cases in the area of visualization since after all it involves visual interpretation by the user.

David Laidlaw

The papers I mentioned at the end of my last answer are examples of highly cited publications without empirical evaluations. I do not know if they are "awesome, cool, impressive, compelling, fascinating, and exciting," but reviewers and citing authors at least found them compelling enough to accept and cite. That suggests that they were judged novel and significant. Empirical evaluations are not always needed.

As far as part two of your question, there are certainly examples of both false positives and false negatives in our review process. I have seen manuscripts that should have been accepted be rejected because a reviewer insisted that a user study was missing or flawed. I have also seen papers be accepted without sufficient evidence of significance. As a field, we can always strive to improve, and improvements in reviewing would be welcome. One major challenge here is that judging design is difficult. Each design must be judged on its merits in the context of all related work. And if that were not challenging enough, the related work is constantly growing and changing, so the evaluation criteria are, too. I do not think that there is a simple answer here; we need to keep discussing and growing as a community. And, we need to avoid making rigid rules.

Krešimir Matković

Someone could argue that peer reviews represent a sufficient evaluation for (some of) the visualization research. Suppose that there is an awesome method and if reviewers like it, we can consider it evaluated. Could you briefly comment on it, is it also a form of evaluation? Further, I think a commercial success of an innovative visualization method which has not been formally evaluated by means of a user study could also be considered as an evaluation. Could you, please, briefly comment on these thoughts.

David Laidlaw

Evidence of the significance of research work can take many forms. We have agreed that empirical evaluations can be part of that evidence. Sometimes, statements of self-evidence can, too. Other types of evidence might include download counts for software, estimates of installed base or number of users, publication of results created using a visualization artifact, commercial success, or awards. Peer review is the way to evaluate whether a visualization artifact is sufficiently novel and significant, but what is presented should be evidence supporting a positive evaluation. The peer reviewers evaluate that evidence. What that evidence comprises is a part of the research design process.

Thomas Wischgoll

I completely agree with David. The paper needs to provide some guidance to the reviewers as well as to how to judge the quality of the results. That is, the purpose the evaluation serves. A paper can provide some other form of evidence to provide a feel for the quality of the results. But the peer-review process is not really a replacement as the reviewers do not have access to the full software, for example, they have to solely rely on what is presented in the paper (and potentially additional material, such as a video).

Krešimir Matković

We have mentioned various forms of evaluation so far—Case studies, user studies, design studies, surveys, etc. It seems that many visualization researchers are not trained in performing studies. Moreover, they do not know which type of study to choose and what is the difference between some of them. As we do not have a common visualization curriculum at universities, is there a way that all visualization researchers have the same understanding of what evaluation means, in particular user studies? Should we try to find new means to promote importance of evaluation to all, or do you think most researchers already know it, or do you, maybe, think that awesome research does not need such an evaluation anyhow?

Thomas Wischgoll

User studies and several other forms of evaluation are probably something the typical visualization researcher is not trained on all that much. This is especially true for the typical scientific visualization researcher. I think in the information visualization community, user studies are significantly more common as the focus and target audience is oftentimes broader, i.e., methods are designed for a larger group of people in information visualization, whereas scientific visualization applications sometimes only have a very limited number of users. But of course, there are exceptions to that statement as well. But to your point, it seems to me that due to the fact that in the information visualization realm user studies are more common, researchers there probably have a little more experience in that area. However, I would assume that many people were not rigorously trained in that area. So, some guidance could be helpful even though there are a number of visualization researchers who attempted to provide such guidance in several publications throughout the recent past.

In several of our studies in the past, albeit not all of them directly related to visualization, we included researchers from psychology and human factors engineering to ensure that the studies follow the necessary scientific rigor on the evaluation part and make sure that the conclusions drawn from the collected data are accurate. In both of those areas, user studies of some form are fairly common and drawing from the experience of those researchers can be very helpful. There is also a lot of existing research on the perceptual side that is relevant to the area of visualization that one can directly tap into and avoid repeating the same type of research or at least use it as a baseline for a formal study.

But to answer your question more directly, I think that most visualization researchers are aware of the need for evaluation and to some extent of the different forms of evaluations, including user studies. However, the degree of what is required to ensure that the results form such a user study are valid may not be as known to everyone in the community. Some of the other chapters in this book try to give an answer to some of those aspects but there probably is a lot more that could be provided. After all, this is a fairly big topic with lots of different avenues that one can take. And, not all of them are valid or appropriate for a given visualization approach.

8.3 Concluding Remarks

The empirical evaluation of the visualization research is far from trivial. As we described above, there are many facets that should be taken care of. We have to select an elevation method, then find appropriate users, correctly execute the evaluation, and present results properly. We agreed that we do need evaluation, and, at the same time, it is clear that it is possible to have a valuable and innovative visualization paper without a user study. Finding the proper evaluation method might be tricky.

As not all members of the visualization community have a proper training in performing studies, many visualization researchers and reviewers often colloquially referred to a controlled laboratory experiment as a user study, while many others consider any empirical study involving the **actual** users is a user study. These two definitions not only are quite different but also have a limited amount of overlapping. A controlled laboratory experiment is typically conducted in the university environment and its participants commonly include a good number of students and university staff. In many applications, a visualization tool or system is designed for a specific group of users who have better knowledge about the data to be visualized and the tasks to be performed. Although it is possible to design a controlled laboratory experiment with domain experts as participants, this approach is not commonly used because (i) it is difficult to design a set of stimuli for complex scenarios, (ii) the variation of users' expertise typically becomes a confounding effect, and (iii) the users may find performing tasks in a controlled setting time-consuming or somehow patronizing. Unlike controlled laboratory experiments or semi-controlled crowdsourcing studies, evaluation with domain experts is difficult to reproduce since others cannot easily replicate the same real-world settings, application-specific tasks, and domain experts with similar knowledge. Nevertheless, the lack of reproducibility is a naturally occurring feature rather than a shortcoming.

In our opinion, we should strive to better understanding of the evaluation in the visualization community. There is a vibrant subgroup of the community which organizes BELIV workshop at the IEEE VIS conference. As we do not have a common visualization curriculum across all universities, we recommend to include evaluation topic in teaching of visualization whenever possible. The current state of user studies in visualization often seems like something that has to be done in order for a paper to get accepted instead of contributing to the merit of the paper. This is definitely wrong. We should all learn when to use a user study and when some other means of evaluation is more appropriate. An unsuitable user study creates more harm than good to a paper. A proper evaluation enriches the paper and helps in its acceptance, for sure.

Our take-home message is, unless your work is really "so awesome, cool, impressive, compelling, fascinating, and exciting that reviewers, colleagues, users are totally convinced just by looking at your work and some examples," consider finding a proper means of evaluation. It will certainly make the paper better. And, even if you write only awesome and fascinating papers, do your homework in study of evaluation methods. We need brilliant minds in evaluation as well. This is probably the only way to ensure an exciting future for visualization research. We did a lot in the last 30 years; we should ensure there will be another fruitful 30 years.

Acknowledgements The authors would like to thank Laura McNamara for numerous discussions. Her input was very valuable and it helped improve the chapter considerably.

References

1. Abdul-Rahman, A., Chen, M., Laidlaw, D.H.: A survey of variables used in empirical studies for visualization. In: Chen, M., Hauser, H., Rheingans, P., Scheuermann, G. (eds.) Foundations of Data Visualization. Springer, Berlin (2019)
2. Chen, M., Ebert, D.S.: An ontological framework for supporting the design and evaluation of visual analytics systems. Comput. Graph. Forum **38**(3), 131–144. https://onlinelibrary.wiley.com/doi/abs/10.1111/cgf.13677 (2019). https://doi.org/10.1111/cgf.13677
3. Demiralp, C., Jackson, C., Karelitz, D., Zhang, S., Laidlaw, D.H.: Cave and fishtank virtual-reality displays: a qualitative and quantitative comparison. IEEE Trans. Vis. Comput. Graph. **12**(3), 323–330 (2006)
4. Gillmann, C., Wischgoll, T., Hamann, B., Hagen, H.: Accurate and reliable extraction of surfaces from image data using a multi-dimensional uncertainty model. Graph. Model. **99**, 13–21. http://www.sciencedirect.com/science/article/pii/S1524070318300365 (2018). https://doi.org/10.1016/j.gmod.2018.07.004
5. Glendenning, K., Wischgoll, T., Harris, J., Vickery, R., Blaha, L.: Parameter space visualization for large-scale datasets using parallel coordinate plots. J. Imaging Sci. Technol. **60**(1), 10,406–1–10,406–8 (2016)
6. Gomez, S.R., Guo, H., Ziemkiewicz, C., Laidlaw, D.H.: An insight- and task-based methodology for evaluating spatiotemporal visual analytics. In: Proceedings of IEEE VAST (2014)
7. Guo, H., Gomez, S.R., Ziemkiewicz, C., Laidlaw, D.H.: A case study using visualization interaction logs and insight metrics to understand how analysts arrive at insights. In: Proceedings of IEEE VAST (2015)
8. Holten, D., Van Wijk, J.J.: Force-directed edge bundling for graph visualization. Comput. Graph. Forum **28**(3), 983–990 (2009). https://doi.org/10.1111/j.1467-8659.2009.01450.x
9. Isenberg, T., Isenberg, P., Chen, J., Sedlmair, M., Möller, T.: A systematic review on the practice of evaluating visualization. IEEE Trans. Vis. Comput. Graph. **19**(12), 2818–2827 (2013). https://doi.org/10.1109/TVCG.2013.126

10. Keefe, D., Acevedo, D., Moscovich, T., Laidlaw, D.H., LaViola, J.: Cavepainting: a fully immersive 3D artistic medium and interactive experience. In: Proceedings of ACM Symposium on Interactive 3D Graphics, pp. 85–93 (2001)
11. Kirby, M., Marmanis, H., Laidlaw, D.H.: Visualizing multivalued data from 2D incompressible flows using concepts from painting. Proc. IEEE Vis. **1999**, 333–340 (1999)
12. Koehler, C., Wischgoll, T., Dong, H., Gaston, Z.: Vortex visualization in ultra low reynolds number insect flight. IEEE Trans. Vis. Comput. Graph. **17**(12), 2071–2079 (2011). https://doi.org/10.1109/TVCG.2011.260
13. Kosara, R., Healey, C.G., Interrante, V., Laidlaw, D.H., Ware, C.: User studies: why, how, and when. Comput. Graph. Appl. **23**(4), 20–25 (2003)
14. Laidlaw, D.H., Kirby, M., Jackson, C., Davidson, J.S., Miller, T., DaSilva, M., Warren, W., Tarr, M.: Comparing 2D vector field visualization methods: a user study. IEEE Trans. Vis. Comput. Graph. **11**(1), 59–70 (2005)
15. Lam, H., Bertini, E., Isenberg, P., Plaisant, C., Carpendale, S.: Empirical studies in information visualization: seven scenarios. IEEE Trans. Vis. Comput. Graph. **18**(9), 1520–1536 (2012). https://doi.org/10.1109/TVCG.2011.279
16. Lam, H., Tory, M., Munzner, T.: Bridging from goals to tasks with design study analysis reports. IEEE Trans. Vis. Comput. Graph. **24**(1), 435–445 (2018). https://doi.org/10.1109/TVCG.2017.2744319
17. Lorensen, W.E., Cline, H.E.: Marching cubes: a high resolution 3D surface construction algorithm. SIGGRAPH Comput. Graph. **21**(4), 163–169 (1987). https://doi.org/10.1145/37402.37422
18. Matkovic, K., Gracanin, D., Jelovic, M., Hauser, H.: Interactive visual steering - rapid visual prototyping of a common rail injection system. IEEE Trans. Vis. Comput. Graph. **14**(6), 1699–1706 (2008). https://doi.org/10.1109/TVCG.2008.145
19. Meyer, M., Sedlmair, M., Munzner, T.: The four-level nested model revisited: blocks and guidelines. In: Proceedings of the 2012 BELIV Workshop: Beyond Time and Errors - Novel Evaluation Methods for Visualization, BELIV '12, pp. 11:1–11:6. ACM, New York, NY, USA (2012). https://doi.org/10.1145/2442576.2442587
20. Munzner, T.: A nested model for visualization design and validation. IEEE Trans. Vis. Comput. Graph. **15**(6), 921–928 (2009). https://doi.org/10.1109/TVCG.2009.111
21. Sedlmair, M., Meyer, M., Munzner, T.: Design study methodology: reflections from the trenches and the stacks. IEEE Trans. Vis. Comput. Graph. **18**(12), 2431–2440 (2012). https://doi.org/10.1109/TVCG.2012.213
22. Smith, G.C.S., Pell, J.P.: Parachute use to prevent death and major trauma related to gravitational challenge: systematic review of randomised controlled trials. BMJ **327**(7429), 1459–1461. https://www.bmj.com/content/327/7429/1459 (2003). https://doi.org/10.1136/bmj.327.7429.1459
23. Upson, C., Faulhaber, T.A., Kamins, D., Laidlaw, D., Schlegel, D., Vroom, J., Gurwitz, R., van Dam, A.: The application visualization system: a computational environment for scientific visualization. IEEE Comput. Graph. Appl. **9**(4), 30–42 (1989). https://doi.org/10.1109/38.31462
24. van Dam, A., Laidlaw, D.H., Simpson, R.M.: Experiments in immersive virtual reality for scientific visualization. Comput. Graph. **26**(4), 535–555 (2002)
25. Weber, G.H., Carpendale, S., Ebert, D., Fisher, B., Hagen, H., Shneiderman, B., Ynnerman, A.: Apply or die: on the role and assessment of application papers in visualization. IEEE Comput. Graph. Appl. **37**(3), 96–104 (2017). https://doi.org/10.1109/MCG.2017.51
26. Wischgoll, T., Glines, M., Whitlock, T., Guthrie, B.R., Mowrey, C.M., Parikh, P.J., Flach, J.: Display infrastructure for virtual environments (dive). J. Imaging Sci. Technol. **61**(6), 60,406–1–60,406–11 (2017)
27. Wischgoll, T., Scheuermann, G.: Detection and visualization of closed streamlines in planar flows. IEEE Trans. Vis. Comput. Graph. **7**(2), 165–172 (2001). https://doi.org/10.1109/2945.928168

28. Ziemkiewicz, C., Chen, M., Laidlaw, D., Preim, B., Weiskopf, D.: Open challenges in empirical visualization research. In: Chen, M., Hauser, H., Rheingans, P., Scheuermann, G. (eds.) Foundations of Data Visualization. Springer, Berlin (2019)

Chapter 9
Evaluation of Visualization Systems with Long-Term Case Studies

Bernhard Preim and Alark Joshi

Abstract New visualization systems need to be evaluated ideally with participants representing the target user group doing real tasks. Most evaluations are short, i.e. participants receive an instruction and use the system only once. These short evaluations have many drawbacks, e.g., participants do not get familiar with the system. Long-term case studies that involve the regular use of a system for at least several weeks lead to more reliable assessments of user acceptance. We discuss strategies to plan and conduct long-term case studies and present examples that demonstrate the feasibility of this evaluation strategy.

9.1 Introduction

Many evaluation concepts have been employed to assess the value of individual visualization techniques or whole visualization systems. Isenberg et al. provided an in-depth analysis of evaluation practice in visualization and considered eight major variants [8]. Regarding their terminology, we focus on *empirical* evaluations that include actual users, instead of hypothetical discussion of usage scenarios or formal analysis based on quality metrics. As Isenberg et al. point out, most of the existing empirical evaluations relate to user preferences and other usability or user experience criteria but do not focus on how visualization systems are actually used to solve complex problems. The case study type of empirical evaluation that we discuss in this chapter is considered as "particularly strong form of evaluation for understanding work practices and visual data analysis" [8]. However, case study-based visualiza-

B. Preim (✉)
Otto-von-Guericke-Universität Magdeburg, Magdeburg, Germany
e-mail: bernhard@isg.cs.uni-magdeburg.de

A. Joshi
University of San Francisco, San Francisco, USA
e-mail: apjoshi@usfca.edu

© Springer Nature Switzerland AG 2020
M. Chen et al. (eds.), *Foundations of Data Visualization*,
https://doi.org/10.1007/978-3-030-34444-3_9

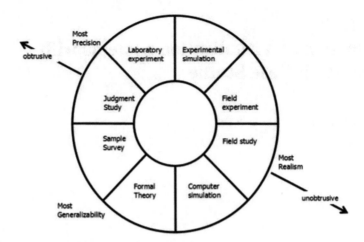

Fig. 9.1 Field studies, an alternative term for long-term case studies, are unobtrusive and yield realistic observations (From: Carpendale et al. [3])

tions are rare compared to the substantial portion of application-oriented visualization research in the visualization community. In this chapter, we discuss the potential and current practice of case study-based evaluation in visualization research. We emphasize one aspect of expressive case study reports, namely the long-term character. Today's complex visualization systems may involve longer learning periods and problem-solving activities that require substantial time. Thus, care is necessary to provide enough time for experts to use the system and for visualization researchers to observe and analyze the usage of the system.

Long-term case studies have their roots in ethnographic research [4], e.g., in cultural anthropology, where researchers live in a different culture, e.g., in an African tribe, take part in the daily activities and carefully document their first-hand experiences (*diary of use*). "The observer tries as much as possible to be unobtrusive," ideally not affecting what is being observed [3]. With ethnographic methods, a few researchers gain insight and in-depth experiences over a long time. *Field tests* and *workplace studies* are alternative names used in human–computer interaction (HCI) [3, 6].

Case study-based methods were introduced in HCI early. A survey by Hughes et al. [7] documents success stories both in academic and commercial settings, where time and budget constraints need to be considered as well. Figure 9.1 puts field studies in the context of other evaluation methods and highlights that they are particularly realistic, but not very precise.

9.1.1 Putting Long-Term Case Studies in the Context of Empirical Evaluation

Long-term case studies are a promising instrument of empirical evaluation and "yields realistic and believable narratives" of real users interacting with a visualization tool [5]. They are motivated by shortcomings of the more frequently used controlled laboratory studies as stated by Shneiderman and Plaisant [25]: "laboratory studies became ever more distant from practical problems and broader goals." Carpendale adds that the use of very small datasets, students as test subjects, and unrealistic tasks lead to the problem that the results of information visualization evaluations are not believable and actually, the developed techniques are not adopted [3]. In particular, systems that require substantial learning effort and are intended to be used for complex problem solving or discovery activities cannot be adequately assessed within one or two hours in a laboratory experiment with well-defined tasks. The simple fact that the evaluation takes place in a laboratory and not in a realistic work context reduces ecological validity, i.e., the amount to which the results can be translated to realistic settings.

Visualizations related to isosurface and volume renderings are often evaluated by means of task-based perceptual experiments typically involving a comparison of methods with respect to shape and depth perception. In the terminology of Carpendale [3], these are referred to as *judgement studies* (recall Fig. 9.1). Similar to (other) laboratory experiments, they favor precision or realism. Although valid tasks and methods are available, the evaluations only explain perceptual aspects at a rather low level (see Preim et al. [19] for a survey and Saalfeld et al. [21] for a tutorial-like paper on how to perform such experiments with a focus on medical visualization). In medical visualization, for example, the actual purpose is to support advanced diagnostics (Is a muscle infiltrated by a tumor and to what extent?) and treatment planning (Is the patient operable? How much tissue needs to be removed? And how to access the pathology?). For understanding such cognitive activities involving problem solving, decision making, and discovery perception-based experiments are not directly relevant. Moreover, almost all these experiments relate to static rendered images, that is, the whole value of interactive exploration, e.g., rotation and clipping, for which 3D visualization techniques are provided, are ignored.

Long-term case studies typically involve a few highly specialized professionals that use a system in their familiar work environment for tasks that are relevant to them, based on data that they have available [2]. The discovery in large scientific, business or finance data, police analysis, and medical research based on large and heterogeneous data are examples of such situations. As we will see in this chapter, there are a number of examples how long-term case studies were used for medical visualization, information visualization, and visual analytics applications.

9.2   Goals and Variants of Long-Term Case Studies

Ethnographics-inspired evaluations were carried out in human–computer interaction
and software engineering as a research method for a deep understanding of processes
and the use of interactive products. A deep involvement in the users' activities can
provide genuine insight into the processes and daily routines of the users.

These observational methods may be applied early in the development process to
analyze current work practices and establish initial requirements [7]. Here, we focus
instead on *evaluative ethnography*, that is, the evaluation of innovative visualization
systems based on a working prototype. Evaluation ethnography includes an assessment of the prototype, the deployment in particular contexts, and workflows and the
extraction of ideas for redesign—three of the five stages of empirical evaluation as
discussed by Lam et al. [16].

9.2.1   Goals

Long-term case study evaluations last at least several weeks and are carefully documented by the users with both verbal notes and screenshots. Regular interviews,
logging protocols, screen capture, and video analysis may be added to *understand*
differences [25]. They may reveal:

- patterns of use, e.g., typical problems as well as actions to tackle them,
- characteristic changes of these patterns over time,
- the social context of system use,
- engagement and motivation,
- the variety of data to be processed and tasks to be solved in practice, and
- unintended usage scenarios.

Such findings may have serious implications for further design which makes these
methods appropriate for *formative* evaluation where the major goal is to refine or add
requirements for the further development of a system. If, for example, some features
are not used at all, they may be removed or at least "hidden" in submenus or dialogs
that rarely appear and thus do not distract.

If certain usage patterns become obvious, the system may be redesigned to provide
guidance, e.g., to support the user along a certain analysis path. As a social aspect,
it may become obvious that domain experts cooperate with others in the analysis
of data. As an example in medicine, the analysis of medical image data is a careful
cooperation between radiology technicians and radiologists and the results of the
diagnostic report are presented to referring physicians from other medical disciplines.
If such a collaborative aspect is identified and analyzed, requirements to directly
support cooperation may arise. In fact, early uses of ethnographic methods were
already focused on analyzing social aspects in office contexts or air-traffic control [1].

Long-term case studies may also reveal *how* engaged users are often despite
struggling with the system, how motivated they are, and how they (and perhaps
their colleagues) trust a system. These user experience (UX)-related properties and

their changes over time are essential for visualization systems to be used in research and industrial practices. The understanding of actual data and tasks often leads to requirements related to the support of more file formats or related to a better support to convert data.

Additionally, unintended usage scenarios are observed that typically involve creative workarounds to achieve a goal, the system was not meant to be used for. As a consequence, a redesign should directly support these usage scenarios. As an example, Whitaker [28] analyzed e-mail use and found that mail systems are not only used for communication (as intended) but also for reminding to activities and as an archive of communication and knowledge.

9.2.2 Multi-dimensional In-depth Long-Term Case Studies

Multi-dimensional in-depth long-term case studies (MILCs) were introduced as a special long-term evaluation technique particularly for InfoVis. This evaluation concept by Shneiderman and Plaisant [25] was introduced to evaluate creativity support tools. The major goal of MILC evaluations is to "study the creative activities that users of information visualization systems engage in." *Multi-dimensional* relates to the integrated use of observations, interviews, and logging protocols. Shneiderman and Plaisant also explain what they consider as long-term: a system used in different stages with a minimum duration of several weeks. The following stages are discriminated:

- the *training stage*, where the users get familiar with the system, optionally a written tutorial to assist independent use,
- an *early use stage*, where the users are visited also with the goal of assisting in using the system and identifying smaller problems that may be solved soon, whereas
- in the *mature use stage*, the system is no longer altered. Thus, changes in usage patterns in the *mature use stage* are not due to changes in the system and may reflect that usage patterns change over time.
- a *final stage* in which the documentation is summarized and a final review is carried out.

The methods of data collection are the same in the early and mature use stages. Only the stable system state makes the difference between the two. Not all authors that base their evaluations on the MILC principles follow all recommendations. Valiati et al. [27], for example, report on three MILC studies, where they have *not* discriminated between early and mature use stages. The system was not improved at all during the whole study. They employed most of the instruments recommended by Shneiderman and Plaisant [25] but did not provide logging functions.

Shneiderman and Plaisant discriminate basically two types of MILCs:

- a moderate MILC as part of a typical research project, where the early use and mature use stages last approximately four weeks and
- long MILCs that may last up to several years where the evaluation is the core activity of a research project.

Most MILC evaluations are moderate variants. In the examples discussed by Valiati et al. [27], the study duration was between six weeks and four months, 5–8 meetings with users were arranged and the overall time of observing users was between 12 and 18 h. This example confirms the recommendations of Shneiderman and Plaisant to combine different instruments, such as observation, interviews, and thinking aloud (recall [25]). They traced the problems identified in the long-term evaluation to the instruments used to detect them: While some problems were explicitly described by the analysts during interviews, a considerable portion were detected based on observation.

9.3 An Overview of Long-Term Evaluations in Visualization Research

In the following, we briefly describe selected examples of long-term case study evaluations. They were chosen, since they rigorously report on goals, preparation, conduction, and analysis. The underlying papers do not introduce a visualization framework, but focus on the evaluation of an already presented system. Thus, the evaluation is not a minor part of a large paper. Among the seven scenarios from Lam et al. [16], they all relate to *visual data analysis*. It seems that long-term case studies are particularly important in this scenario. In the scientific literature, there are more long-term case study evaluations of visualization, but they are described in considerable less detail.

9.3.1 Evaluating the Rank-by-Feature Framework

Seo et al. have developed a comprehensive visual analytics framework that enables the efficient analysis of high-dimensional data [23]. Many metrics (*interestingness measures*) are involved to rank individual features and pairs of features to direct the further analysis to potentially interesting aspects, e.g., features, where the distribution strongly deviates from a normal distribution or pairs of features where a strong linear or quadratic correlation exists. As a general unsupervised learning method, hierarchical clustering with an interactive dendrogram visualization is provided. The system was primarily used for analyzing gene expression data and was initially presented along with informal evaluations including feedback from domain experts (Fig. 9.2).

To get a deeper understanding, if and how the rank-by-feature framework change the researchers exploration process, a MILC evaluation was performed [24]. Six participants were recruited that had used the framework and published scientific results obtained with it. These researchers were from different fields (including statisticians, biologists, metereologists) and were not involved in the design and development of the tool.

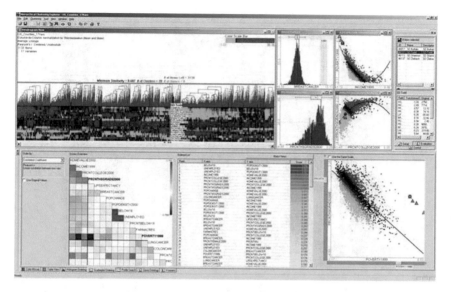

Fig. 9.2 Rank-by-feature framework with hierarchical clustering (top left), a matrix view depicting correlations between dimensions (bottom left), an ordered list with most interesting feature combinations (bottom middle) as well as histograms and scatterplots for selected dimensions and combinations thereof (From: [23])

9.3.2   Evaluating the Social Action Tool

Few users employed a social network analysis tool with graph-based visualizations and statistics related to graph-based data [18]. The long-term case study was performed according to the MILC variant (recall [25]), where the early and mature use stages lasted four weeks. The evaluation was started with a 2 h training session and a documentation was provided to further support the autonomous use of the system.

9.3.3   Evaluating the Jigsaw Analysis Tool

Another prominent visual analytics tool analyzed with ethnographic methods is JIG-SAW, a tool that enables the analysis of large document collections [14, 26]. Clustering is provided where the similarity of documents is analyzed depending on the co-occurrence of words. Documents may be also sorted according to different criteria. Thus, document views, list views, and cluster views are essential components of the systems (see Fig. 9.3).

The evaluations with three intelligence analysis experts (two from academia, one from industry) lasted between two and fourteen months. Interviews (45–60 min.) were audio-recorded, fully transcribed, and carefully analyzed with specialized soft-

Fig. 9.3 Jigsaw system used with a multi-monitor setup. The top view provides a list visualization with connections between people where selected people are highlighted (From: [26])

ware to understand core themes [15]. The prepared questions of the semi-structured interviews relate to specific tasks for which Jigsaw is used, the goals of the analysis, the data to be used, features considered essential or superfluous. The analysis with one expert revealed that he employed mostly documents related to a narrow time frame since otherwise, the resulting visualizations are overwhelming. It turned out that a feature was missing that allowed users to select/deselect documents for the current analysis. Mostly, the analysis of documents served to understand whether there are relations between two persons and, if so, to better understand what type of relation they have. Graph views that provide a visual interpretation of the data were new to them and appreciated.

The long-term case study provided many insights in the learning process required to use the Jigsaw system and in unexpected pattern of system use.

9.3.4 Evaluating the Impact of a Medical Visualization Tool

Working in the medical domain requires long-term immersion into a medical facility that frequently leads to a deeper understanding of the primary pipeline for the treatment of a patient. This includes the processes followed at the facility as well as all the individuals involved in the processes. In previous work, Joshi et al. [13] worked closely with neurosurgeons to understand challenges with respect to image-guided surgery. Neurosurgeons, radiologists, neurologists, and technicians are all involved in process of surgical planning and the actual surgery. The researchers identified challenges associated with data representation of all the modalities being used for image-guided surgery such as CT, MRI, EEG electrode strips, and in some cases,

Fig. 9.4 A figure showing context (EEG electrodes) around the position of the surgical instrument. (From: [13])

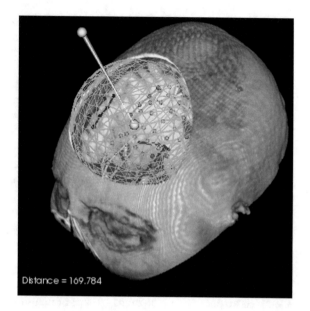

Distance = 169.784

PET scans and DTI imagery. They developed a system that allows contextual representation of the data during surgery and evaluated it with neurosurgeons and residents [12, 13] (Fig. 9.4).

Due to the embedded nature of the researchers involved in the project, other problems related to occlusion in vascular neurosurgery too were identified and addressed [10]. These techniques were incorporated into existing image-guided surgery software and were evaluated over a long-period of time for ease-of-use and adoption. Technicians and surgeons continued to use the technique via the image-guided navigation system.

Expert analysis provided crucial insight into use cases and usability of the system. As the research team continued to work with the surgeons and operating room technicians, other challenges with respect to the ambient lighting in the operating room were identified and light-sensitive solutions [9] were designed and deployed in the operating room.

These solutions to their problems were identified, resolved, deployed, and evaluated over a two-year period to iteratively improve the image-guided surgery system.

9.3.5 Generalized Experiences

A common result of all long-term evaluations was that experts always started their analysis with clear analytical questions in mind. The visualization researchers have not observed pure exploration activities without any hypothesis. The initial use of the system for the selected data may lead, of course, to interesting or even surprising situations, that stimulate follow-up questions, e.g., to understand a phenomenon in more detail or to confirm a pattern. The analytical questions are very specific for the particular domain but as Valiati et al. [27] point out, most of them can be mapped to rather general visualization tasks, such as gaining an overview, searching for a particular configuration, and comparison that were performed at different abstraction levels. This generalization may help to translate the experiences to other areas and enable other researchers to *reproduce* these experiences or find out that the results cannot be confirmed eventually leading to more reliable knowledge about analytical patterns and appropriate computer support. All long-term case studies discussed in this section relied on very few experts. The three evaluations described by Valiati et al. had one expert only. The rank-by-feature evaluation had six experts, the largest number, we found in such an evaluation.

The overall assessment of long-term evaluations revealed a number of tasks that were not supported well at least by early information visualization systems [27]. Users want to:

- document and record (intermediate) results for themselves or discussion with others,
- emphasize or comment on items, groups of items or relations,
- to verify observations derived from data visualizations with statistical methods

9.4 Planning, Conducting, and Reporting

A long-term case study obviously requires careful planning and sufficient time. It is likely that we rarely see this type of evaluation since it does not nicely fit in the tight schedule of paper publishing where the implementation is often finished only a few weeks before the deadline. The most important aspect is the recruiting of experts to use the system for a longer time. These experts need to either be the target users or be representative, in particular they should have approximately the qualification and experience, of the target users. Sometimes, the few top level experts for a special domain are not available and are replaced by users with a little lower experience. However, if the system is intended for users with long-term experience and responsibility to make decisions, students or junior researchers in this area are not representative enough.

Once the users are selected, the evaluation and documentation procedures organizational issues should be discussed (see hints in [25]). The software needs to be prepared carefully, including a short tutorial/documentation to enable autonomous use, logging capabilities, and testing. Since realistic tasks should be investigated, the selection of specific goals, tasks, and data is the responsibility of the domain expert. However, discussions between visualization researcher and the domain expert are required to ensure that the data and tasks are representative.

This involves a considerable effort on the side of the domain expert and consequently, publications in their scientific domain are a typical result [9, 11].

9.4.1 Reporting

Reporting on a case study requires considerable thoughts as well according to Isenberg et al. [8], the following aspects are crucial in any type of reporting on empirical evaluation:

- be specific about your domain experts (age, gender, qualification, experience in the domain, and with similar software, …),
- be specific about the nature of your relation to them, e.g., Are they co-authors of the paper? Are they independent or part of the same institution/project?
- be careful with definitive statements and try to include proper statements of uncertainty when justified.

In addition, we recommend additional components for reporting based on Valiati et al. [27] who described three MILC evaluations in a standardized manner.

- description of the data used by the experts, e.g., number of dimensions, number of datasets, and size of a document collection
- analytics questions that the experts tried to answer
- severe usability problems that may have avoided that experts could analyze the data in the way they originally wanted to perform.

9.5 Challenges and Limitations

A major challenge of long-term evaluations is that very self-disciplined users are needed that are willing and able to document over a longer time why they used the system, what they considered satisfying, surprising or frustrating. Users often stop the evaluation earlier than expected [2, 17]. Participants of long-term evaluations are not only few, but often also more tech-savvy than average users leading to a selection bias that further reduces the generalizability of the results.

Another limitation is that due to the small number of test persons, there is more randomness involved than in a laboratory study, i.e., it is a bit by chance how the system is actually used and which data are used. The environment in which case study work is performed is realistic, but not controllable. Thus, statistical analysis is typically not meaningful. "The outcomes should not be too generalized" as Elmqvist et al. argue [5]. Since only a few users are involved, long-term case studies do not help to characterize the use of system for a diverse set of users that differ, e.g., in their spatial ability.

To ease the burden on the target users of the system, developers could consider automating the data collection process through system logs and infrequent face-to-face meetings with the users. This would provide insight into whether a deployed system is being used as well as identify pain points for users that may be preventing them from using the system.

9.5.1 Combinations with Other Methods

Long-term case studies have a number of advantages that were stated in the introduction and motivate this chapter. However, as Carpendale [3] points out, no single evaluation method can fully characterize the value of interactive visualization systems. Long-term case studies enable realistic observations, but they are not precise. Even the MILC variant (recall [25]) that combines a number of methods within a case study evaluation remains limited. Therefore, combinations with other methods are relevant.

Instead of discussing all possible combinations, we will focus on one combination that is particularly relevant for visual analytics systems that often aim at discovery processes. The combination with an explicit recording of *insights* is a natural choice and provides a clear focus for long-term case studies. The number and quality of such insights, e.g., whether insights are surprising and can be verified, is considered as an evaluation measure in *insight-based evaluations* [22]. The original insight-based evaluations were laboratory experiments where analysts should freely use the system (after appropriate training) to find interesting relations. Seo et al. [24] combined the MILC evaluation with insight-based analysis of the rank-by-feature framework. This combination is promising since discovery processes often are not very effective when restricted to a limited amount of time. This combination, however, does not solve the major problem of long-term case studies, namely that they comprise only a very few participants. Therefore, Seo et al. [24] added a broader survey where they asked a larger number of users, again authors of publications that employ their tool, to take part in an interview [24]. This interview cannot provide such a rich description of system use as in the long-term evaluation, but since much more participants are involved, more reliable and generalizable statements about usage patterns and usefulness can be derived.

The conduction of insight-based long-term case study evaluations has to consider many aspects (see [3] for a discussion). A crucial question is when analysts are interviewed with respect to what they have learned about the domain using a visual analytics system and a selection of datasets. Insights may occur suddenly, but also hours or even days after a system was used. Of course, the insights that were gained strongly depend on the domain knowledge of the analyst, her motivation, and creativity. Preim et al. [20] provide a discussion of the evaluation practice in medical visualization, where long-term case studies and their combination with other methods are discussed.

9.6 Conclusions

Long-term case study is a viable empirical evaluation method for visualization systems that enables an understanding of cognitive activities, such as problem solving and decision making. Long-term case studies in visualization research have some unique aspects compared to applications in human–computer interaction, e.g., discovery processes in visual analytics applications. Thus, we discussed primarily such visualization examples and hope to stimulate further attempts in this direction. This qualitative and observational evaluation method overcomes many limitations of laboratory-based studies and enables a deep understanding of system use. It can be adapted to different time-frames and budgets ranging from several weeks to a few years. The observation of users doing real work in their familiar (work) context is a key aspect. The MILC variant described by Shneiderman and Plaisant [25] provides guidance how to perform such evaluations in an informative manner. Since only a few users are involved and the working environment cannot be controlled, long-term case studies are limited. A combination with other evaluation methods, e.g., questionnaires, allows to derive quantitative assessments. Long-term case studies were successfully used in a number of InfoVis and visual analytics applications. In other areas, particularly, in scientific visualization applications, the method is underutilized but promising as well.

Long-term case studies evolve into continuous use of a deployed system only if the researchers are immersed and have clearly addressed an existing problem in the workflow of the target users. The maintenance and iterative development of the system in conjunction with the end users results in successful outcomes. If you would like your system to be used for a long period of time, you have to be willing to maintain and support it for that same duration as well.

References

1. Bentley, R., Hughes, J.A., Randall, D., Rodden, T., Sawyer, P., Shapiro, D., Sommerville, I.: Ethnographically-informed systems design for air traffic control. In: Proceedings of the 1992 ACM Conference on Computer-Supported Cooperative Work, pp. 123–129 (1992)
2. Breakwell, G.M., Hammond, S., Fife-Scha, C.: Research Methods in Psychology. Sage Publications, Thousand Oaks (1995)
3. Carpendale, S.: Evaluating information visualizations. Information Visualization, pp. 19–45. Springer, Berlin (2008)
4. Crabtree, A., Rodden, T., Tolmie, P., Button, G.: Ethnography considered harmful. In: Proceedings of the ACM SIGCHI Conference on Human Factors in Computing Systems, pp. 879–888 (2009)
5. Elmqvist, N., Yi, J.S.: Patterns for visualization evaluation. Inf. Vis. **14**(3), 250–269 (2015)
6. González, V., Kobsa, A.: A workplace study of the adoption of information visualization systems. Proceedings of I-KNOW, vol. 3, pp. 92–102 (2003)
7. Hughes, J., King, V., Rodden, T., Andersen, H.: The role of ethnography in interactive systems design. Interactions **2**(2), 56–65 (1995)
8. Isenberg, T., Isenberg, P., Chen, J., Sedlmair, M., Möller, T.: A systematic review on the practice of evaluating visualization. IEEE Trans. Vis. Comput. Graph. **19**(12), 2818–2827 (2013)
9. Joshi, A., Papanastassiou, A., Vives, K., Spencer, D., Staib, L., Papademetris, X.: Light-sensitive visualization of multimodal data for neurosurgical applications. In: IEEE International Symposium on Biomedical Imaging (ISBI) (2010)
10. Joshi, A., Qian, X., Dione, D.P., Bulsara, K.R., Breuer, C.K., Sinusas, A.J., Papademetris, X.: Effective visualization of complex vascular structures using a non-parametric vessel detection method. IEEE Trans. Vis. Comput. Graph. (VIS 2008), **14**(6), (2008)

11. Joshi, A., Scheinost, D., Okuda, H., Murphy, I., Staib, L.H., Papademetris, X.: Unified framework for development, deployment and testing of image analysis algorithms. In: MICCAI Workshop on Systems and Architectures for Computer Assisted Interventions (2009)
12. Joshi, A., Scheinost, D., Spann, M., Papademetris, X.: Evaluation of multi-viewport based visualization for electrode navigation during stereotactic image guided neurosurgery. In: International Brain Mapping and Interoperative Surgical Planning Society's 6th World Congress for Brain Mapping and Image Guided Therapy (2009)
13. Joshi, A., Scheinost, D., Vives, K.P., Spencer, D.D., Staib, L.H., Papademetris, X.: Novel interaction techniques for neurosurgical planning and stereotactic navigation. IEEE Trans. Vis. Comput. Graph. (VIS 2008), **14**(6), (2008)
14. Kang, Y.A., Gorg, C., Stasko, J.: Evaluating visual analytics systems for investigative analysis: deriving design principles from a case study. In: Proceedings of IEEE Visual Analytics Science and Technology, pp. 139–146 (2009)
15. Kang, Y.A., Stasko, J.: Examining the use of a visual analytics system for sensemaking tasks: case studies with domain experts. IEEE Trans. Vis. Comput. Graph. **18**(12), 2869–2878 (2012)
16. Lam, H., Bertini, E., Isenberg, P., Plaisant, C., Carpendale, S.: Empirical studies in information visualization: seven scenarios. IEEE Trans. Vis. Comput. Graph. **18**(9), 1520–1536 (2012)
17. Ohly, S., Sonnentag, S., Niessen, C., Zapf, D.: Diary studies in organizational research: an introduction and some practical recommendations. J. Pers. Psychol. **9**, 79–93 (2010)
18. Perer, A., Shneiderman, B.: Integrating statistics and visualization: case studies of gaining clarity during exploratory data analysis. In: Proceedings of the ACM SIGCHI Conference on Human Factors in Computing Systems, pp. 265–274 (2008)
19. Preim, B., Baer, A., Cunningham, D., Isenberg, T., Ropinski, T.: A survey of perceptually motivated 3d visualization of medical image data. Comput. Graph. Forum **35**(3), 501–525 (2016)
20. Preim, B., Isenberg, P., Ropinski, T.: A critical analysis of the evaluation practice in medical visualization. In: Proceedings of the EG Workshop on Visual Computing in Biology and Medicine (2018)
21. Saalfeld, P., Luz, M., Berg, P., Preim, B., Saalfeld, S.: Guidelines for quantitative evaluation of medical visualizations on the example of 3d aneurysm surface comparisons. Comput. Graph. Forum **37**, (2018)
22. Saraiya, P., North, C., Lam, V., Duca, K.A.: An insight-based longitudinal study of visual analytics. IEEE Trans. Vis. Comput. Graph. **12**(6), 1511–1522 (2006)
23. Seo, J., Shneiderman, B.: A rank-by-feature framework for unsupervised multidimensional data exploration using low dimensional projections. In: Proceedings of IEEE Symposium on Information Visualization, pp. 65–72 (2004)
24. Seo, J., Shneiderman, B.: Knowledge discovery in high-dimensional data: case studies and a user survey for the rank-by-feature framework. IEEE Trans. Vis. Comput. Graph. **12**(3), 311–322 (2006)
25. Shneiderman, B., Plaisant, C.: Strategies for evaluating information visualization tools: multidimensional in-depth long-term case studies. In: Proceedings of the Workshop on BEyond Time and Errors: Novel Evaluation Methods for Information Visualization (2006)
26. Stasko, J.T., Görg, C., Liu, Z., Singhal, K.: Jigsaw: supporting investigative analysis through interactive visualization. In: Proceedings of the IEEE Symposium on Visual Analytics Science and Technology, pp. 131–138 (2007)
27. Valiati, E.R., Freitas, C.M., Pimenta, M.S.: Using multi-dimensional in-depth long-term case studies for information visualization evaluation. In: Proceedings of the Workshop on BEyond Time and Errors: Novel Evaluation Methods for Information Visualization, p. 9 (2008)
28. Whittaker, S., Sidner, C.L.: Email overload: exploring personal information management of email. In: Proceedings of the ACM SIGCHI Conference on Human Factors in Computing Systems, pp. 276–283 (1996)

Chapter 10
Vis4Vis: Visualization for (Empirical) Visualization Research

Daniel Weiskopf

Abstract Appropriate evaluation is a key component in visualization research. It is typically based on empirical studies that assess visualization components or complete systems. While such studies often include the user of the visualization, empirical research is not necessarily restricted to user studies but may also address the technical performance of a visualization system such as its computational speed or memory consumption. Any such empirical experiment faces the issue that the underlying visualization is becoming increasingly sophisticated, leading to an increasingly difficult evaluation in complex environments. Therefore, many of the established methods of empirical studies can no longer capture the full complexity of the evaluation. One promising solution is the use of data-rich observations that we can acquire during studies to obtain more reliable interpretations of empirical research. For example, we have been witnessing an increasing availability and use of physiological sensor information from eye tracking, electrodermal activity sensors, electroencephalography, etc. Other examples are various kinds of logs of user activities such as mouse, keyboard, or touch interaction. Such data-rich empirical studies promise to be especially useful for studies in the wild and similar scenarios outside of the controlled laboratory environment. However, with the growing availability of large, complex, time-dependent, heterogeneous, and unstructured observational data, we are facing the new challenge of how we can analyze such data. This challenge can be addressed by establishing the subfield of *visualization for visualization (Vis4Vis)*: visualization as a means of analyzing and communicating data from empirical studies to advance visualization research.

D. Weiskopf (✉)
University of Stuttgart, Stuttgart, Germany
e-mail: weiskopf@visus.uni-stuttgart.de

© Springer Nature Switzerland AG 2020
M. Chen et al. (eds.), *Foundations of Data Visualization*,
https://doi.org/10.1007/978-3-030-34444-3_10

10.1 Introduction

This position statement primarily focuses on empirical studies with user involvement but also touches other empirical studies that may collect data from technical performance benchmarks to assess the computational characteristics of a visualization system.

I argue that we need to establish a new subfield to address the challenges of empirical evaluation in visualization research:

We need *visualization for visualization (Vis4Vis).*

The underlying problem is the difficulty in performing an appropriate evaluation for complex visualization systems. For these, many of the traditional approaches to empirical research adopted from other fields cannot be used directly. Other chapters of this book [22] discuss various aspects of the underlying problems, methodological challenges, and possible solutions.

I argue that one promising route is to use as much information as possible from empirical studies. Unfortunately, many of the traditional methods for user studies and other empirical research in visualization come from other fields and earlier times in which there was much less data accessible from studies. One example of such data that is still underutilized in visualization research is gaze data from eye tracking experiments. Section 10.3 discusses examples of eye tracking in visualization research in more detail. However, there are many other potential sources of sensor data that could be collected. Several of these examples rely on physiological sensors, often in the context of work on human–computer interaction (HCI): electroencephalography (EEG) [3] and, in general, the use of brain–computer interfaces (BCIs) and EEG for interaction [37], pervasive BCI [68], near-infrared spectroscopy (NIRS) [39, 80], functional magnetic resonance imaging (fMRI) [28], or the combination of several physiological sensors to characterize emotions [84] or investigate interfaces [71].

However, data is not restricted to coming from physiological sensors. For example, logging user activities with the visualization interface, based on recording mouse, keyboard, touch, or other ways of interaction, can provide a detailed and rich source of highly relevant information [83]. Other examples are video and audio recordings during user studies that can serve as a basis for think-aloud protocol analysis [32].

Overall, technological advances for various kinds of sensors and other data sources have made it easy and cost-effective to capture largely increasing amounts of data for empirical visualization research. And with further progress in technology, in particular, for non-stationary or wearable devices for visualization and user studies, we will see even more diverse types of user studies in visualization research. A recent trend in the visualization community addresses immersive analytics [62], which will lead to the problem of evaluating visualizations in the context of virtual reality or augmented reality.

With the challenges of empirical research for complex visualizations on the one hand, and opportunities that come with advanced data acquisition on the other hand, we will have to rethink how we can conduct, evaluate, and report empirical studies. With this text, I focus on the issues related to data analysis for the evaluation and reporting of the results of studies based on large, complex, time-dependent, hetero-geneous, and unstructured observational data. I argue that visual data analysis and communication are a promising approach to address these issues. Accordingly, I will discuss opportunities and open questions for visualization research. My proposal for the need for *Vis4Vis*, especially in the context of empirical visualization research, extends my position statement that I gave as part of the panel discussion at the 2016 Workshop on Beyond Time And Errors: Novel Evaluation Methods For Visualization (BELIV).[1]

10.2   Background of Empirical Studies

The relevance of empirical studies for evaluation, especially user-oriented evaluation, is well accepted by the visualization research community. Other chapters of this book [22] discuss several aspects of empirical studies in more detail, and I refer to them for background reading.

In general, there are many well-established approaches to empirical studies for visualization and visual analytics [21, 70, 85]. Tory [82] provides a recent overview and categorization of user study approaches, covering various quantitative and qual-itative methods. Freitas et al. [35] discuss a user-centered perspective on evaluation. There are also examples in which different types of study methods are combined, including the combination of usability metrics and eye tracking [25].

Evaluation methodology is the special focus of the series of BELIV Workshops, which investigate approaches beyond the traditional user performance measures of completion time and accuracy. Therefore, many BELIV Workshop papers address topics relevant to this text. For example, Elmqvist and Yi [31] describe a collection of patterns for evaluation, Ellis and Dix [30] provide an explorative analysis of user studies, Lam and Munzner [57] discuss quantitative empirical studies in the context of meta-analysis, and Anderson [2] employs cognitive measures for evaluation.

However, the above papers do not focus on empirical studies that use rich sets of observations, whereas Kurzhals et al. [51, 52] consider this approach as critical for future and improved evaluation methods for visual analytics. They especially focus on the combination of eye tracking information with traditional task performance indicators, but they also discuss the issue of data fusion integrating further time-oriented data acquired during an empirical study. One example is the combination

---

[1]Panel "On the Future of Evaluation and BELIV" with panelists Daniel Weiskopf, Laura McNa-mara, Mark Whiting, Niklas Elmqvist, and Tamara Munzner, BELIV 2016 (Workshop on Beyond Time And Errors: Novel Evaluation Methods For Visualization) at IEEE VIS 2016. https://beliv-workshop.github.io/2016/schedule.html.

of eye tracking and interaction logs [9]. Kurzhals et al. [51] call for exploratory data analysis and hypothesis building to address the difficult analysis questions that come with complex data. In follow-up work, Kurzhals et al. [50] adopt the perspective of analysis tasks on eye tracking data, with a respective overview of such tasks.

A further step in the direction of integrating different data sources from empirical research into an interactive visual analysis approach was taken by Blascheck et al. [10, 11]: they describe how visual analytics methods can be used to evaluate visual analytics systems, for example, by including think-aloud protocol analysis, eye tracking information, or interaction data from the same experiment. Blascheck et al. [7] enrich this approach by integrating visual data analysis and coding of user behavior.

I argue to follow-up and extend this direction of advanced visualization methods for analyzing complex and rich data sources. This will become particularly relevant for studies that address more complex research questions than in traditional, quite focused, and restricted laboratory studies. A trend in HCI and other communities tries to address realistic scenarios by adopting research in the wild [27], following early work on cognition in the wild from the perspective of anthropology [43, 59, 81]. A related evaluation need has been identified in the visualization community by Lam et al. [56] and Isenberg et al. [44]. They discuss scenarios that go beyond traditional user experience, user performance, or (technical) algorithm performance, for example, how we can evaluate communication through visualization, visual data analysis and reasoning, or collaborative visual data analysis. I am convinced that the visualization of data-rich recordings will be especially useful for empirical research in such areas.

10.3   Example: Eye Tracking Studies and Evaluation

Let us use eye tracking studies as one example of experimental research with data-rich observations. Gaze is a highly relevant source of data for empirical visualization research because it provides quite accurate and fast information that can be useful to understand attention, reading patterns, and the like. Even though there is not always a direct interpretation of eye tracking data [48], most studies can be set up in a way that eye tracking provides informative feedback if it is used with the right study design and interpretation of results [36]. Eye tracking might even be an alternative way to measure indicators of insight [67]. Background on eye tracking is described in the books by Duchowski [29] and Holmqvist et al. [41].

This section focuses on eye tracking for user studies and how we can visually analyze gaze information acquired in such studies. There are other, yet related applications of eye tracking: For example, gaze can serve as a basis for interaction techniques [45], eye movements can be employed for activity recognition [16, 33], eye tracking can help identify tasks and abilities of users of information visualizations [79], and it can be used to improve interactive visualization by recommendations built on inferred user interest [76, 78] and by adaptive interfaces based on the recognition of user tasks and intent [77].

Now, let us focus on eye tracking in empirical visualization research. Extending the fundamental visualization pipeline [23, 38], the process of acquisition and visual analysis of eye tracking data can be described by the pipeline of Fig. 10.1, as defined by Kurzhals et al. [50]. The study data consists of gaze information and—potentially—further complementary data. These are processed and annotated before the mapping to the visualization is computed. The overarching process consists of two interlinked loops: a foraging loop to investigate and explore the study observables, and a sensemaking loop for the interpretation of the data [69]. This interpretation may lead to confirming, rejecting, or building new hypotheses.

Figure 10.1 shows that data-rich information from eye tracking leads to a quite complex data analysis problem. General, rather high-level analysis tasks include compare, relate, and detect [50]. There are a number of specific questions such as: on which parameters or data are these tasks performed (independent or dependent variables), do we want to define derived variables from raw data (other types of independent or dependent variables), which visualization techniques support these tasks and data types, what are the eventual research questions that should be answered by the analysis?

There is a comprehensive overview of visualization techniques for eye tracking data [12, 13], along with a taxonomy that incorporates types of data, stimuli, and visualization techniques. Alternatively, Andrienko et al. [4] provide a critical assessment and review of geo-inspired visual analytics techniques from the perspective of eye tracking analysis. These overview and review papers are a good starting point for choosing appropriate visualization techniques, depending on the visual analysis problem; see center part of Fig. 10.1.

Overall, there has been quite some progress recently in novel and improved visualization techniques to support the evaluation of eye tracking studies. In particular, there are techniques that allow researchers to combine spatiotemporal gaze analysis [55]

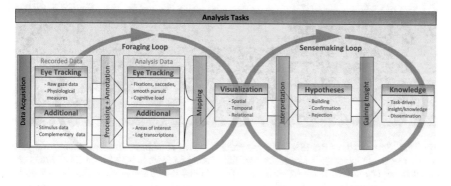

Fig. 10.1 Schematic pipeline for the visual analysis of eye tracking data. All stages (data acquisition, processing, mapping, interpretation, and gaining insight) are influenced by the analysis task. Figure reprinted by permission from Springer: book chapter by Kurzhals et al. [50] © 2017

with the integrated interpretation of scanpaths and areas of interest (AOIs) [53] (see Fig. 10.2 for an example), visually compare scanpaths [49], examine large sets of gaze trajectories by bundling [42], analyze time-dependent AOIs for long-timespan studies [64], work with fixation metrics for the large-scale analysis of information visualizations [20], show gaze and stimulus simultaneously in a volume representation [14], or relate gaze to data of interest in a visualization [46].

There are many examples of the usefulness of such visual data analysis for eye tracking experiments. Typically, visual data analysis is a critical component in pilot studies that can then inform the design of the study process and statistical evaluation. I just want to briefly sketch a few typical examples of how visualization supported our own previous work on eye tracking evaluation of visualization techniques. One example is an eye tracking study that compares parallel coordinates and scatterplots [66]. Here, the visualization of scanpaths, attention, and AOIs for pilot studies helped us formulate hypotheses that eventually led to an advanced computational description of transitions between AOIs that could be used for statistical testing of complex reading behavior. Similarly, for an eye tracking study on transportation maps [65], visualization allowed us to define a new numerical indicator for geodesic distance plots that served as a basis for statistical inference on reading behaviors. Finally, Burch et al. [17] showcased many different types of visualization techniques and discussed how they could be used to identify qualitative findings in eye tracking

Fig. 10.2 Screenshot of the *ISeeCube* system [53], which combines visual spatiotemporal gaze analysis with AOI-oriented analysis. The spatiotemporal analysis is based on a space-time cube visualization (**A**) that includes selected scanpaths (**B**) and the results of clustering controlled by user-specified parameters (**C**). The AOI-oriented analysis is supported by hierarchical clustering and scarfplots of AOI sequences (**D**) and a detailed view of a selected AOI (**E**). The timeline of the video stimulus allows for temporal navigation (**F**). The screenshot was taken when using *ISeeCube* [53] implemented by Kurzhals. Image © 2019 Daniel Weiskopf

data from a study on tree visualization techniques [19]: visualization allowed us to identify reading strategies, reasons for the bad performance of radial tree layouts, and spatiotemporal characteristics of the eye tracking information.

Despite the advances in visual analysis and the above success stories, Kurzhals et al. [52] pointed out a number of open issues related to evaluating visualization and visual analytics with eye tracking: we are still missing sufficient methods for scanpath comparison, fusion of different data sources (e.g., gaze with interaction logs or EEG), and practical tools and working analysis systems. Furthermore, Kurzhals et al. see the need of linking to cognitive models and translational evaluation of human cognition, which asks for building an interdisciplinary community that combines expertise in computer, cognitive, and social sciences. I think that these issues still remain as challenges today. In particular, the combination of data from different sources is a key aspect that needs to be addressed further. There is a need to reach out beyond eye tracking alone and include various other types of data that we can access during studies.

Another challenge is scalability, especially if we want to address long-timespan studies and/or studies with large numbers of participants, leading to a big data visual analytics problem for eye tracking [8]. This problem will also arise when visualization is evaluated with pervasive eye tracking [24], unconstrained mobile eye tracking, or in-the-wild research, typically with mobile eye tracking glasses. The analysis becomes challenging here because each study participant will see individual stimuli, which makes it hard to register or align gaze data between participants and relate them to the semantics of objects from the stimuli. In fact, the data analysis has to include much analysis for time-varying image data acquired by the world camera of the eye tracking glasses. There are some first attempts in this direction [54] that combine computer-based image analysis with visual interaction, but we are still far from a simple, reliable, and time-efficient analysis process.

Up to now, the discussion has focused on eye tracking as an element of methods for quantitative research. However, for a more comprehensive evaluation approach, qualitative methods should also be considered—typically leading to a combination in the form of mixed methods [47]. I see an integration of data-rich research methods (often the quantitative ones, especially when based on physiological sensors like eye tracking) with data-poor research methods (often the qualitative ones) as another area where visualization can play an important role. An example of this research direction is the triangulation of different approaches (here, gaze combined with think-aloud protocol analysis and interaction logs) by Blascheck et al. [10, 11]. Taking this approach further, visual analysis and coding of participants' behavior and actions are possible [7], integrating data-rich gaze information in the form of word-sized graphics [5] with other sources of information from experiments.

10.4   Generalized Problem Characterization

The above discussion was centered around the specific example of eye tracking studies and the evaluation of the results of such studies. Many of the basic challenges already occur in this context of eye tracking and carry over to other types of studies. This section extends the discussion to a generalized view on visual data analysis for empirical visualization research.

10.4.1   Data and Visualization Types

The choice of visualization technique largely depends on the type of data that needs to be analyzed. In general, observational data will be large, complex, time-dependent, heterogeneous, and unstructured, coming from different types of sensors or information sources. However, in general, we can assume that observational data can be assigned some time stamp, i.e., data even from different sources can be eventually registered along the timeline (even though it might be difficult technically). In other words, the underlying data model is that of a time-dependent data set with different types of time-varying data attributes.

The actual data attributes can be of largely varying type, and they may not be sampled at the same timepoints or same frequency. Some might not even be sampled at points in time, but spread across the timeline or even be associated with the full trial (i.e., the full timeline). There is a large set of potential variables that could be acquired as raw data during the experiments; see the respective survey by Rahman et al. [1]. Typical types of time-series data consist of multidimensional data, i.e., multiple real-valued fields, or multiple categorical (nominal) data attributes (e.g., categories of events from user logs). Other types of much larger data sources include videos (images) and audio that may, for example, be recorded for protocol analysis or mobile eye tracking. Data may also include information about technical or algorithmic measures of performance [15, 58, 72]. For any kind of such data, we may also obtain measures of reliability or uncertainty, which is relevant for many types of sensor data.

The characterization of data does not stop at the stage of the original or raw data. In fact, many examples of visual analysis work on derived data that might be more informative than raw data. For the example of eye tracking in Fig. 10.1, the 'analysis data' is typically derived data. Preferably, the derived data is fully automatically computed from the original sources, but there might be cases where user intervention might be required, for example, for the visual-interactive annotation of data.

The choice of visualization technique(s) depends on the type of data to be analyzed. A general strategy is to use multiple coordinated views to support several data attributes [73]. More integrated visual representations may lead to better results but typically require a specific visual design. To address the complexity of the data analysis problem and facilitate scalability to large data, interactive visualization is routinely combined with automatic data analysis—such as statistical methods, unsupervised, or supervised learning—in a visual analytics setup. Finally, the choice of visualization may also depend on the independent variables, for example, whether we have to analyze data for individual participants or groups of participants, or whether we need comparative visualization to show differences with respect to independent variables.

10.4.2   Analysis and Dissemination Goals

Of course, the choice of visualization technique also depends on the goals of the analysis. Typical analysis tasks include outlier detection, summarization, or grouping. A related perspective on data analysis goals for knowledge discovery in databases (KDD) is provided by Fayyad et al. [34]. Where possible, automatic data analysis or statistical techniques are employed to support the task, but as discussed above, the typical approach will follow the combination with interactive visualization. In particular, the visual analysis should also include the original input data or stimuli. The analysis of qualitative aspects of studies is especially challenging [26]; a general approach is based on coding such qualitative study data [74].

A fundamental issue of any visual data analysis is the question of reliability: interactive data exploration might lead to different findings, depending on the interaction steps taken by the analyst. This issue is present for the analysis of study data as well; after all, we want reliable and robust results from studies. Therefore, interactive visualization is typically accompanied by statistical analysis to obtain more controlled answers, yet based on hypotheses informed by visualization. The sensemaking loop of Fig. 10.1 indicates hypothesis building and testing for the example of eye tracking experiments; however, the general structure of the sensemaking loop extends to any kind of experimental evaluation and could include statistical testing.

Another issue is related to properly planning the setup of the studies. Their quality critically depends on an appropriate choice of stimuli or other input shown to the participants. Therefore, the generation of input data is of high relevance to support informative results of studies or facilitate benchmarking. A promising approach employs generative data models to do so [75].

Finally, the goal of visualization does not stop at data analysis. In fact, visualization is equally relevant for disseminating results of studies after interpretation and insight generation in the sensemaking loop of Fig. 10.1. Therefore, visualization approaches for dissemination [6] and storytelling [60, 61] are required.

10.5   Future Research Perspectives and Call for Action

Based on the specific observations and experiences with eye-tracking-based empirical visualization research (Sect. 10.3) and the generalized problem characterization

(Sect. 10.4), I have identified the following, quite subjective recommendations for future research directions and a call for action.

Let us be our own domain experts: visualization for visualization (*Vis4Vis*)!

I argue that we should prominently position visualization research as an application domain for visualization. So far (at the time of writing in 2018), the call for papers and keywords in the paper submission systems of the main conferences of the visualization community (IEEE VIS, EuroVis, PacificVis) specifically ask for application or design study papers, but they do not explicitly consider visualization research—even in cases where they list many other research areas. Furthermore, the call for papers and submission keywords typically contain empirical research, especially user studies, but they focus on actual studies and not on methods that support the evaluation of studies. The series of BELIV Workshops is a good example of a venue that specifically asks for the development of research methods and, thus, implicitly supports the topic of *Vis4Vis*. Similarly, the series of Workshops on Eye Tracking and Visualization (ETVIS)[2] [18] facilitates such research, yet restricted to eye tracking.

To advance our field, a more prominent integration of *Vis4Vis* in the main conferences would be helpful. Being our own domain experts offers several benefits. First, we have an intrinsic and tight link to assessing whether our visual data analysis methods work well or how they need to be improved, leading to short development cycles; therefore, we can expect a fast development of useful visualization techniques that may even carry over to applications beyond those for empirical visualization research. Second, we will benefit from improved ways of evaluating our empirical studies, leading to a better understanding of visualization. Finally, since other disciplines such as HCI are facing similar evaluation challenges, there is a potential impact of improved data analysis for empirical research outside the visualization community.

Data-driven research for the next generation of empirical studies in visualization!

I am convinced that the integration of as-much-as-possible data acquired during studies is a viable way to conduct advanced empirical visualization studies that may support in-the-wild experiments, unconstrained settings, and individual participants and group work alike. Therefore, in the sense of *Vis4Vis*, we are facing the challenge of data fusion and combined visual analysis of massive, often messy sensor and other study data. This, in particular, may include various kinds of physiological sensor, image/video, and audio data. However, with the recent progress in machine learning,

---

[2]ETVIS: Workshop on Eye Tracking and Visualization. https://www.etvis.org.

especially deep neural networks, there is a great potential that we will be able to work with data-rich experiments, with a strong emphasis on data-driven research. In fact, the combination of machine learning with visual analytics is a most promising approach to address these hard analysis problems, for example, in combination with video visual analytics [40]. In this context, it will be critical to keep the original data as long as possible in the analysis pipeline in order to be able to obtain reliable results. Furthermore, it is equally important to obtain reliable and controlled results for data analysis by complementing visual analysis with rigorous statistical testing.

New ways of reporting, privacy preservation, and open science!

With extended or new approaches to visual data analysis, we are also facing the issue of how we can report findings from empirical research. One part of this issue is the concise presentation of results, for example, in a research article. Here, traditional styles of reporting (by using established statistical descriptions) no longer work, but it is not yet clear how the wide variety of more complex analysis results could be summarized in a brief, comprehensible, and replicable way. Here, visualization can play an important role in the sense of using it for storytelling of the scientific data, but respective methods are yet to be developed.

Another part of this issue is related to how we should communicate the massive data potentially acquired during studies. The straightforward approach is to provide the complete set of research data along with the publication, for example, in repositories that guarantee reliable and long-term access of open research data. However, raw data alone is not useful, and even if meta-information is provided, it might still be hard to fully replicate previous studies if they come with complex data. Therefore, it might become relevant to even provide visual analysis tools and descriptions thereof along with the research data. Alternatively, our community could establish a set of tools on which the reproducibility of studies could rely, adopting similar ideas from eye tracking research [63]. The issues of both storytelling and open science are connected to the development of visual data analysis methods in the sense of *Vis4Vis*.

Furthermore, with open empirical data, we have to carefully consider issues related to the privacy of participants and research ethics. With data-rich empirical data combined from different types of sensors, we might acquire enough information that could lead to a breach of anonymity if the data is published in original raw format, i.e., there is an intrinsic conflict between open science and privacy preservation. However, visualization has the potential to help here if it is extended toward novel privacy-preserving visualizations integrated into the research process. The outcome could be privacy-preserving, modified versions of the original data that could still be shared as open research data—with sufficient details to support the reproducibility of the relevant research results.

> **Best practices for the next generation of evaluation methods!**

The three areas of recommendations and future research directions mentioned above will have to be complemented by the adoption of the visualization techniques in the processes and reporting of empirical visualization research. To this end, I see an ongoing process of identifying best practices for novel evaluation approaches and establishing new standards of empirical research.

10.6 Conclusion

It is obvious that visualization for visualization (*Vis4Vis*) is not the only answer to the challenges that we are facing in improving our set of methods for empirical visualization research. For example, many more of these challenges are discussed by Ziemkiewicz et al. [86]. However, I am convinced that there is room for more advanced visualization methods for data analysis and reporting to be used in the context of studies within the visualization community, eventually improving our approach to empirical research.

Acknowledgements Funded by the Deutsche Forschungsgemeinschaft (DFG, German Research Foundation)—Project-ID 251654672—TRR 161 (Project B01 and Task Force TF-B). I thank the participants of the Dagstuhl Seminar 18041 ("Foundations of Data Visualization") for fruitful discussions. Special thanks to Kuno Kurzhals for the many discussions on eye tracking and visualization. The screenshot in Fig. 10.2 was taken from his *ISeeCube* implementation [53].

References

1. Abdul-Rahman, A., Chen, M., Laidlaw, D.: A survey of variables used in empirical studies for visualization. In: Chen, M., Hauser, H., Rheingans, P., Scheuermann, G. (eds.) Foundations of Data Visualization, pp. 155–173. Springer, Berlin (2020)
2. Anderson, E.W.: Evaluating visualization using cognitive measures. In: Proceedings of the Workshop on Beyond Time and Errors: Novel Evaluation Methods for Visualization (BELIV), pp. 1–4 (2012)
3. Anderson, E.W., Potter, K.C., Matzen, L.E., Shepherd, J.F., Preston, G.A., Silva, C.T.: A user study of visualization effectiveness using EEG and cognitive load. Comput. Graph. Forum **30**(3), 791–800 (2011)
4. Andrienko, G.L., Andrienko, N.V., Burch, M., Weiskopf, D.: Visual analytics methodology for eye movement studies. IEEE Trans. Vis. Comput. Graph. **18**(12), 2889–2898 (2012)
5. Beck, F., Blascheck, T., Ertl, T., Weiskopf, D.: Exploring word-sized graphics for visualizing eye tracking data within transcribed experiment recordings. In: Burch, M., Chuang, L., Fisher, B., Schmidt, A., Weiskopf, D. (eds.) Eye Tracking and Visualization: Foundations, Techniques, and Applications, pp. 113–128. Springer, Berlin (2016)
6. Beck, F., Koch, S., Weiskopf, D.: Visual analysis and dissemination of scientific literature collections with SurVis. IEEE Trans. Vis. Comput. Graph. **22**(1), 180–189 (2016)

7. Blascheck, T., Beck, F., Baltes, S., Ertl, T., Weiskopf, D.: Visual analysis and coding of data-rich user behavior. In: Proceedings of the IEEE Conference on Visual Analytics Science and Technology, pp. 141–150 (2016)
8. Blascheck, T., Burch, M., Raschke, M., Weiskopf, D.: Challenges and perspectives in big eye-movement data visual analytics. In: Proceedings of the IEEE International Symposium on Big Data Visual Analytics, pp. 1–8 (2015)
9. Blascheck, T., Ertl, T.: Towards analyzing eye tracking data for evaluating interactive visualization systems. In: Proceedings of the Workshop on Beyond Time and Errors: Novel Evaluation Methods for Visualization (BELIV), pp. 70–77 (2014)
10. Blascheck, T., John, M., Koch, S., Bruder, L., Ertl, T.: Triangulating user behavior using eye movement, interaction, and think aloud data. In: Proceedings of the ACM Symposium on Eye Tracking Research & Applications, pp. 175–182 (2016)
11. Blascheck, T., John, M., Kurzhals, K., Koch, S., Ertl, T.: VA^2: a visual analytics approach for evaluating visual analytics applications. IEEE Trans. Vis. Comput. Graph. 22(1), 61–70 (2016)
12. Blascheck, T., Kurzhals, K., Raschke, M., Burch, M., Weiskopf, D., Ertl, T.: State-of-the-art of visualization for eye tracking data. In: EuroVis – STARs, pp. 63–82 (2014)
13. Blascheck, T., Kurzhals, K., Raschke, M., Burch, M., Weiskopf, D., Ertl, T.: Visualization of eye tracking data: a taxonomy and survey. Comput. Graph. Forum 36(8), 260–284 (2017)
14. Bruder, V., Kurzhals, K., Frey, S., Weiskopf, D., Ertl, T.: Space-time volume visualization of gaze and stimulus. In: Proceedings of the ACM Symposium on Eye Tracking Research & Applications, pp. 12:1–12:9 (2019)
15. Bruder, V., Müller, C., Frey, S., Ertl, T.: On evaluating runtime performance of interactive visualizations. IEEE Trans. Vis. Comput. Graph. (2019). https://doi.org/10.1109/TVCG.2019.2898435
16. Bulling, A., Ward, J.A., Gellersen, H., Troster, G.: Eye movement analysis for activity recognition using electrooculography. IEEE Trans. Pattern Anal. Mach. Intell. 33(4), 741–753 (2011)
17. Burch, M., Andrienko, G.L., Andrienko, N.V., Höferlin, M., Raschke, M., Weiskopf, D.: Visual task solution strategies in tree diagrams. In: Proceedings of the IEEE Pacific Visualization Symposium, pp. 169–176 (2013)
18. Burch, M., Chuang, L., Fisher, B., Schmidt, A., Weiskopf, D. (eds.): Eye Tracking and Visualization: Foundations, Techniques, and Applications. Springer, Berlin (2016)
19. Burch, M., Konevtsova, N., Heinrich, J., Höferlin, M., Weiskopf, D.: Evaluation of traditional, orthogonal, and radial tree diagrams by an eye tracking study. IEEE Trans. Vis. Comput. Graph. 17(12), 2440–2448 (2011)
20. Bylinskii, Z., Borkin, M.A.: Eye fixation metrics for large scale analysis of information visualizations. In: Burch, M., Chuang, L., Fisher, B., Schmidt, A., Weiskopf, D. (eds.) Eye Tracking and Visualization: Foundations, Techniques, and Applications, pp. 235–255. Springer, Berlin (2016)
21. Carpendale, S.: Evaluating information visualizations. In: Kerren, A., Stasko, J.T., Fekete, J.-D., North, C. (eds.) Information Visualization: Human-Centered Issues and Perspectives, pp. 19–45. Springer, Berlin (2008)
22. Chen, M., Hauser, H., Rheingans, P., Scheuermann, G. (eds.): Foundations of Data Visualization. Springer, Berlin (2020)
23. Chi, E.H.: A taxonomy of visualization techniques using the data state reference model. In: Proceedings of the IEEE Symposium on Information Visualization, pp. 69–75 (2000)
24. Chuang, L., Duchowski, A., Qvarfordt, P., Weiskopf, D.: Ubiquitous gaze sensing and interaction (Dagstuhl Seminar 18252). Dagstuhl Rep. 8(6), 77–148 (2019)
25. Çöltekin, A., Heil, B., Garlandini, S., Fabrikant, S.I.: Evaluating the effectiveness of interactive map interface designs: a case study integrating usability metrics with eye-movement analysis. Cartogr. Geogr. Inf. Sci. 36(1), 5–17 (2009)
26. Corbin, J., Strauss, A.: Basics of Qualitative Research: Techniques and Procedures for Developing Grounded Theory, 4th edn. SAGE Publications, Thousand Oaks (2015)
27. Crabtree, A., Chamberlain, A., Grinter, R.E., Jones, M., Rodden, T., Rogers, Y.: Introduction to the special issue of 'the turn to the wild'. ACM Trans. Comput. Hum. Interact. 20(3), 13:1–13:4 (2013)

28. Cui, X., Bray, S., Bryant, D.M., Glover, G.H., Reiss, A.L.: A quantitative comparison of NIRS and fMRI across multiple cognitive tasks. Neuroimage **54**(4), 2808–2821 (2011)
29. Duchowski, A.: Eye Tracking Methodology: Theory and Practice, 2nd edn. Springer, Berlin (2007)
30. Ellis, G., Dix, A.J.: An explorative analysis of user evaluation studies in information visualisation. In: Proceedings of the Workshop on Beyond Time And Errors: Novel Evaluation Methods for Visualization (BELIV), pp. 1–7 (2006)
31. Elmqvist, N., Yi, J.S.: Patterns for visualization evaluation. In: Proceedings of the Workshop on Beyond Time And Errors: Novel Evaluation Methods for Visualization (BELIV), pp. 12:1–12:8 (2012)
32. Ericsson, K.A., Simon, H.A.: Protocol Analysis: Verbal Reports as Data, revised edn. MIT Press, Cambridge (1993)
33. Fathi, A., Li, Y., Rehg, J.M.: Learning to recognize daily actions using gaze. In: Proceedings of the European Conference on Computer Vision, pp. 314–327. Springer, Berlin (2012)
34. Fayyad, U., Piatetsky-Shapiro, G., Smyth, P.: The KDD process for extracting useful knowledge from volumes of data. Commun. ACM **39**(11), 27–34 (1996)
35. Freitas, C.M.D.S., Pimenta, M.S., Scapin, D.L.: User-centered evaluation of information visualization techniques: making the HCI-InfoVis connection explicit. In: Huang, W. (ed.) Handbook of Human Centric Visualization, pp. 315–336. Springer, Berlin (2014)
36. Goldberg, J.H., Helfman, J.I.: Comparing information graphics: a critical look at eye tracking. In: Proceedings of the Workshop on Beyond Time And Errors: Novel Evaluation Methods for Visualization (BELIV), pp. 71–78 (2010)
37. Gürkök, H., Nijholt, A.: Brain-computer interfaces for multimodal interaction: a survey and principles. Int. J. Hum. Comput. Interact. **28**(5), 292–307 (2012)
38. Haber, R.B., McNabb, D.A.: Visualization idioms: A conceptual model for visualization systems. In: Nielson, G.M., Shriver, B.D., Rosenblum, L.J. (eds.) Visualization in Scientific Computing, pp. 74–93. IEEE Computer Society Press, Washington, D. C. (1990)
39. Hirshfield, L.M., Gulotta, R., Hirshfield, S., Hincks, S., Russell, M., Ward, R., Williams, T., Jacob, R.: This is your brain on interfaces: enhancing usability testing with functional near-infrared spectroscopy. In: Proceedings of the SIGCHI Conference on Human Factors in Computing Systems, pp. 373–382 (2011)
40. Höferlin, B., Höferlin, M., Heidemann, G., Weiskopf, D.: Scalable video visual analytics. Inf. Vis. **14**(1), 10–26 (2015)
41. Holmqvist, K., Nyström, M., Andersson, R., Dewhurst, R., Jarodzka, H., Van de Weijer, J.: Eye Tracking: A Comprehensive Guide to Methods and Measures. Oxford University Press, Oxford (2011)
42. Hurter, C., Ersoy, O., Fabrikant, S., Klein, T., Telea, A.: Bundled visualization of dynamic graph and trail data. IEEE Trans. Vis. Comput. Graph. **20**(8), 1141–1157 (2013)
43. Hutchins, E.: Cognition in the Wild. MIT Press, Cambridge (1995)
44. Isenberg, T., Isenberg, P., Chen, J., Sedlmair, M., Möller, T.: A systematic review on the practice of evaluating visualization. IEEE Trans. Vis. Comput. Graph. **19**(12), 2818–2827 (2013)
45. Jacob, R.J.K., Karn, K.S.: Eye tracking in human-computer interaction and usability research: ready to deliver the promises. In: Hyönä, J., Radach, R., Deubel, H. (eds.) The Mind's Eye: Cognitive and Applied Aspects of Eye Movement Research, pp. 573–605. Elsevier, Amsterdam (2003)
46. Jianu, R., Alam, S.S.: A data model and task space for data of interest (DOI) eye-tracking analyses. IEEE Trans. Vis. Comput. Graph. **24**(3), 1232–1245 (2018)
47. Johnson, R.B., Onwuegbuzie, A.J., Turner, L.A.: Toward a definition of mixed methods research. J. Mix. Methods Res. **1**(2), 112–133 (2007)
48. Kim, S.H., Dong, Z., Xian, H., Upatising, B., Yi, J.S.: Does an eye tracker tell the truth about visualizations?: Findings while investigating visualizations for decision making. IEEE Trans. Vis. Comput. Graph. **18**(12), 2421–2430 (2012)
49. Koch, M., Kurzhals, K., Weiskopf, D.: Image-based scanpath comparison with slit-scan visualization. In: Proceedings of the ACM Symposium on Eye Tracking Research & Applications, pp. 55:1–55:5 (2018)

50. Kurzhals, K., Burch, M., Blascheck, T., Andrienko, G., Andrienko, N., Weiskopf, D.: A task-based view on the visual analysis of eye-tracking data. In: Burch, M., Chuang, L., Fisher, B., Schmidt, A., Weiskopf, D. (eds.) Eye Tracking and Visualization: Foundations, Techniques, and Applications, pp. 3–22. Springer, Berlin (2016)
51. Kurzhals, K., Fisher, B.D., Burch, M., Weiskopf, D.: Evaluating visual analytics with eye tracking. In: Proceedings of the Workshop on Beyond Time And Errors: Novel Evaluation Methods for Visualization (BELIV), pp. 61–69 (2014)
52. Kurzhals, K., Fisher, B.D., Burch, M., Weiskopf, D.: Eye tracking evaluation of visual analytics. Inf. Vis. **15**(4), 340–358 (2016)
53. Kurzhals, K., Heimerl, F., Weiskopf, D.: ISeeCube: visual analysis of gaze data for video. In: Proceedings of the ACM Symposium on Eye Tracking Research & Applications, pp. 43–50 (2014)
54. Kurzhals, K., Hlawatsch, M., Seeger, C., Weiskopf, D.: Visual analytics for mobile eye tracking. IEEE Trans. Vis. Comput. Graph. **23**(1), 301–310 (2017)
55. Kurzhals, K., Weiskopf, D.: Space-time visual analytics of eye-tracking data for dynamic stimuli. IEEE Trans. Vis. Comput. Graph. **19**(12), 2129–2138 (2013)
56. Lam, H., Bertini, E., Isenberg, P., Plaisant, C., Carpendale, S.: Empirical studies in information visualization: seven scenarios. IEEE Trans. Vis. Comput. Graph. **18**(9), 1520–1536 (2012)
57. Lam, H., Munzner, T.: Increasing the utility of quantitative empirical studies for meta-analysis. In: Proceedings of the Workshop on Beyond Time And Errors: Novel Evaluation Methods for Visualization (BELIV) (2008). Article No. 2
58. Larsen, M., Harrison, C., Kress, J., Pugmire, D., Meredith, J.S., Childs, H.: Performance modeling of in situ rendering. In: Proceedings of the International Conference for High Performance Computing, Networking, Storage and Analysis, pp. 276–287 (2016)
59. Lave, J.: Cognition in Practice. Cambridge University Press, Cambridge (1988)
60. Lee, B., Henry Riche, N., Isenberg, P., Carpendale, S.: More than telling a story: transforming data into visually shared stories. IEEE Comput. Graph. Appl. **35**(5), 84–90 (2015)
61. Ma, K.-L., Liao, I., Frazier, J., Hauser, H., Kostis, H.: Scientific storytelling using visualization. IEEE Comput. Graph. Appl. **32**(1), 12–19 (2012)
62. Marriott, K., Schreiber, F., Dwyer, T., Klein, K., Henry Riche, N., Itoh, T., Stuerzlinger, W., Thomas, B.H. (eds.): Immersive Analytics. Springer, Berlin (2018)
63. Munz, T., Chuang, L., Pannasch, S., Weiskopf, D.: VisME: visual microsaccades explorer. J. Eye Mov. Res. **12**(6) (2019). https://doi.org/10.16910/jemr.12.6.5
64. Muthumanickam, P.K., Vrotsou, K., Nordman, A., Johansson, J., Cooper, M.D.: Identification of temporally varying areas of interest in long-duration eye-tracking data sets. IEEE Trans. Vis. Comput. Graph. **25**(1), 87–97 (2019)
65. Netzel, R., Ohlhausen, B., Kurzhals, K., Woods, R., Burch, M., Weiskopf, D.: User performance and reading strategies for metro maps: an eye tracking study. Spat. Cogn. Comput. **17**(1–2), 39–64 (2017)
66. Netzel, R., Vuong, J., Engelke, U., O'Donoghue, S.I., Weiskopf, D., Heinrich, J.: Comparative eye-tracking evaluation of scatterplots and parallel coordinates. Vis. Inform. **1**(2), 118–131 (2017)
67. North, C.: Toward measuring visualization insight. IEEE Comput. Graph. Appl. **26**(3), 6–9 (2006)
68. Peck, E.M., Solovey, E.T., Chauncey, K., Sassaroli, A., Fantini, S., Jacob, R.J.K., Girouard, A., Hirshfield, L.M.: Your brain, your computer, and you. Computer **43**(12), 86–89 (2010)
69. Pirolli, P., Card, S.: The sensemaking process and leverage points for analyst technology as identified through cognitive task analysis. In: Proceedings of the International Conference on Intelligence Analysis, vol. 5, pp. 2–4 (2005)
70. Plaisant, C.: The challenge of information visualization evaluation. In: Proceedings of the Working Conference on Advanced Visual Interfaces, pp. 109–116 (2004)
71. Prendinger, H., Mori, J., Ishizuka, M.: Using human physiology to evaluate subtle expressivity of a virtual quizmaster in a mathematical game. Int. J. Hum. Comput. Stud. **62**(2), 231–245 (2005)

72. Rizzi, S., Hereld, M., Insley, J., Papka, M.E., Uram, T., Vishwanath, V.: Performance modeling of vl3 volume rendering on GPU-based clusters. In: Proceedings of the Eurographics Symposium on Parallel Graphics and Visualization, pp. 65–72 (2014)
73. Roberts, J.C.: State of the art: coordinated multiple views in exploratory visualization. In: Proceedings of the International Conference on Coordinated and Multiple Views in Exploratory Visualization, pp. 61–71 (2007)
74. Saldana, J.: The Coding Manual for Qualitative Researchers, 3rd edn. SAGE Publications, Thousand Oaks (2015)
75. Schulz, C., Nocaj, A., El-Assady, M., Frey, S., Hlawatsch, M., Hund, M., Karch, G.K., Netzel, R., Schätzle, C., Butt, M., Keim, D.A., Ertl, T., Brandes, U., Weiskopf, D.: Generative data models for validation and evaluation of visualization techniques. In: Proceedings of the Workshop on Beyond Time and Errors: Novel Evaluation Methods for Visualization (BELIV), pp. 112–124 (2016)
76. Shao, L., Silva, N., Eggeling, E., Schreck, T.: Visual exploration of large scatter plot matrices by pattern recommendation based on eye tracking. In: Proceedings of the ACM Workshop on Exploratory Search and Interactive Data Analytics, pp. 9–16 (2017)
77. Silva, N., Blascheck, T., Jianu, R., Rodrigues, N., Weiskopf, D., Raubal, M., Schreck, T.: Eye tracking support for visual analytics systems: foundations, current applications, and research challenges. In: Proceedings of the ACM Symposium on Eye Tracking Research & Applications, pp. 11:1–11:10 (2019)
78. Silva, N., Schreck, T., Veas, E., Sabol, V., Eggeling, E., Fellner, D.W.: Leveraging eye-gaze and time-series features to predict user interests and build a recommendation model for visual analysis. In: Proceedings of the ACM Symposium on Eye Tracking Research & Applications (2018). Article No. 13
79. Steichen, B., Carenini, G., Conati, C.: User-adaptive information visualization: using eye gaze data to infer visualization tasks and user cognitive abilities. In: Proceedings of the ACM International Conference on Intelligent User Interfaces, pp. 317–328 (2013)
80. Strait, M., Canning, C., Scheutz, M.: Reliability of NIRS-based BCIs: A placebo-controlled replication and reanalysis of Brainput. In: CHI '14 Extended Abstracts on Human Factors in Computing Systems, pp. 619–630 (2014)
81. Suchman, L.A.: Plans and Situated Actions: The Problem of Human-Machine Communication. Cambridge University Press, Cambridge (1987)
82. Tory, M.: User studies in visualization: a reflection on methods. In: Huang, W. (ed.) Handbook of Human Centric Visualization, pp. 411–426. Springer, Berlin (2014)
83. Vuillemot, R., Boy, J., Tabard, A., Perin, C., Fekete, J.-D. (eds.): Proceedings of LIVVIL: Logging Interactive Visualizations and Visualizing Interaction Logs (2016). Workshop at IEEE VIS 2016, https://hal.inria.fr/hal-01535913/file/proceedings.pdf
84. Wagner, J., Kim, J., André, E.: From physiological signals to emotions: implementing and comparing selected methods for feature extraction and classification. In: Proceedings of the IEEE International Conference on Multimedia and Expo, pp. 940–943 (2005)
85. van Wijk, J.J.: Evaluation: a challenge for visual analytics. IEEE Comput. **46**(7), 56–60 (2013)
86. Ziemkiewicz, C., Chen, M., Laidlaw, D., Preim, B., Weiskopf, D.: Open challenges in empirical visualization research. In: Chen, M., Hauser, H., Rheingans, P., Scheuermann, G. (eds.) Foundations of Data Visualization, pp. 237–246. Springer, Berlin (2020)

Chapter 11
"Isms" in Visualization

Min Chen and Darren J. Edwards

Abstract In visualization, there are many different wisdoms and opinions about why visualization works, what makes a good visualization, and how to design and evaluate visualization. Collectively these wisdoms and options have shaped a landscape of the schools of thought in the field of visualization. In this chapter, we examine various schools of thought in visualization, juxtaposing them with schools of thought in computer science and psychology. We deliberate the possibility that some schools of thought in computer science and psychology may have influenced those in visualization. Based on our observation of the development of schools of thought in the discipline of psychology, we believe that it is the empirical evidence that informs the development of theories, which are often embedded in some schools of thought. Meanwhile, empirical studies have a crucial role in visualization to inform and validate postulated theories.

11.1 Introduction

The field of visualization does not really have isms, but is not short of schools of thought. In 2012, a VisWeek panel, entitled *Quality of Visualization: the Bake Off* [15], presented four different approaches to evaluating the quality of visualization, which were referred to as "four schools of thought". Four established visualization scientists, Kelly Gaither, Eduard Gröller, Penny Rheingans, and Matthew Ward, were asked to articulate these approaches. Despite that they held a broader

M. Chen (✉)
University of Oxford, Oxford, UK
e-mail: min.chen@oerc.ox.ac.uk

D. J. Edwards
Swansea University, Swansea, UK
e-mail: d.j.edwards@swansea.ac.uk

© Springer Nature Switzerland AG 2020
M. Chen et al. (eds.), *Foundations of Data Visualization*,
https://doi.org/10.1007/978-3-030-34444-3_11

view than the school of thought that each was championing, they presented exquisite and persuasive cases for the four schools.

In the order when the position statements were presented in the panel, these four schools of thought are:

V1$_A$ **School of A**, where A for algorithms or automation. Gröller argued: *"Let us reduce the quality of a visualization to the quality of the involved algorithm". "An optimization process should automatically figure out which algorithms and parameter settings best fulfil the user defined declarations and constraints"* [15].

V1$_B$ **School of E**, where E for experiments or empirical studies. Rheingans reasoned: *"A little empirical evidence never hurts". "Promising results from empirical studies seem to signal that a new tool might be a winner"* [15].

V1$_C$ **School of M**, where M for metrics or measurements. Ward articulated: *"As the field of visualization evolves, more and more measures have been proposed as a means of comparing alternate visualizations or even [measuring] the effectiveness of a single visualization"* [15].

V1$_D$ **School of R**, where R for real users or real-world applications. Kelly Gaither asserted: *"Our success is measured in 'Aha' moments, and these moments are precious and rare". "In my world, visualizations are never produced in isolation or the absence of domain knowledge"* [15].

Here we use labels in the form X_i (e.g., **V1$_A$** and C_B) to tag a school of thought. The main label indicates a fundamental question that is a bone of contention, while the subscript identifies a specific school of thought in the context of this question. More labels will gradually be introduced in this chapter.

In many scientific and scholarly subjects, different schools of thought were formulated by those who share some common beliefs or some opinions with a set of common characteristics. To denote such schools of thought, sets of principles, belief systems, doctrines, ideologies, or spiritual currents, as well as the related bodies of teaching, the suffix "-ism" is typically used. It derives from the Ancient Greek suffix "-ισμὸς", meaning "taking side with". Among many uses of "ism" as a suffix, all [*stem*]-ism words in this chapter fall into the category of words referring to a belief in [*stem*] or a doctrine or principle of [*stem*].

For example, in psychology, there are structuralism, functionalism, pragmatism, behaviourism, gestaltism, associationism, and cognitivism (see Sect. 11.4). In philosophy, there are numerous -isms. The wiktionary page, *Glossary of Philosophical Isms* [76], lists several hundreds of isms, reflecting the long history of scholarly investigations into and discourses on many aspects of our world and our mind. Such diversity reflects the advancement and maturity of a discipline.

In the following sections, we first continue our discussion on a number of major clusters of opinions, which can be considered as schools of thought. We then examine schools of thought in computer science and psychology, which are the two disciplines that the subject of visualization is most closely related. Finally, we offer our observations and concluding remarks.

11.2   Schools of Thought in Visualization

In Sect. 11.1, we have already encountered four schools of thought in visualization. The discourses in several chapters in this book reflect some viewpoints of these schools of thought. For example, one may observe a discussion on $V1_B$ versus $V1_D$ in Chap. 8 [46], an evaluation methodology based on $V1_D$ in Chap. 9 [57] and a technical approach based on $V1_C$ in Chap. 10 [74].

In this section, we discuss three more fundamental questions that have been the bones of contention in visualization.

11.2.1   What is Visualization for?

One fundamental question that everyone in the field of visualization cannot help ask is "what is visualization really for"? In a 2014 article published in a philosophy venue[14], Chen et al. gathered some twenty different statements offering answers to this question, including five statements in Scott Owen's original 1999 collection [49]. Recently, Streeb compiled a comprehensive collection of some 120 statements [63]. Here we broadly divided these statements into five schools of thought, for which we improvise some "ism" terms.

$V2_A$ **Insightism.** Many visualization researchers and practitioners argued that the main purpose of visualization is for gaining insight from data. For example, McCormick et al. stated in 1987 [47]: *"The goal of visualization in computing is to gain insight by using our visual machinery"*. Earnshaw and Wiseman stated in 1992 [19]: *"Visualization is concerned with exploring data and information in such a way as to gain understanding and insight into the data"*. Similar statements can easily be found in numerous written documents.

Some statements presented stronger arguments, making visualization as the main source of insight, such as the statement by Hearst [31] *"Visualization has been shown to be successful at providing insight about data for a wide range of tasks"*. Others presented weaker arguments, designating visualization to an assisting role, such as the statement by Thomas and Cook [65]: *"People use visual analytics tools and techniques to synthesize information and derive insight from massive, dynamic, ambiguous, and often conflicting information"*. Note that this statement implies three sources of insight, i.e., "people" as the main source, "visualization", and "analytics" as the assisting tools.

The spectrum from strong insightism to weak insightism partly depends on the interpretation of the word "insight". In Gaither's statement for the school of R, $V1_D$, an insight is considered as a deep understanding of a complex problem, or an "Aha" moment in a complex situation. Many others define an insight in visualization as a correct conclusion inferred from viewing visualization. For example, Gomez et al. define gaining insight as tasks in the forms of "who+when+where $\rightarrow$ what", "when+where+what $\rightarrow$ who", and so on [30]. The strong insightism usually correlates with a broad or "weak" definition of insight, while the weak insightism usually correlates with a narrow or "strong" definition of insight.

V2$_B$ Cognitivism. Using the term borrowed from psychology, we outline a school of thought that focuses on the perceptual and cognitive benefits of visualization. Perhaps the strongest statement is that by Spence [61], who asserted: *"Visualization is solely a human cognitive activity and has nothing to do with computers"*. A number of visualization researchers offered answers to the question what visualization is for by articulating that visualization can enable *seeing the unseen* [47], *maximizing human understanding* [49], *amplifying cognition* [12], and *helping think* [24].

While the fundamental idea of cognitivism in visualization is not in any way the exactly same as that of cognitivism in psychology, this school of thought does reflect the essence of the fundamental idea in psychology, i.e., cognition impacts the behaviour of visualization.

V2$_C$ Communicationism. Most people appreciate that visualization can provide effective aid to information communication and knowledge dissemination. Tableau, a major provider of visualization technology, stated at its Web site [64]: *"Data visualization is another form of visual art that grabs our interest and keeps our eyes on the message"*. *"Data visualization helps to tell stories by curating data into a form easier to understand, highlighting the trends and outliers. A good visualization tells a story, removing the noise from data and highlighting the useful information"*.

In her book *Effective Data Visualization* [21], Evergreen offered an animated answer to the question of why we visualize: *"Seriously, that's the most important question to ask when creating a data visualization. It's the first thing I ask a client who sends me data for redesign. And it's the primary reason we visualize: Because we have a point to communicate to the world. We have a compelling finding to share, a big idea revealed in our analysis that needs to say to people. A point"*.

V2$_D$ Economism. Some visualization researchers and practitioners attempted to answer the "what for" question from some more tangible benefits of visualization, avoiding hinging an answer on a less observable and measurable benefit such as insight. For example, Bertin referred to the benefit of external memorization in his book *Semiology of Graphics* [6] as *"the artificial memory that best supports our natural means of perception"*. Friedhoff and Kiely highlighted the benefit of saving time in [27]: *"If the information is rendered graphically"*, researchers *"can assimilate it at a much faster rate"*. Ware offered a similar statement in his book *Information Visualization: Perception for Design* [72]: "One of the greatest benefits of data visualization is the sheer quantity of information that can be rapidly interpreted if it is presented well". Tufte asserted the relative benefit of visualization in comparison with statistics in his book *The Visual Display of Quantitative Information* [68]: *"Indeed graphics can be more precise and revealing than conventional statistical computations"*.

Chen et al. gave a somehow "economism" definition of visualization [14]: *"Visualization (or more precisely, computer-supported data visualization) is a study of transformation from data to visual representations in order to facilitate effective and efficient cognitive processes in performing tasks involving data. The fundamental measure for effectiveness is correctness and that for efficiency is the time required for accomplishing a task"*. A few years later, Chen and Golan (an economist) proposed an information-theoretic metric for analysing the cost-benefit ratio of human- and machine-centric data intelligence processes [16]. In the context of visualization, the metric defines the benefit as the amount of information (Shannon entropy) in the original data subtracted by the amount of information in a visualization image and further subtracted by the potential informative distortion that is mainly caused by information loss in visualization and may also be due cognitive biases, but can be alleviated by human knowledge. They considered energy is the fundamental measure of the cost, which can be approximated by time and monetary measurements.

V2$_E$ **Pragmatism.** Visualization researchers and practitioners are in general open-minded as to what is visualization is really for. While the question is yet to be convincingly answered, the majority in the community focus on the utility of visualization in different application contexts. This approach echoes the school of thought of pragmatism. One example is the list of functions summarized by Marty [45]: *"answer a question"*, *"pose new questions"*, *"explore and discover"*, *"support decisions"*, *"communicate information"*, *"increase efficiency"*, and*"inspire"*. Another list by Chen et al. [14] include functions *"making observation"*, *"facilitating external memorization"*, *"stimulating hypotheses and other thoughts"*, *"evaluating hypothesis"*, and *"disseminating knowledge"*.

Many taxonomies of visualization (e.g., [11, 54, 78]) and many surveys on visualization topics (e.g., [4, 9, 34, 39, 40, 58, 67]) include "tasks" as one of the main dimensions for categorization, reflecting the typical view of pragmatism.

Having schools of thought is not in any way suggesting that the visualization community is divided. Many in the community often embrace different schools of thought. For example, the aforementioned statement by Thomas and Cook [65] is an instance of weaker insightism but it also captures a sense of pragmatism. In the book, there is also a statement: *"Visual representations and interaction technologies provide the mechanism for allowing the user to see and understand large volumes of information at once"*. This captures senses of cognitivism and economism. van Wijk presented a visualization pipeline from data to knowledge (insight) [70], while proposed an economic model for measuring the gained knowledge (insight) as well as the cost of visualization processes. This exemplifies the views of both insightism and economism. Stasko [62] made perhaps the broadest argument about the value of visualization, including contributing factors of time, insight, essence, and confidence. These four factors correspond to the arguments of economism, insightism, communicationism, and cognitivism, respectively.

11.2.2 Faithfulness and Integrity Versus Embellishment and Distortion

In the field of visualization, many passionately argue that data visualization must be faithful to the data being depicted and should not be embellished with chartjunks. Tufte made a powerful argument for "graphical integrity" in his book *The Visual Display of Quantitative Information* [68] and coined the term *lie factor* to indicate the level of deviation of a visualization image from its source data, and the term *chartjunk* to describe decorative visual features in a visualization image.

This school of thought is widely endorsed by visualization researchers and practitioners. There are countless online blogs that repeat and reinforce Tufte's views on graphical integrity and chartjunks. Pandey et al. presented an empirical study, confirming that several types of distortion in visualization can have deceptive effects on viewers [50]. Kindlmann and Scheidegger presented an algebraic framework for defining three principles that formalize the notion of graphical integrity [38].

While hardly anyone in the visualization community would support any practice intended to deceive viewers, there have been many visualization techniques that inherently cause distortion to the original data. These include logarithmic plots, metro maps, magic lenses, focus+context visual designs, colour and opacity transfer functions, illustrative deformation, and so on. There might just be a hidden gap between theory and practice or between idealism and pragmatism, until the debate about chartjunks brought the bone of contention to the fore.

The debate started with a paper by Bateman et al. [3], which reported an empirical study showing that visual embellishment could aid memorization of the data depicted. Another paper by Hullman et al. [32] proposed a possible explanation that *"introducing cognitive difficulties to visualization"* *"can improve a user's understanding of important information"*. Since the finding and the explanation represented a major departure from the widely endorsed views on chartjunks, the works stimulated many discussions in the community (e.g., [23, 25]).

In general, the question about distortion differs from that about chartjunks, though most of those who are against distortion are likely also against chartjunks. Here we treat these two questions separately in our definitions of the following four schools of thought.

$V3_A$ **Essentialism.** Do not introduce any visual embellishment that is unnecessary for comprehending the data depicted.

$V3_B$ **Decorationism.** Visual embellishment can be used in visualization and can bring benefit.

$V3_C$ **Isomorphism.** Do not introduce any distortion that is inconsistent with the source data.

$V3_D$ **Polymorphism.** Distortion can be featured in visualization and can bring benefit.

In fact, if one reads carefully some original discourses, one may find that the gaps between essentialism and decorationism and between isomorphism and polymorphism are not totally unbridgeable. On a case by case basis, most people with different schools of thought can often agree on whether an embellishment is unnecessary or not, or whether a distortion is inconsistent or not. For example, Few, who has been a champion against chartjunks, stated impartially: "*Embellishments can at times, when properly chosen and designed, represent information redundantly in useful ways, ...*" [25].

There are many other questions that are bones of contention and other clusters of opinions that can be characterized as schools of thought. For example, the necessity and usefulness of many techniques (e.g., animation, 3D visual designs, virtual reality, and so on) often attract different opinions in a manner similar to the contention between essentialism and decorationism or between isomorphism and polymorphism.

11.2.3 Human-Centric Processes Versus Machine-Centric Processes

In the field of visualization, regardless whether one is in favour of any particular school of thought in terms $V1_A$–$V1_D$, $V2_A$–$V2_E$, or $V3_A$–$V3_D$, everyone holds a view that human intelligence is necessary in any reasonably complex or mission-critical data intelligence processes. Here *data intelligence* is an encompassing term for processes such as statistical inference, computational analysis, data visualization, human–computer interaction, machine learning, business intelligence, simulation, prediction, and decision-making.

Around 2004, a new area *visual analytics* [65] emerged in the field of visualization to develop data intelligence workflows that always have humans in the loop.

This view may not be shared by many researchers and practitioners in data mining, machine learning, and some other machine-centric aspects of data intelligence. For example, in a textbook on data mining and machine learning [77], Witten et al. wrote "*Economists, statisticians, forecasters, and communication engineers have long worked with the idea that patterns in data can be sought automatically, identified, validated, and used for prediction. ... as the world grows in complexity, overwhelming us with the data it generates, data mining becomes our only hope for elucidating hidden patterns. ... It can lead to new insights, ...*".

In a textbook on machine learning [59], Rothman wrote "*In May 2017, Google revealed AutoML, automated machine learning system that could create an artificial intelligence solution without the assistance of a human engineer. IBM Cloud and Amazon Web Services (AWS) offer machine learning solutions that do not require AI developers*".

Today, the latter view is widely believed, which can be evidenced by many nonfictional scientific writings, such as Fry's book *Hello World: How to be Human in the Age of Machine* [28], and Frank's book *What to do when Machines do Everything* [26].

Here, we can clearly see two schools of thought about whether humans should have a significant role in any reasonably complex or mission-critical data intelligence processes:

V4$_A$ **Mechanism.** Most, if not all, data intelligence processes can be automated using data mining and machine learning techniques. The amount of data available in this era of "big data" makes automation both necessary and feasible.

V4$_B$ **Anti-mechanism.** Any reasonably complex data intelligence workflow should always have humans in the loop, where humans' analytical capability can be enhanced using interactive visualization techniques.

While many of those who hold the view of **V4$_A$** may also make an economic case for automation, this is rarely the viewpoint disagreed by the school of thought **V4$_B$**. The fundamental difference between the two schools of thought is about whether or not it is possible for machine-centric processes to replace humans in all or most data intelligence workflows. Meanwhile, almost all of those who hold the view of **V4$_B$** do not actually oppose any kind of automation regardless, especially since most visualization images are generated automatically in practice. We use the term "anti-mechanism" here to indicate its association with one of the schools of thought in computer science. **V4$_A$** and **V4$_B$** reflect two polarized viewpoints about the possibility. In practice, most researchers and practitioners in visualization as well as in computer science often have to make use of both human- and machine-centric processes in data intelligence workflows that are "currently" considered to be complex. In the next section, we continue this line of discussion in the context of computer science.

11.3   Isms in Computer Science

The discipline of computer science and engineering, where the subject of visualization resides mainly, has inherited a wide range of mathematical concepts and methods but has displayed very limited interest in most philosophical schools of thoughts in mathematics, except on the topic of machine intelligence. Johnson-Laird first outlined four postulations [36], which were discussed in detail in Penrose's book *Shadows of the Mind* [53]. Here we list these four postulations by quoting Penrose' text [53] with "ism" tags found in the literature.

C$_A$ *"All thinking is computation; in particular, feelings of conscious awareness are evoked merely by the carrying out of appropriate computations".* [**strong AI, hard AI, functionalism, mechanism, computationalism**].

C$_B$ *"Awareness is a feature of the brain's physical action; and whereas any physical action can be simulated computationally, computational simulation cannot by itself evoke awareness".* [**weak AI, soft AI**].

C$_C$ *"Appropriate physical action of the brain evokes awareness, but this physical action cannot even be properly simulated computationally".* [**anti-mechanism**].

▶ How to prove $a \times b = b \times a$?

▶ 5×3:

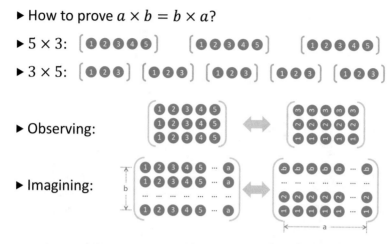

▶ 3×5:

▶ Observing:

▶ Imagining:

Fig. 11.1 This example shows that a human mathematician can make observation of the configuration of 5×3 and another configuration of 3×5 and ascertain that $5 \times 3 = 3 \times 5$. The mathematician can then make further observations for different values of a and b or imagine how the 5×3 and 3×5 configurations may be extended to different values of a and b. The combined effort of observation and imagination enables the mathematician to conclude that $a \times b = b \times a$. This figure was redrawn based on an illustration in [53]

$\mathbf{C_D}$ *"Awareness cannot be explained by physical, computational, or any other scientific terms"*. [**mysticism**].

In terms of human and machine intelligence, these four postulations exemplify four different schools of thought. Penrose has been the most prominent champion for the postulation of $\mathbf{C_C}$ through his two books [52, 53]. Interestingly Penrose started with his reasoning in [53] using two examples of visualization as shown in Figs. 11.1 and 11.2. He pointed out that one can visually inspect the patterns shown in these two examples and conclude that the proof can be extrapolated to the general formulations as mentioned in the captions of Figs. 11.1 and 11.2. Penrose then offered a proof for the postulation [53]:

> Human mathematicians are not using a knowably sound algorithm in order to ascertain mathematical truth.

by following the reasoning strategy that Gödel used to prove his Incompleteness Theorems [7, 29], and Turing used to prove his theorem on the Halting Problem [18, 69]. This proof provided a basis for his School of Thought $\mathbf{C_C}$.

In the literature, a number of authors have provided critical comments on Penrose's conclusion, while making cases for the school of thought $\mathbf{C_A}$, including, for instance, the critiques by Sloman [60], LaForte et al. [41], and Berto [7].

Penrose's mathematical and algorithmic reasoning can be traced back to Lucas's article [42], where he opened his discourse with:

> Gödel's Theorem seems to me to prove that Mechanism is false, that is, that minds cannot be explained as machines.

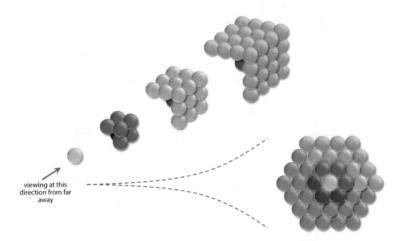

Fig. 11.2 Hexagonal numbers are numbers that can be arranged as hexagonal arrays: 1, 7, 19, 37, 61, 91, 127, etc. To prove a postulation that the sum of successive hexagonal numbers is a cube (e.g., $1 + 7 + 19 = 3^3$ and $1 + 7 + 19 + 37 = 4^3$), a mathematician can visually observe the relationship between the illustrated 2D and 3D configurations and then imagine their extensions. This figure was redrawn based on an illustration in [53]

Mainzer, who is a philosopher of science and a leading thinker on complex systems, related the discourse further back to the schools of thought in philosophy [43]:

> In the history of philosophy and science, there have been many different suggestions like Democritus, Lamettrie, et al., proposed to reduce mind to atomic interactions. Idealists like Plato, Penrose, et al. emphasized that mind is completely independent of matter and brain. For Descartes, Eccles, et al. mind and matter are separate substances interacting with each other. Leibniz believed in a metaphysical parallelism of mind and matter because they cannot interact physically. According to Leibniz, mind and matter are supposed to exist in "pre-established harmony" like two synchronized clocks. Modern philosophers of mind like Searle defended a kind of evolutionary naturalism. Searle argues that mind is characterized by intentional mental states which are intrinsic features of the human brain's biochemistry and which is therefore cannot be simulated by computers.

While many in the field of visualization and beyond may not have been following these discourses, some of the schools of thought in visualization have exhibited some alignments with these four schools of thought in computer science. For example, the school of thought $V4_A$, mechanism, may be related to C_A, while the school of thought $V4_B$, anti-mechanism, may be related to C_C or C_D. Those who sit between $V4_A$ and $V4_B$ may align with C_B.

While we do not know whether those arguing for the benefits of visualization believe that aspects of human mind may not be computational, we can reasonably assume that many of them at least hold a view that many aspects of human mind cannot be simulated by computational algorithms available today and in the near future. For example, many statements featuring insightism or cognitivism exhibit likely support for C_C, or at least a time-limited notion of C_B.

11.4   Isms in Psychology

The discipline of psychology saw the formation of many schools of thought. As early as 1927, Jastrow wrote perhaps the first survey on "isms" in psychology [35]. A good number of books on the philosophy of psychology provide a large collection of readings on the topic, which include volumes edited by Block [8] and Bermudez [5]; and books authored by Margolis [44], Botterill and Carruthers [10], Walsh et al. [71], and Weiskopf [73].

In this section, we first list a number of isms (in alphabetic order) that have been frequently mentioned in the literature, and we then describe these isms following a chronological order.

PS_A **Associationism.** Mental connections between events and ideas (H. Ebbinghaus 1850–1909).

PS_B **Behaviourism.** Study of observable emitted behaviour (I. Pavlov 1849–1936).

PS_C **Cognitivism.** Understanding how people think as state transitions (J. Piaget 1896–1980).

PS_D **Functionalism.** Mental operations and practical use of consciousness (J. R. Angell 1869–1949).

PS_E **Gestaltism.** Study of holistic concepts, not merely as sums of parts (M. Wertheimer 1880–1943).

PS_F **Pragmatism.** Knowledge is validated by its usefulness (W. James 1842–1910).

PS_G **Structuralism.** Analysis of consciousness into constituent components (E. Titchener 1867–1927).

Before getting into specific ontologies of psychology, it is important to understand something about philosophy of science in psychology at various times in history, in the form of epidemiological beliefs about truth, opinion, and knowledge, and how this shaped the ontologies of the day and have led to the wide range of "isms" that we find today.

Objective verification through empirical verification came about very slowly in psychology, but had its roots dating back as far as John Locke (1632–1704). Locke had promoted the idea that objective verification through our sensory experience should be sought in order to establish knowledge about the world around us; i.e., knowledge must be based on sensory experience. This forms the basis of the scientific method we have today in relation to prior reasoning, hypothesis testing and a means for falsification using validated forms of objective measurement.

In psychology, this empirical methodology to establishing knowledge was slow to be accepted in the mainstream world view, in comparison with other sciences such as physics, biology, and chemistry. As a result of this slow adoption, early schools of thought (or "isms") in psychology were based on subjective methodology rather than objective empiricism. One example of such a school of thought is structuralism.

In the early 1900s, **structuralism**, which was developed by Wilhelm Wundt and Edward Bradford Titchener [66] and inspired by methodological advances in the fields of chemistry and physics, sought to identify and catalogue complex mental operations using introspective methods. It was perhaps the first serious attempt to formulate a school of thought in psychology, and it was believed that this should be conducted through trained introspection. However, this soon became impossible, as just for sensation 40,000 elements were discovered. In addition, criticisms from psychologists such as William James suggested that introspective methods would only lead to distorted perceptions of these sensations and biased by this subjective approach, for which he coined the term *the psychologist's fallacy* [33].

The competing perspective of **functionalism** was founded by William James [33], which utilized the idea of the practical use of consciousness, and that mental states are constituted solely by their functional role and causal relations, sensory inputs and behavioural outputs, i.e., their function [13]. This is a school of thought about the nature of mental states rather than the properties of these states, which would be the structuralist approach. It also assumes that psycho-physiological mental states should be recognized by what they do (what they transform) rather than what they are made of [1]. For example, when setting a mousetrap, the individual's mental state can be identified as "something that kills mice" rather than identifying sensation properties, or the particular approach to kill mice. However, functionalism still lacked a formal empirical approach, so was superseded by approaches which involved a more empirical approach with the rise of positivism, post-positivism, and critical rationalism.

Gestaltism, which is an early form of cognitivism, was inspired by physics and founded by Max Wertheimer, Kurt Koffka, and Wolfgang Kohler, employed third-person phenomenological inquiry to discover principles of perceptual organization of holistic concepts and is identified as independent from the sum of its parts. In this way, Gestalt psychology is an attempt to discover the perceptual laws which allow for meaningful perceptions from the regularities in the environment form a chaotic world. The main assumption made in Gestaltism is that the mind forms a perceptual global whole from these chaotic environmental regularities and has self-organizing tendencies [75]. A simple illustration of this is when Wertheimer suggests that: "I stand at the window and see a house, trees, sky. Theoretically, I might say there were 327 brightnesses and nuances of colour. Do I have '327'? No. I have sky, house, and trees". This is a clear illustration that the holistic concept of those environmental regularities becomes a whole, which is unique from the sum of its parts to form meaningful perception.

Behaviourism, adopting the epidemiological approach of positivism and falsifiability advocated by philosophers such as John Locke in its methodology revolutionized the way that psychological experimentation would take place based on gaining knowledge in science through empirical investigation. Behaviourists attempted to explain the psychological phenomenon through empirically defined objective phenomena in the form of overt stimuli–responses which could be objectively measured. In this way, they focused on what people and animals do in response to different environmental situations [51].

Behaviourism developed in stages, starting from basic association learning, of the studies conducted by Pavlov such as the pairing association of a bell with food which led to a dog salivating in the learning phase, to the learned association (conditioned response) of the dog salivating to the stimuli of the bell alone—a process called classical conditioning [51]. Skinner then further developed this model through operant conditioning which demonstrated that behaviours which led to some form of pleasant outcome (positive reinforcement) was likely to be repeated, while behaviours which led to less pleasant outcomes or painful outcomes were more likely to be avoided (negative reinforcement) [22].

Cognitivism, developed when linguist Noam Chomsky criticized Skinner's explanation of operant conditioning to adequately explain the emergence of language. In Chomsky's book *Syntactic Structures*, he suggests that for a language to emerge, then an innate universal grammar was necessary in the form of a transformational generative grammar (TGG) [17]. As this was difficult for behaviourists at the time to explain, the cognitive revolution was born in the 1960s and still dominates mainstream psychology today.

Ulric Neisser [48] used the term *Cognitive Psychology* for the first time to describe a person as a dynamic information-processing system, where mental operations can be given in computational terms. Cognitive psychology relates to sensory input, and how it is elaborated, stored in mental representations, recovered from memory and used in cognitive tasks, rather than focusing on just behavioural outputs.

Though early cognitive psychology work began as far back as Hermann Ebbinghaus mapped out the learning and forgetting curves in experimental studies of memory in 1885 [20]. It also developed and continued to develop when Piaget [55] explored cognitive development and the four stages of cognitive development, which included (1) sensory motor stage, (2) preoperational stage, (3) concrete operational stage, and (4) the formal operational stage. Each of these stages added growing cognitive complexity as the child grew older and were able to complete more complex cognitive tasks. However, cognitivism did not really develop until Baddeley and Hitch [2] produced the working memory model (WMM) which specified a central executive, visuospatial sketchpad and articulatory-phonological loop. This model was the first to incorporate information theory into its account, whereby through identifying these memory limitations, and this led researchers to make accurate predictions about behavioural performances under these types of conditions.

It should also be noted that in psychology there has been a debate on the nature of behaviour, and on whether it originates from nature or nurture. The idea of **Nativism** dates as far back as the philosopher Immanuel Kant in the eighteenth century who argued in his critique of pure reason [37] that the human mind knows objects in innate, a priori ways. More recently in psychology, nurture theorists such as behaviourists have long argued that behaviour (and psychology) is subject to the learning reinforcement contingencies in the environment. However, cognitivists, though they accept learning, also account for innate components, such as Chomsky's TGG [17] and other cognitive linguists such as Pinker [56].

11.5 Conclusions

In this chapter, we have used the notion of "schools of thought" to frame different views in the field of visualization. While the definitions of these schools of thought unavoidably "discretize" the otherwise complex and intermingled spectra of opinions, our intention is to bring out different views for further discussion, evaluation, validation, or falsification.

Some schools of thought visualization can be related to schools of thought in psychology and computer science. We have already seen the mentioning of cognitivism in visualization ($V2_B$) and psychology (PS_C), and mentioning of mechanism and anti-mechanism in visualization ($V4_A$ and $V4_B$) and computer science (C_A and C_C).

During the 2012 VisWeek panel [15], the convener of the panel related the four schools of thought mentioned in Sect. 11.1 to schools of thought in psychology as:

- School of A (algorithms or automation) $V1_A \rightarrow$ functionalism PS_D;
- School of E (experiments or empirical studies) $V1_B \rightarrow$ behaviourism PS_B;
- School of M (metrics or measurements) $V1_C \rightarrow$ structuralism PS_G;
- School of R (Real users or Real world applications) $V1_D \rightarrow$ pragmatism PS_F.

Although the suggestion by the panel convener was meant to be provocative as a tradition of the panel discussions in IEEE VIS (VisWeek) conferences, the mappings indicate that visualization researchers and practitioners have been thinking deeply about many aspects of visualization in ways similar to many pioneers in other scholarly subjects.

When a scientific or scholarly subject reaches a certain level of maturity, the scientists or scholars will naturally attempt to make abstraction and generalization from empirical evidence and practical experience. It would be wrong if the scientists or scholars did not do that. As an inherent and integral part of the processes for abstraction and generalization, there will be different viewpoints, different abstract theories and models, different postulations, and so on.

As a scientific discipline, there is no reason for the field of visualization to be afraid of different schools of thought. In particular, empirical studies will have a significant role in the evolution of schools of thought, including their creation, betterment, convergence, divergence, and obsolesce. Empirical studies are important means for stimulating new postulations, evidencing various viewpoints, and validating abstract theories and models. Meanwhile, for researchers who are interested in theoretical research and empirical studies, having schools of thought is no doubt a blessing. Meanwhile, as researchers, practitioners, authors, and reviewers, we must respect schools of thought that we do not agree. We must learn to judge the novelty, rigour, and significance of a scientific contribution not based on whether or not this fits with our own school of thought.

Acknowledgements The authors would like to thank Professor Hans-Christian Hege, Zuse Institute Berlin (ZIB), Germany, for some insightful discussions during the Dagstuhl Seminar 18041 on Foundations of Data Visualization in January 2018. The authors also appreciate very much the comments and suggestions made by Professor Helwig Hauser, University of Bergen, Norway, and have revised the early version accordingly.

References

1. Angell, J.R.: The province of functional psychology. Psychol. Rev. **14**(2), 61 (1907)
2. Baddeley, A.D., Hitch, G.: Working memory. Psychol. Learn. Motiv. **8**, 47–89 (1974)
3. Bateman, S., Mandryk, R.L., Gutwin, C., Genest, A., McDine, D., Brooks, C.: Useful junk? The effects of visual embellishment on comprehension and memorability of charts. In: Proceedings of the ACM SIGCHI Conference on Human Factors in Computing Systems, pp. 2573–2582 (2010)
4. Beck, F., Vehlow, C., Weiskopf, D.: Visualizing group structures in graphs: a survey. Comput. Graph. Forum **36**(6), 201–225 (2017)
5. Bermudez, J.L.: Philosophy of Psychology: Contemporary Readings. Routledge, Abingdon (2006)
6. Bertin, J.: Semiology of Graphics: Diagrams, Networks, Maps. Esri Press, Redlands (1983)
7. Berto, F.: There's Something About Gödel: The Complete Guide to the Incompleteness Theorem. Wiley-Blackwell, New York (2009)
8. Block, N. (ed.): Readings in Philosophy of Psychology, vol. I. Harvard University Press, Cambridge (1980)
9. Borgo, R., Micallef, L., Bach, B., McGee, F., Lee, B.: Information visualization evaluation using crowdsourcing. Comput. Graph. Forum **37**(3), 573–595 (2018)
10. Botterill, G., Carruthers, P.: The Philosophy of Psychology. Cambridge University Press, Cambridge (1999)
11. Buja, A., Cook, D., Swayne, D.F.: Interactive high-dimensional data visualization. J. Comput. Graph. Stat. **5**, 78–99 (1996)
12. Card, S., Mackinlay, J., Shneiderman, B.: Readings in Information Visualization: Using Vision to Think. Morgan Kaufmann Publishers, Burlington (1999)
13. Carr, H.A.: Psychology: A Study of Mental Activity. American Psychological Association, Washington (1925)
14. Chen, M., Floridi, L., Borgo, R.: What is visualization really for? In: The Philosophy of Information Quality, Springer Synthese Library, vol. 358, pp. 75–93 (2014)
15. Chen, M., Gaither, K., Gröller, E., Rheingans, P., Ward, M.: Quality of visualization: the bake off. In: IEEE VisWeek Conference: Panel (2012)
16. Chen, M., Golan, A.: What may visualization processes optimize? IEEE Trans. Vis. Comput. Graph. **22**(12), 2619–2632 (2016)
17. Chomsky, N.: Syntactic Structures. Mouton & Co., The Hague (1957)
18. Davis, M. (ed.): The Undecidable: Basic Papers on Undecidable Propositions Unsolvable Problems, and Computable Functions. Raven Press, New York (1965)
19. Earnshaw, R.A., Wiseman, N.: An introduction to scientific visualization. In: Scientific Visualization, Techniques and Applications. Springer, Berlin (1992)
20. Ebbinghaus, H.: Memory: a contribution to experimental psychology. Ann. Neurosci. **20**(4), 155–156 (2013)
21. Evergreen, S.D.H.: Effective Data Visualization: The Right Chart for the Right Data. SAGE Publications, Thousand Oaks (2016)
22. Ferster, C.B., Skinner, B.F.: Schedules of Reinforcement. Appleton Century Crofts, New York (1957)
23. Few, S.: Benefitting infovis with visual difficulties? Provocation without a cause. http://www.perceptualedge.com/articles/visual_business_intelligence/visual_difficulties.pdf (2011)
24. Few, S.: Now You See It. Analytics Press, Oakland (2009)

25. Few, S.: The chartjunk debate: a close examination of recent findings. http://www.perceptualedge.com/articles/visual_business_intelligence/the_chartjunk_debate.pdf (2011)
26. Frank, M., Roehrig, P., Pring, B.: What to do When Machines do Everything: How to Get Ahead in a World of AI, Algorithms, Bots, and Big Data. Wiley, New York (2017)
27. Friedhoff, R.M., Kiley, T.: The eye of the beholder. Comput. Graph. World **13**(8), 46 (1990)
28. Fry, H.: Hello World: How to be Human in the Age of the Machine. Doubleday, New York (2018)
29. Gödel, K.: ÃIJber formal unentscheidbare sätze der principia mathematica und verwandter systeme, i. Monatshefte für Mathematik und Physik **38**(1), 173–198 (1931)
30. Gomez, S.R., Guo, H., Ziemkiewicz, C., Laidlaw, D.H.: An insight- and task-based methodology for evaluating spatiotemporal visual analytics. In: Proceedings of the IEEE Conference on Visual Analytics Science and Technology, pp. 63–72 (2014)
31. Hearst, M.A.: Search User Interfaces. Cambridge University Press, Cambridge (2009)
32. Hullman, J., Adar, E., Shah, P.: Benefitting infovis with visual difficulties. IEEE Trans. Vis. Comput. Graph. **17**(12), 2213–2222 (2011)
33. James, W.: The Principles of Psychology, vol. 2. Henry Holt and Company, New York (1890)
34. Jänicke, S., Franzini, G., Cheema, M.F., Scheuermann., G.: Visual text analysis in digital humanities. Comput. Graph. Forum **36**(6), 226–250 (2017)
35. Jastrow, J.: Concepts and "isms" in psychology. Am. J. Psychol. **39**, 1–6 (1927)
36. Johnson-Laird, P.N.: How could consciousness arise from the computations of the brain? In: Blakemore, C., Greenfield, S. (eds.) Thoughts on Intelligence, Identity and Consciousness (1987)
37. Kant, I.: Critique of Pure Reason. Cambridge University Press, Cambridge (1998)
38. Kindlmann, G., Scheidegger, C.: An algebraic process for visualization design. IEEE Trans. Vis. Comput. Graph. **20**(12), 2181–2190 (2014)
39. Kreiser, J., Meuschke, M., Mistelbauer, G., Preim, B., Ropinski, T.: A survey of flattening-based medical visualization techniques. Comput. Graph. Forum **37**(3), 597–624 (2018)
40. Kucher, K., Paradis, C., Kerren, A.: The state of the art in sentiment visualization. Comput. Graph. Forum **27**(1), 71–96 (2018)
41. LaForte, G., Hayes, P.J., Ford, K.M.: Why Gödel's theorem cannot refute computationalism. Artif. Intell. **104**, 265–286 (1998)
42. Lucus, J.R.: Minds, machines, and Gödel. Philosophy **36**, 112–127 (1961)
43. Mainzer, K.: Thinking in Complexity, The Complex Dynamics of Matter, Mind, and Mankind, 3rd. edn. Springer, Berlin (1997)
44. Margolis, J.: Philosophy of Psychology. Prentice-hall, Upper Saddle River (1984)
45. Marty, R.: Applied Security Visualization. Addison-Wesley, Boston (2009)
46. Matković, K., Wischgoll, T., Laidlaw, D.H.: Empirical evaluations with domain experts. In: Chen, M., Hauser, H., Rheingans, P., Scheuermann, G. (eds.) Foundations of Data Visualization. Springer, Berlin (2019)
47. McCormick, B.H., DeFanti, T.A., Brown, M.D.: Visualization in scientific computing. ACM SIGGRAPH Comput. Graph. **21**(6) (1987)
48. Neisser, U.: Cognitive Psychology. Appleton-Century-Crofts, New York (1967)
49. Owen, G.S.: HyperVis – teaching scientific visualization using hypermedia. Technical Report, ACM SIGGRAPH Education Committee. http://www.siggraph.org/education/materials/HyperVis/hypervis.htm (1999)
50. Pandey, A.V., Rall, K., Satterthwaite, M.L., Nov, O., Bertini, E.: How deceptive are deceptive visualizations? An empirical analysis of common distortion techniques. In: Proceedings of the 33rd Annual ACM Conference on Human Factors in Computing Systems, pp. 1469–1478 (2015)
51. Pavlov, A.P.: Le crétacé inférieur de la russie. Nouveaux Mémoires de la Société impériale des Naturalistes de Moscou **21**, 1–87 (1901)
52. Penrose, R.: Emperor's New Mind. Oxford University Press, Oxford (1989)
53. Penrose, R.: Shadows of the Mind. Vintage, New York (1995)

54. Pfitzner, D., Hobbs, V., Powers, D.: A unified taxonomic framework for information visualiza-
    tion. In: Proceedings of the Asia-Pacific Symposium on Information Visualisation, pp. 57–66
    (2003)
55. Piaget, J.: Origins of Intelligence in the Child. Routledge & Kegan Paul, Abingdon (1936)
56. Pinker, S.: The Language Instinct: How the Mind Creates Language. Penguin (2003)
57. Preim, B., Joshi, A.: Evaluation of visualization systems with long-term case studies. In: Chen,
    M., Hauser, H., Rheingans, P., Scheuermann, G. (eds.) Foundations of Data Visualization.
    Springer, Berlin (2019)
58. Pretorius, A.J., Khan, I.A., Errington, R.J.: A survey of visualisation for live cell imaging.
    Comput. Graph. Forum **36**(1), 46–63 (2017)
59. Rothman, D.: Artificial Intelligence by Example. Packt Publishing, Birmingham (2018)
60. Sloman, A.: The Emperor's New Mind concerning computers, minds and the laws of physics.
    Bull. Lond. Math. Soc. **24**, 87–96 (1992)
61. Spence, R.: Information Visualization: Design for Interaction. Pearson, London (2007)
62. Stasko, J.T.: Value-driven evaluation of visualizations. In: Proceedings of the 5th Workshop on
    Beyond Time and Errors: Novel Evaluation Methods for Visualization (BELIV (2014)
63. Streeb, D., El-Assady, M., Keim, D.A., Chen, M.: Why visualize? Untangling a large network
    of arguments. IEEE Trans. Vis. Comput. Graph. https://doi.org/10.1109/TVCG.2019.2940026
    (Early access in 2019)
64. Tableau: Data visualization beginner's guide: a definition, examples, and learning resources
    (Accessed in April 2019). https://www.tableau.com/learn/articles/data-visualization
65. Thomas, J.J., Cook, K.A. (eds.): Illuminating the Path: The Research and Development Agenda
    for Visual Analytics. IEEE Computer Society, Washington, D. C. (2005)
66. Titchener, E.B.: The 'type-theory' of the simple reaction. Mind **18**, 236–241 (1896)
67. Tominski, C., Gladisch, S., Kister, U., Dachselt, R., Schumann., H.: Interactive lenses for
    visualization: an extended survey. Comput. Graph. Forum **36**(6), 173–200 (2017)
68. Tufte, E.R.: The Visual Display of Quantitative Information, 2nd edn. Graphics Press, Cheshire
    (2001)
69. Turing, A.: On computable numbers, with an application to the entscheidungsproble. Proc.
    Lond. Math. Soc. **42**, 230–265 (1937)
70. van Wijk, J.J.: The value of visualization. In: Proceedings of the IEEE Visualization, pp. 79–86
    (2005)
71. Walsh, R.T.G., Teo, T., Baydala, A.: A Critical History and Philosophy of Psychology: Diversity
    of Context, Thought, and Practice. Cambridge University Press, Cambridge (2014)
72. Ware, C.: Information Visualization: Perception for Design, 2nd edn. Morgan Kaufmann,
    Burlington (2004)
73. Weiskopf, D.: An Introduction to the Philosophy of Psychology. Cambridge University Press,
    Cambridge (2015)
74. Weiskopf, D.: Vis4Vis: visualization for (empirical) visualization research. In: Chen, M.,
    Hauser, H., Rheingans, P., Scheuermann, G. (eds.) Foundations of Data Visualization. Springer,
    Berlin (2019)
75. Wertheimer, M.: Laws of Organization in Perceptual Forms. A Source Book of Gestalt Psy-
    chology. Routledge & Kegan Paul, Abingdon (1923)
76. wiktionary.org: Appendix: glossary of philosophical isms (Accessed in April 2019). https://
    en.wiktionary.org/wiki/Appendix:Glossary_of_philosophical_isms
77. Witten, I.H., Frank, E., Hall, M.A.: Data Mining: Practical Machine Learning Tools and Tech-
    niques, 3rd edn. Morgan Kaufmann, Burlington (2010)
78. Zhou, M.X., Feiner, S.K.: Visual task characterization for automated visual discourse synthesis.
    In: Proceedings of the SIGCHI Conference on Human Factors in Computing Systems, pp. 392–
    399 (1998)

Chapter 12
Open Challenges in Empirical Visualization Research

Caroline Ziemkiewicz, Min Chen, David H. Laidlaw, Bernhard Preim
and Daniel Weiskopf

Abstract In recent years, empirical studies have increasingly been seen as a core part of visualization research, and user evaluations have proliferated. It is broadly understood that new techniques and applications must be formally validated in order to be seen as meaningful contributions. However, these efforts continue to face the numerous challenges involved in validating complex software techniques that exist in a wide variety of use contexts. The authors, who represent perspectives from across visualization research and applications, discuss the leading challenges that must be addressed for empirical research to have the greatest possible impact on visualization in the years to come. These include challenges in developing research questions and hypotheses, designing effective experiments and qualitative methods, and executing studies in specialized domains. We discuss those challenges that have not yet been solved and possible approaches to addressing them. This chapter provides an informal survey and proposes a road map for moving forward to a more cohesive and grounded use of empirical studies in visualization research.

C. Ziemkiewicz (✉)
Forrester Research, Inc, Cambridge, MA, USA
e-mail: cziemkiewicz@forrester.com

M. Chen
University of Oxford, Oxford, UK

D. H. Laidlaw
Brown University, Providence, RI, USA

B. Preim
Otto-von-Guericke-Universität Magdeburg, Magdeburg, Germany

D. Weiskopf
University of Stuttgart, Stuttgart, Germany

© Springer Nature Switzerland AG 2020
M. Chen et al. (eds.), *Foundations of Data Visualization*,
https://doi.org/10.1007/978-3-030-34444-3_12

12.1 Introduction

The visualization field has long had a complex relationship with empirical validation. In the 2000's, as it was quickly growing from a niche graphics subfield into a major research area of its own, there was a proliferation of reports and panels on the major unsolved challenges in visualization. A common theme in these challenges was a need to reliably prove the value of visualization. For example, Keim et al. [16] noted the issue of user acceptability; if domain users did not see how visualization could help them, they would not adopt it, and so there was no way to test whether visualization helped them. In a report directed at national funding agencies in the USA, Johnson et al. [14] cite similar challenges in demonstrating value and involving domain scientists in research. This example suggests some of the practical context behind this push for validation. As visualization researchers sought support for their work, it was necessary to find objective metrics that could show the value of their methods.

With this backdrop, evaluation in visualization has historically focused on user studies that either measure the effectiveness of visualization versus a traditional method or the relative effectiveness of two or more visualization techniques. However, while our overall approach to empirical research has remained much the same, the context around it has changed dramatically. Visualization has been adopted widely in commercial and government settings. As "big data" became a household term, the value of visualization came to be broadly understood as an efficient interface between people and information. Empirical visualization research has yielded general guidelines that are familiar to commercial designers outside the research community.

In this new context, it may be necessary to revisit the role of empirical research in visualization. In a world where visualization is assumed to have value, demonstrating that a visualization is usable may no longer be sufficient validation. In a moment like this, it is worthwhile to look back at what challenges have been addressed and which remain open. In 2004, Plaisant [22] identified what were then the major challenges in visualization evaluation. In some cases, the visualization community has made substantial progress in these challenges: for example, building task taxonomies [2, 5, 25], adding to the variety of evaluation approaches [12, 21, 27], and using contests to develop benchmark problems and datasets [23].

However, there are other challenges named in 2004 that remain unsolved today. Even as visualization grows more popular, the core problem of motivating domain users to buy into research continues to be a limitation. As a community, while we have developed more techniques for evaluation, we have not consistently established best practices for either research methodology or experimental stimulus design. Researchers continue to face challenges in controlling the experimental parameters in study design. It is possible that many of these issues could be addressed by a greater understanding of related psychological fields, but incorporating that understanding is a nontrivial exercise. Each of these challenges faces unique obstacles, but there is promising work that points to possible ways of addressing them.

12.2   Challenge 1: Motivating Domain Experts

12.2.1   Current Challenges

The difficulty of substantially involving domain experts in research has perhaps been the most cited challenge in visualization evaluation, and remains as relevant today as ever. Empirical research strongly benefits from realistic assessments of current technology as well as from realistic evaluations of research prototypes. In the early stages of development, the observation of experts solving real problems is essential to understand workflows, processes, constraints, and non-routine factors that would not be detected with questionnaires or interviews carried out at places distant from the working environment.

Similarly, empirical evaluations of a prototype strongly benefit from a high degree of realism. If they are carried out at the workplace of the domain experts and serve to solve real tasks, domain experts are fully motivated. In contrast, if artificial or archival data are used, the motivation is lower. Additionally, many visualization methods are aimed at niche user groups with advanced training, such as scientists, medical professionals, and analysts. The tasks these users engage in are frequently complex and involve learning, problem solving, or decision making. However, many empirical studies use simplified low-level tasks, such as basic perceptual tasks, searching for data, navigation, or routine activity. Evaluations in the developer's laboratory using such abstracted low-level tasks are much simpler to carry out, but often the results have at best a very indirect relation to the true activities of users.

Apart from designing abstracted studies outside of the domain context, another common approach to this problem is to involve domain experts briefly at key points in the process. For example, a researcher might develop a tool, then have a domain expert to evaluate it using an interview or other form of qualitative feedback. While this can be a way to work around the domain expert's schedule, it asks the user to evaluate something for which they have no prior context. A pair of surveys of evaluation methods used in visualization papers argues for a systematic lack of process evaluation methods such as requirements gathering and analysis of user workflows [13, 18]. Without this key context, tools are likely to be disconnected from the user's work context, and the value of their feedback may be limited.

12.2.2   Possible Approaches

One of the reasons for the systematic lack of process evaluation methods is that they are difficult to publish except as part of a lengthy design study. An approach to address this problem may be to create venues for such papers, for example, by designing a workshop around them or by introducing a new paper type. Another possibility would be to investigate methods that combine controlled and uncontrolled empirical methods; for example, contextual inquiry, and observational studies in a laboratory environment [18]. Ultimately, as Sedlmair et al. [26] point out, adoption of a system in the field is a problem to approach at the organizational level, not on the level of individual end users. Visualization researchers who wish to motivate domain experts must learn to observe and integrate with the experts' environment and work context.

12.3   Challenge 2: Systematic Lack of Research Methodology Skills

12.3.1   Current Challenges

A major factor that limits the effectiveness of empirical research in visualization is that visualization researchers, especially those from a computer science background, are not guaranteed to be trained in basic human subjects research methodology. In the field of psychology, from which visualization researchers often borrow approaches, there is no shortage of researchers who have been trained to design and conduct controlled empirical studies and formal qualitative research. In the field of visualization, the number of researchers who have had direct experience in designing and conducting empirical studies is significantly smaller. Computer science education does not prioritize these skills, as evidenced by the fact that user-centered design and research methods are not included as part of the core computer science curriculum [15]. This lack of skilled resources means that it is impossible to conduct large numbers of high-quality studies on any given topic, leaving many core research questions unanswered.

This skill deficit is reflected in visualization research in a number of ways. One of the most widespread is a lack of detailed and consistent statistical reporting of empirical results [6, 7, 13]. Researchers who present studies without using the appropriate statistical tests, making corrections for multiple comparisons, or reporting effect sizes not only limit the impact of their own work but make it difficult or impossible to produce meta-analyses and surveys. Moreover, it is still not uncommon to see papers with evaluations that consist only of unstructured feedback from a small number of experts. Contributing to this problem is a broad lack of knowledge about qualitative methods that leads to confusion between qualitative research and informal feedback-gathering [13].

This problem affects all of visualization but can be especially difficult in scientific visualization (SciVis), where researchers are less likely to come from a human–computer interaction (HCI) background. SciVis research often requires specialized algorithmic knowledge, and the social context of computer science education frequently puts distance between these "hard" algorithmic skills and the "soft" skills of user research. SciVis researchers face the additional challenge of balancing collaboration. Information visualization researchers dealing with generic or broadly understood data may forego domain collaborators in favor of psychologists or HCI experts, but SciVis researchers almost always need to collaborate with experts from a scientific domain. Coordinating multiple collaborations, especially among in-demand experts, carries significant risks. As a result, teams including SciVis researchers, domain experts, and empirical research specialists remain relatively rare.

12.3.2   Possible Approaches

While visualization researchers understand the value of collaboration with experts in empirical methods, this does not always translate into active participation in such collaborations. Providing specific funding incentives has the potential to push these partnerships forward. As an example, cooperation between visual analytics and data analysis in Germany was initiated by a national research priority program on Scalable Visual Analytics which encouraged collaborations between both fields and between funded projects [17]. To address the skills gap within the community, one possibility is to revise standards for computer science curricula to include user-centered research as a core topic [15]. A more immediate action could be to compile a community portal to collect resources on empirical methods, similar to efforts such as The Fluid Project [1] but tailored to the specific needs of visualization researchers.

12.4   Challenge 3: Data Collection and Generation

12.4.1   Current Challenges

Although visualization researchers have made considerable progress in recent years in developing formal taxonomies and models of evaluation tasks, there has been less emphasis on developing repeatable approaches to data generation. In a field where the nature of the data can considerably change the effectiveness of the method being tested, unrealistic data is a serious threat to ecological validity. Examples include data at a scale much smaller than would be encountered in real tasks, data that lacks the errors and inconsistencies common to real datasets, and data with strong statistical patterns that might not normally be present. While benchmark datasets are useful for comparison, they often do not capture these real-world data challenges.

At the same time, real-world datasets can be difficult to collect and use for a variety of reasons, such as privacy, size, protection of proprietary information, or legal restrictions on dissemination [26]. A common approach in such situations are to build sanitized datasets by removing or perturbing sensitive information. However, security research has shown that even sophisticated privacy-preserving data mining methods can be vulnerable to re-identification, especially in cases where multiple data sources can be combined [20]. Even in cases where real data can be used as-is, it can be difficult to generalize evaluation results from a single dataset. Finding multiple datasets that represent a realistic range of conditions only compounds the problem.

Generative data models can be an effective approach to this problem, but they require careful design to avoid biases [24]. A generative data model can be used to automate the generation of multiple datasets with desired properties, which can address the issue of testing against multiple valid datasets to support generalization. There are a number of significant technical challenges associated with such models; many involve complex simulations, and as most models are developed for one-off cases, standardized techniques and replication are rare. Moreover, interactions between a generative model and a visualization technique can be difficult to predict. There is no guarantee that a model that produces data with desirable characteristics will still have the same characteristics after being processed as part of a given visualization algorithm.

12.4.2 Possible Approaches

The type of formalization that has been applied to tasks and visual representations in recent years has helped to produce more rigorous and controlled experiments. However, the way we describe data is still most often in the terms used by Jacques Bertin fifty years ago [3]. More specialized typologies of data that take into account contemporary concerns such as scale, heterogeneity, and uncertainty could go a long way toward defining a design space in which datasets used in experiments can vary. Generative models have the potential to address many problems in data collection, but the field will advance more quickly if designers of generative models adopt open practices and make models available for replication and benchmarking. As more such models are made available, it will be possible to identify best practices and guidelines for further development [24].

12.5 Challenge 4: Experimental Design Space and Tradeoffs

12.5.1 Current Challenges

At the core of many of the challenges in visualization evaluation is that it involves the combination of two highly complex systems: the human user and the data visualization system. In such a situation, the number of experimental variables that must be controlled can quickly become unmanageable. The skills deficit discussed in Sect. 12.3 compounds this problem, as there is a lack of institutional knowledge about how to balance tradeoffs and control variables in experimental design. This leads to a number of issues affecting the ecological validity of experiments as well as the ability of other researchers to evaluate and make use of experimental results.

One common problem is when the assumption that a system should be evaluated in one experiment leads to overstuffed design. In some ways, this problem has been exacerbated by the increased push for no system to go unevaluated. While this is an admirable goal, in practice, treating evaluation as a box that must be checked often leads to user studies that either lack a clear hypothesis or attempt to test too many hypotheses at once. Such user studies often suffer from a mismatch in validation method to type of contribution; for example, a paper whose primary contribution is a novel visual encoding does not necessarily require a task-based evaluation, provided the authors make no claims about improving performance on that specific task [19]. Nonetheless, user studies remain common in such situations, often using ad hoc tasks that have not been rigorously designed.

Knowledge of appropriate design space tradeoffs also affects the quality of reviewing. Lack of familiarity with empirical methods is one issue, but partial familiarity can cause its own share of problems. A reviewer with knowledge of only one method may apply the rigor metrics of that method to an unrelated one, leading to inappropriate evaluations [6]. For example, a researcher who uses qualitative methods may receive criticism for not including statistical analyses suited for quantitative methods. A better understanding of the experimental design space, and an acknowledgement that no one study can cover it exhaustively, remains elusive.

12.5.2   Possible Approaches

In psychology and related disciplines, it is common to publish a series of related studies in a single publication, which each experiment building on the knowledge gained in the previous one. Such a structure allows researchers to produce more tightly controlled individual study designs while still approaching a larger research question. While linked studies of this type are sometimes seen in visualization perception research [4, 11, 29], it may also be a useful method for technique or system evaluation. In this model, user evaluations may even be published separately from the system itself, which in many cases may require more limited validation methods. In order to improve the control of variables in study design, one possibility is to publish and promote evaluation checklists, a method that has been used effectively in other domains [10].

12.6 Challenge 5: Engagement with Relevant Psychology Fields

12.6.1 Current Challenges

As in other human-centered computer science disciplines, understanding visualization depends heavily on understanding the people who use it. Psychology is a key component of any empirical research in visualization. Yet explicit engagement with psychology research remains infrequent outside of a few specific areas, such as research on color scale design [28]. This can lead to findings that are divorced from important context. An experiment on how well a user can remember information in a particular visual representation must take into account the expected performance of visual working memory in general; a field study observing adoption of a system cannot be generalized without a working knowledge of how quickly new technology is usually adopted in workplaces. Perhaps the most pervasive example of such issues is the widespread assumption that the effectiveness of a visual representation can be generalized between users without taking into account natural variation in spatial ability and other cognitive factors [30].

This lack of engagement with psychology also causes issues when it leads to ignorance of challenges in psychological research that affect visualization researchers as well. The difficulty of integrating knowledge gained in increasingly specialized subfields was named "The Grand Challenge" of psychology by Axel Cleeremans of Université libre de Bruxelles in 2010 [8]. Clearly this is a concern for visualization as well, as community discussion at the 2018 IEEE VIS Conference centered around the problem of unifying the diverging fields of scientific visualization, information visualization, and visual analytics. Visualization researchers are also just beginning to take notice of the replication crisis in psychology [9], but have yet to adopt the reforms made by psychologists in its wake.

These challenges themselves can create pitfalls for outside researchers looking to make use of psychological findings or methods. The complexity of psychology's many disparate fields, and lack of communication between these fields [8], can obscure important connections and make it difficult to know where to start looking for answers. Visualization researchers very often know that psychology is important to their work, but without clear goals and an understanding of the research space, it is rare for sustained productive conversation to happen between the two disciplines.

12.6.2 Possible Approaches

In order to engage more fully with psychology research, it may be necessary to modernize our research practices to meet the changes made by psychologists in recent years. For example, adopting open science practices, especially sharing data and code (where possible), would be a positive step for the visualization field on its own. But it could also foster collaboration by making materials and tools available to psychology researchers themselves. In some cases, these experts may have visualization needs that our community is unaware of, and a greater degree of communication may help reveal them. This can be a challenging process, as publication cultures and research goals will vary across fields. Work to identify common ground and mutual goals will be a necessary first step. We can also learn from psychology now by addressing some of the known issues that affect both fields; for example, submitting research reports ahead of performing experiments in order to reduce positive effect bias.

12.7   Conclusion and Next Steps

In this chapter, we have discussed five key challenges in empirical visualization research in detail and proposed possible approaches to addressing them. By doing so, we hope to build on the successes of the past in developing a research agenda for the future. It is vital to note the areas in which we have made progress as well as those where challenges remain. Empirical studies in visualization have advanced in many ways over the past decade, as has visualization itself. But even as the value of visualization becomes more broadly accepted, the current evaluation paradigm more often than not focuses on testing whether a visualization is generally effective or not. By addressing these challenges, we hope to make space for research that goes beyond this paradigm to answer more specific, contextualized, and meaningful questions that drive the future of visualization research.

References

1. Administrator, F.P.: Fluid project wiki. https://wiki.fluidproject.org
2. Amar, R., Eagan, J., Stasko, J.: Low-level components of analytic activity in information visualization. In: IEEE Symposium on Information Visualization, 2005. INFOVIS 2005, pp. 111–117. IEEE (2005)
3. Bertin, J., Berg, W.J., Wainer, H.: Semiology of Graphics: Diagrams, Networks, Maps. University of Wisconsin Press, Madison (1983)
4. Bezerianos, A., Isenberg, P.: Perception of visual variables on tiled wall-sized displays for information visualization applications. IEEE Trans. Vis. Comput. Graph. **18**(12), 2516–2525 (2012)
5. Brehmer, M., Munzner, T.: A multi-level typology of abstract visualization tasks. IEEE Trans. Vis. Comput. Graph. **19**(12), 2376–2385 (2013)
6. Carpendale, S.: Evaluating information visualizations. Information Visualization, pp. 19–45. Springer, Berlin (2008)
7. Chen, C., Yu, Y.: Empirical studies of information visualization: a meta-analysis. Int. J. Hum.-Comput. Stud. **53**(5), 851–866 (2000)
8. Cleeremans, A.: The grand challenge for psychology. APS Observer **23**(8) (2010)
9. Collaboration, O.S., et al.: Estimating the reproducibility of psychological science. Science **349**(6251), aac4716 (2015)
10. Crisan, A., Elliott, M.: How to evaluate an evaluation study? comparing and contrasting practices in vis with those of other disciplines. In: Proceedings of the 2018 Workshop on Beyond Time and Errors: Novel Evaluation Methods for Information Visualization. IEEE (2018)

11. Heer, J., Bostock, M.: Crowdsourcing graphical perception: using mechanical turk to assess visualization design. In: Proceedings of the SIGCHI Conference on Human Factors in Computing Systems, pp. 203–212. ACM (2010)
12. Isenberg, P., Zuk, T., Collins, C., Carpendale, S.: Grounded evaluation of information visualizations. In: Proceedings of the 2008 Workshop on Beyond Time and Errors: Novel Evaluation Methods for Information Visualization, p. 6. ACM (2008)
13. Isenberg, T., Isenberg, P., Chen, J., Sedlmair, M., Möller, T.: A systematic review on the practice of evaluating visualization. IEEE Trans. Vis. Comput. Graph. **19**(12), 2818–2827 (2013)
14. Johnson, C., Moorhead, R., Munzner, T., Pfister, H., Rheingans, P., Yoo, T.S.: NIH-NSF visualization research challenges report. Institute of Electrical and Electronics Engineers (2005)
15. Joint Task Force on Computing Curricula, A.f.C.M.A., Society, I.C.: Computer Science Curricula 2013: Curriculum Guidelines for Undergraduate Degree Programs in Computer Science, p. 999133. ACM, New York (2013)
16. Keim, D.A., Mansmann, F., Schneidewind, J., Ziegler, H.: Challenges in visual data analysis. In: 10th International Conference on Information Visualization, IV 2006, pp. 9–16. IEEE (2006)
17. Konstanz, U.: Scalable visual analytics: Interactive visual analysis systems of complex information spaces. http://www.visualanalytics.de/node/2
18. Lam, H., Bertini, E., Isenberg, P., Plaisant, C., Carpendale, S.: Empirical studies in information visualization: seven scenarios. IEEE Trans. Vis. Comput. Graph. **18**(9), 1520–1536 (2012)
19. Munzner, T.: A nested process model for visualization design and validation. IEEE Trans. Vis. Comput. Graph. **15**(6), 921–928 (2009)
20. Narayanan, A., Shmatikov, V.: Myths and fallacies of personally identifiable information. Commun. ACM **53**(6), 24–26 (2010)
21. North, C.: Toward measuring visualization insight. IEEE Comput. Graph. Appl. **26**(3), 6–9 (2006)
22. Plaisant, C.: The challenge of information visualization evaluation. In: Proceedings of the Working Conference on Advanced Visual Interfaces, pp. 109–116. ACM (2004)
23. Plaisant, C., Fekete, J.D., Grinstein, G.: Promoting insight-based evaluation of visualizations: from contest to benchmark repository. IEEE Trans. Vis. Comput. Graph. **14**(1), 120–134 (2008)
24. Schulz, C., Nocaj, A., El-Assady, M., Frey, S., Hlawatsch, M., Hund, M., Karch, G., Netzel, R., Schätzle, C., Butt, M., et al.: Generative data models for validation and evaluation of visualization techniques. In: Proceedings of the Sixth Workshop on Beyond Time and Errors on Novel Evaluation Methods for Visualization, pp. 112–124. ACM (2016)
25. Schulz, H.J., Nocke, T., Heitzler, M., Schumann, H.: A design space of visualization tasks. IEEE Trans. Vis. Comput. Graph. **19**(12), 2366–2375 (2013)
26. Sedlmair, M., Isenberg, P., Baur, D., Butz, A.: Information visualization evaluation in large companies: Challenges, experiences and recommendations. Inf. Vis. **10**(3), 248–266 (2011)
27. Shneiderman, B., Plaisant, C.: Strategies for evaluating information visualization tools: multi-dimensional in-depth long-term case studies. In: Proceedings of the 2006 AVI Workshop on Beyond Time and Errors: Novel Evaluation Methods for Information Visualization, pp. 1–7. ACM (2006)
28. Silva, S., Santos, B.S., Madeira, J.: Using color in visualization: a survey. Comput. Graph. **35**(2), 320–333 (2011)
29. Tory, M., Kirkpatrick, A.E., Atkins, M.S., Moller, T.: Visualization task performance with 2d, 3d, and combination displays. IEEE Trans. Vis. Comput. Graph. **12**(1), 2–13 (2006)
30. Ziemkiewicz, C., Ottley, A., Crouser, R.J., Chauncey, K., Su, S.L., Chang, R.: Understanding visualization by understanding individual users. IEEE Comput. Graph. Appl. **32**(6), 88–94 (2012)

Part III
Collaboration with Domain Experts

Visualization requires three components to work: data, tasks, and an audience. Any foundation on data visualization will deal with all three. Furthermore, as in other computer science disciplines, there is basic and applied visualization research. Basic visualization research is driven by generalized visualization tasks, evaluation issues, or theoretical questions about visualization. Applied visualization starts from an application (most likely outside visualization) including data, task, and audience, and tries to find the best visualization solution for the given case. Of course, applied visualization research may (and should) lead to new visualization tasks which triggers new basic visualization research. Also, basic visualization research offers new possibilities for applications.

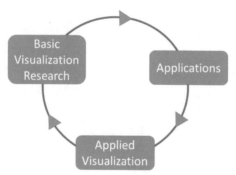

In addition, current and past experience shows that there is a substantial difference between visualizations for domain experts knowing data, underlying model, and assumptions as well as tasks very well, and visualizations for a broad audience with very different backgrounds and much less or even no knowledge about data, models, and tasks. The latter case is covered in another part, so this part of the book focuses only on domain experts. As domains, the collaboration models, data, and tasks vary substantially, this part starts with a chapter that contains seven successful case studies. Visualization experts describe in these cases how they approached the cooperation, what was important, and which lessons they took out of these projects. The domains cover biological, medical, and engineering examples which are the most often seen domains in cooperations between visualization experts and domain experts. While these cooperations mainly concern the cooperation within academia, the second chapter describes experiences and advice from industry practioneers on the collaboration between university and commercial companies. The final chapter of this part takes a somewhat more abstract point of view. It looks at the process of actually selecting a domain expert as collaboration partner, and how to create impact by the research. The authors indicate clearly that just starting a collaboration by chance might work, but more mindful strategies provide better chances to lead to success. This includes some thoughts on measures of success where it becomes clear that this is, at least partly, a subjective question that any visualization researcher (and domain expert) has to define for him- or herself.

Chapter 13
Case Studies for Working with Domain Experts

Johanna Beyer, Charles Hansen, Mario Hlawitschka, Ingrid Hotz,
Barbora Kozlikova, Gerik Scheuermann, Markus Stommel, Marc Streit,
Johannes Waschke, Thomas Wischgoll, and Yong Wan

Abstract The collaboration with domain experts concentrates always on an application domain where the experts work. Usually, they provide the data and directions
of research that require visualization support. This chapter presents seven successful cases of such collaborations. The domain varies from biology and medicine to
mechanical engineering. There are examples of long time cooperation as well as
smaller short-term projects. The description concentrates on the process, output,
and especially on the lessons learnt from these cooperations. The scientific work is
described to understand the context and goals of the cooperation, but many details
can only be found in the references. The reason for this unusual writing is the wish on
the one hand to describe various aspects of collaboration with domain experts which
is an important part of the foundations of data visualization. On the other hand, the
text should not become lengthy and filled with too many details of individual cases
that can be found elsewhere.

J. Beyer
Harvard University, Cambridge, MA, USA

C. Hansen · Y. Wan
University of Utah, Salt Lake City, UT, USA

M. Hlawitschka · J. Waschke
Leipzig University of Applied Sciences, Leipzig, Germany

I. Hotz
Linköping University, Norrköping, Sweden

B. Kozlikova
Masaryk University, Brno, Czech Republic

G. Scheuermann (✉)
Leipzig University, Leipzig, Germany
e-mail: scheuermann@informatik.uni-leipzig.de

M. Stommel
Leibniz Institute of Polymer Research, Dresden, Germany

M. Streit
Johannes Kepler University Linz, Linz, Austria

T. Wischgoll
Wright State University, Dayton, OH, USA

© Springer Nature Switzerland AG 2020
M. Chen et al. (eds.), *Foundations of Data Visualization*,
https://doi.org/10.1007/978-3-030-34444-3_13

255

13.1   Case Study: FluoRender

Yong Wan and Charles Hansen

FluoRender is a software package for visualizing and analyzing 3D and 4D (3D over time) fluorescence microscopy data. FluoRender has become an established system with many features driven by collaborations with biologists delivering visualization, segmentation, measurement, and tracking functions with an emphasis on accuracy, interactivity, and intuitiveness. Originally developed for the Zebrafish community, it has extended to other biological applications. FluoRender has been deployed as a standard tool in research laboratories both domestically and internationally, facilitating research on cell movements, neuronal circuitry, and tissue development during conventional analysis of wild-type and mutant embryos of popular model species. The increasing popularity and growing user base of FluoRender have given rise to new visualization and analysis challenges from both general and specialized workflows. Close collaborations between biologists and computer scientists have provided a systematic insight into the workflows in real-world biological research. A data analysis workflow is indeed a far cry from a rigid pipeline; it has to be highly adaptable and easily customizable for varying data analysis needs. In practice, user interactions and decisions are involved through the entire data analysis workflow. Interactive visualization and analysis functions work hand in hand, allowing exploring and iterative investigations, as well as progressive improvement to the refined results that lead to biological discoveries.

One example of the expanded user base is our collaboration with Professor Gabrielle Kardon and her interest in developing a mouse atlas [27]. Such atlases are important for understanding normal anatomy and the development and function of structures, and for determining the etiology of congenital abnormalities. Although the focus of FluoRender was the analysis and visualization of confocal microscopy data, the atlas required not only volume rendering and segmentation but also polygonal modeling, for muscles and bones, as well as advanced texturing which captured the anisotropy of muscle tissue reflecting what was seen in the confocal scan Fig. 13.1.

13.1.1   Lessons Learned

In the development of FluoRender, we learned several lessons from the integral collaboration with biologists. The first lesson was communication. Domain scientists use terminology from their domain. Visualization scientists use terminology from the visualization and computer science domain. It is critical that communications find the common ground so that there are no misunderstandings. This takes time for a detailed explanation from the visualization researchers of what their ideas and methods accomplish and how they are accomplished. It also is incumbent on domain scientists to explain their ideas and methods in terms that are easily understood. While the visualization scientists may not have the same detailed knowledge that the domain scientists have of a particular biological or domain process, the visualization scientists should have sufficient application knowledge to understand the biology and processes being investigated. Once this is accomplished, collaborative research is enhanced, and advances are more easily made.

Another lesson we have learned is that all participating collaborators have the science to accomplish. Of course, the domain scientists have science research questions they seek to answer in their particular domain. They are seeking answers to questions and testing hypotheses in their particular biological domain. At the same time, the visualization scientists should be advancing the field of visualization. This delicate balance of both advancing the field and providing solutions to needs and problems in biology is critical to a successful collaboration. The visualization researchers should not be simply serving the needs of the domain collaborators but should focus on both advancing visualization and providing solutions to requirements from the domain scientists.

Do not simply ask domain scientists what needs solving or which desired features are missing in a visualization system. By understanding the domain workflow, productive progress can be made. This often requires working with collaborators in their laboratory and having collaborators spend time in the visualization laboratory. It is important to not simply meet and discuss the domain problem but to actually work with collaborators in their research setting and observe what data analysis tasks are easily accomplished while others can be improved using an updated workflow. This leads to better understanding of the practical domain problems in greater detail.

Lastly, it is important to be creative. By providing a creative solution to biological problems, advances in both biology and visualization science can be made. Such

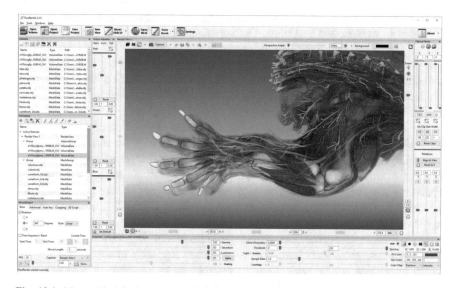

Fig. 13.1 Mouse hind limb atlas, Y. Wan, C. Hansen, SCI Institute and A. Kelsey Lewis and G. Kardon, Human Genetics, University of Utah

Fig. 13.2 Developing
forelimb of a healthy mouse
strain (top) compared to that
of a mutant mouse strain
with a stiff, abnormal gait
(bottom). Lateral triceps in
brachialis muscles (purple),
other types of muscle (red),
and tendons (green). Note
that in the top image
(wild-type mouse), the
lateral triceps and brachialis
muscles are distinct, while in
the bottom image (mutant
mouse), the two muscles are
fused, limiting the forelimb's
function

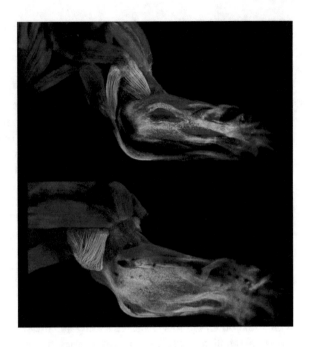

creative solutions should be enjoyable and fun for all sides of the collaboration. Success can be measured by typical means such as publications in domain journals or in visualization journals. It can also be measured by recognition of those in the respective fields. For example, our collaboration with the Kardon laboratory was recognized by the director of NIH in 2017 [9]. Dr. Francis Collins recognized the advances in muscle and soft tissue development research by Professor Kardon in finding that when two muscles are fused and indistinct, forelimb function is limited. This is due to a mutation in a gene called Tbx3. It is already known that the mutation of Tbx3 in human is the cause of a rare condition called ulnar–mammary syndrome (UMS). However, because of the lack of detailed examination and visualization on human patients, muscle anomalies of the UMS patients were overlooked in the original research. Researchers initially declined the idea of fused muscles in UMS patients because of the anatomical differences between mice and humans. Interestingly, at Kardon's urging, a similar pattern of missing muscles was confirmed in the re-examination of a UMS patient. This research demonstrates the astounding similarity between human and mouse genetics, which provides an excellent application stage for visualization tools, as such details in Fig. 13.2 can be prohibitive to obtain for human patients. Further improvements to the clarity are achieved by coloring muscles using the interactive segmentation tools in FluoRender, which are like 3D paintbrushes. They have to be intuitive and enjoyable to use for researchers, as the operations can be repeated from several tens to hundreds of samples in an investigation. The director of NIH also recognized FluoRender [9]:

... there's one more NIH connection to this work. Kardon's team produced this image, featured in the University of Utah's 2016 Research as Art competition, using a free software program, called FluoRender that was developed by another NIH-supported team at the University of Utah. FluoRender enables researchers to take a series of 2D photos from a scanning confocal microscope and turn them into amazingly informative 3D imagery.

13.2 Case Study: Connectomics

Johanna Beyer

This chapter describes an ongoing collaboration between the Visual Computing Group at Harvard University and neuroscientists working in the field of connectomics at the Harvard Center for Brain Science. The collaboration focuses on the visualization and analysis of large-scale connectomics data and has spanned over the last eight years.

13.2.1 Domain Problem

Connectomics aims to reconstruct the detailed neural connectivity in the mammalian brain, containing billions of interconnected nerve cells, at the resolution of individual connections (i.e., synapses). Determining this "wiring diagram" or so-called *connectome* is one of the grand challenges of modern neuroscience and will allow scientists to better understand how the brain functions and develops, and how mental illnesses and neural pathologies manifest themselves on the connectivity level. Recent advances in high-resolution electron microscopy and sample preparation have made it possible to acquire data at the speed and resolution necessary to reconstruct the brain's connectivity at the level of individual synapses. However, the acquired image stacks are typically hundreds of terabytes to petabytes in size, exhibit severe noise and imaging artifacts, and can contain tens of thousands of complex neural structures. A lot of effort has gone into developing novel methods for data acquisition, volume registration, and (semi-)automatic segmentation, resulting in large labeled volumes of brain tissue. The main goal of our collaboration was to enable the next logical step: supporting scalable and interactive volume exploration and visual analysis of the collected data.

13.2.2 Process and Output

This case study encompasses several sub-projects that were all developed within the same collaboration over the last eight years. Projects always started with initial meetings and interviews with the neuroscientists. Most neuroscientists we talked to,

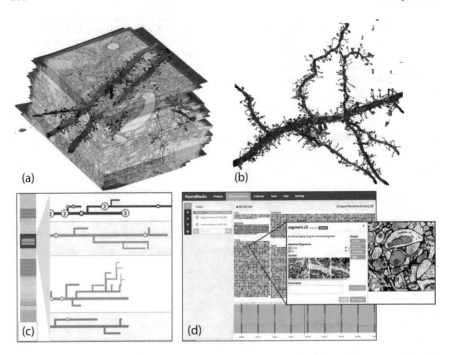

Fig. 13.3 *Exploration, visualization, and analysis of connectomics data.* **a** Volume rendering of a segmented terabyte electron microscopy volume. **b** A segmented dendrite. **c** Connectivity analysis with *Neurolines*. **d** Visual tracking of a segmentation project with a pop-up for visual proofreading

although in the same group, often had very different visualization and analysis tasks and requirements. Therefore, the initial project phase always focused on finding the main collaborator for the next project and quickly coming up with an initial prototype for further discussion. Throughout the development and implementation of the project, we kept a tight feedback loop with the domain scientists, to ensure that our project (a) solved an actual problem and (b) solved the problem that was relevant to our collaborators. To achieve this, we held regular in-person meetings, video conferences, and visited their laboratories to observe our collaborators at their routine data collection and processing tasks. After finishing each software prototype, we made sure to demonstrate it to our collaborators and encouraged them to use the software on their own. We evaluated the usefulness of our projects based on expert feedback and specific use cases that were developed together with our collaborators.

In the initial phase of our collaboration, we focused on developing basic visualization and data management infrastructure for large-scale segmented (i.e., labeled) neuroscience data sets. Having built that initial framework, in later years, we shifted our focus on visual analysis and integrating domain knowledge into our data exploration framework. Figure 13.3 shows some of the different projects we have developed over time.

Our first major project was a scalable volume rendering framework [12] for exploring petascale microscopy data streams. In a second step, we extended the framework to support interactive volume visualization of labeled volumes [4]. Using this framework, scientists could interactively explore their raw image data, as well as their segmentation data. However, a more in-depth quantitative evaluation was still difficult. To allow our collaborators to explore the data based on their domain knowledge, we developed ConnectomeExplorer [3], a tool for interactive domain-specific queries of neuroscience data. These queries allowed the first glimpse into how different segmented neural structures were connected. As connectomics is ultimately interested in the neural connectivity of the brain, our next project Neurolines [1] solely focused on visualizing and exploring the connectivity of axons and dendrites. In this project, we went beyond the initial volumetric data visualization and focused on an abstract 2D view that enabled users to focus on connectivity rather than the detailed 3D morphology of their data. Therefore, we abstracted the topology of 3D brain tissue data into a multi-scale, relative distance-preserving subway map visualization where each neurite is represented as a tree structure based on its real, but adaptively simplified, anatomy, and its branches. During the development of those projects, it became clear to us that the major bottleneck of our collaborators was not the analysis of their data, but the actual segmentation process, and tracking the segmentation status of a volume over time. Therefore, we developed tools for visual proofreading of segmentations [13], as well as for visual segmentation tracking and management [2].

What this list of different projects demonstrates is that the challenges and goals of domain experts often evolve and change over time. Therefore, it is vital to meet with them regularly, observe how they work, and to make an effort to understand their current set of challenges. That includes not just the challenges that are stated explicitly but also the implicit challenges that scientists might not even think of mentioning.

13.2.3 Lessons Learned

Collaborating with domain scientists has its own set of challenges; however, it is also incredibly rewarding. Here are some lessons we learned during our collaboration with neuroscientists:

Understand the domain problem. While this hint seems obvious, make sure to meet with domain scientists regularly. Visit their laboratory, follow them around, and observe their work. What do they spend the most time on, what are the difficult and/or annoying tasks? Domain scientists typically do not have a background in visualization, so they might not know where and how visualization could be most useful. For example, our collaborators would have never thought about a visual tool for tracking the segmentation process over time, even though they considered this as one of the most time-consuming and difficult tasks in their everyday work.

Define your roles and expectations. Make sure that everyone on the project is on the same page regarding each other's roles and responsibilities. Is the end goal of the project a scientific publication, a useful software framework, or, ideally, both? Make sure to address these concerns early on.

Have a user adoption strategy. We have observed a much better adoption of Web-based systems as compared to stand-alone applications that require local installation of software. Neuroscientists typically do not have a desktop PC but prefer laptops, as they tend to move around a lot between wet laboratories and other laboratory spaces, which makes adoption of software that requires specific hardware (e.g., GPUs, large monitors) a lot more difficult.

Go where the domain challenge takes you, not where your previous research has positioned you. Sometimes, you will discover interesting problems and challenges that you did not expect or foresee and that will require you to branch out into a different area or sub-area of research. Yet, if those are the most pressing challenges of your collaborators, embrace them and do not shy away from them. Start with the nail (i.e., the domain problem), then find the hammer, not the other way around.

13.3   Case Study: Mechanical Part Design

Gerik Scheuermann, Markus Stommel, and Ingrid Hotz

Component design is a major task in mechanical engineering. There is a well-defined workflow including structural mechanics simulation and analysis using the finite element method. We looked for possibilities to use tensor visualization of stress fields to leverage the full tensor information for the design. This is in strong contrast to the usual reduction to scalar fields like the von Mises stress which is done in all engineering post-processing tools. We tested nearly all available tensor visualization techniques until we finally found a way to show directional information using tensor lines that actually led to better component design. We describe the problem, the cooperation process, the success of the method, and the learned lessons.

13.3.1   Domain Problem

In this case study, we look at a standard problem in mechanical engineering. Engineers have to design a mechanical part with defined functional and qualitative properties which can be produced by standard methods. This is an important part of the product development process in mechanical engineering and follows a clearly defined workflow. First, a manual sketch is created, followed by a 2D, and finally a 3D CAD model. This model is the basis for a virtual mechanical test using the finite element method (FEM). The FEM result is interpreted using visualization. If the result is sound, a rapid prototype will undergo physical tests before a classical prototype is finally tested. If the virtual test shows problems or has to be further optimized, the design is altered and undergoes again a FEM analysis.

As an example, we use a brake lever of a bike which is currently made from metal and shall be replaced by a plastic component. The example is still fictitious in the sense that the actual manufacturer is not involved but realistic enough to show the potential in engineering terms. The specific engineering goal is to optimize the plastic rib support structures where the baseline is a textbook design using straight ribs. Rib structures are the most often used reinforcement structure for injection-molded plastic parts. Their design means the definition of position, number and shape of the ribs while considering given boundary conditions that follow from the manufacturing process or the part appearance. Even though there are algorithmic optimization methods for some design steps, rib design is still a manual process driven by the engineer's experience. Therefore, this is a typical example of trying to support a domain expert's daily task by enforcing his/her intuition through visualization.

13.3.2   Process and Output

The modern product design process in mechanical engineering is a completely virtual process that leads to a physical prototype by 3D printing. It consists of a number of improvements cycles. Each cycle contains 3D CAD design, FEM, and analysis using visualization. The design criteria include part stiffness, maximum stress peaks, weight, geometrical, or functional boundary conditions and also practical aspects of manufacturing. For material stressing, this comparison is performed so far on the basis of a couple of scalar key metrics.

The idea of this case study (for more details see [19]) is to use the complete FEM result instead, especially all stress tensor information to obtain an optimal design. The visualization partners offered the domain experts (i.e., the engineers) a framework of almost all tensor visualization methods ever invented in visualization research. Especially, we tested multiple linked views and linking-and-brushing for stress tensors, as well as several different tensor line methods. The engineers tested these methods and discussed their meaning in mechanics with visualization researchers. The question was always what does this visualization mean for the mechanics and how can the visual information be used to improve the design. After a number of visualization methods that did not deliver insight into the design process, we finally ended up with tensor lines and fabric textures [14]. Here, we showed planar cuts through the stress field. Thicker and thinner lines showed the eigen directions. This led to an intuitive design of rib support structures; see Fig. 13.4.

This intuitive design meant to follow the tensor lines of the stress field for rib support design. In the first step, we compared the standard textbook design with three different rib designs following different tensor lines. In the second step, we verified the results experimentally by 3D-printed brake levers. The results confirmed the hypothesis that tensor lines are good guidelines for rib support structures and lead to stiffer designs without additional material or production costs. As can be seen in Fig. 13.5, all three test designs performed substantially better than the textbook design. Honestly, we did not try any other designs, so basically every informed rib design was better than the standard.

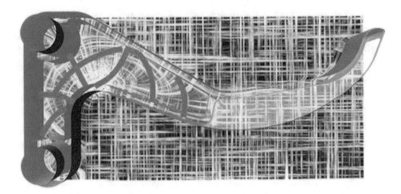

Fig. 13.4 Tensor fabric of the stress tensor in a planar cut through the CAD model. The engineer draws some lines manually to design rib support structures aligned with the tensor lines. These manual lines were the basis for the first new design. Two other drawings based on the same visualization led to the other two designs in our case study

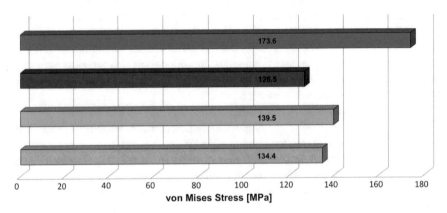

Fig. 13.5 Maximal von Mises stress in the four different rib structure designs. The top blue bar is the textbook result. It shows substantially higher maximal stress, i.e., the design is much worse than the three new designs

For the engineers, a major outcome has been a hypothesis that can substantially support the design process of technical parts. The results of the finite element simulations and the experiments give evidence that tensor lines are valuable for the design of rib patterns.

13.3.3 Lessons Learned

In this case study, it can be clearly seen that close cooperation between domain experts and visualization scientists is needed for success. The engineers do not know about many modern visualization methods, so they have no access to them or no understanding of them. The visualization scientists do not have enough understanding of the creative tasks of the domain expert and his/her thinking that leads from the problem over the visual input to better solutions like a better design in this case. Also, it is difficult and sometimes nearly impossible to find the best visual metaphors for insight into the domain expert side without classical try and error. Testing the different methods, discussing their meaning in terms of the problem at hand is the key. In this example, the engineers need to derive insight into the stress transport in the part, and even more important, an idea how to place the rib support structures from the visualization. Therefore, the lessons are as follows:

- Try to present more data than before.
- Look for information that helps the domain expert with his/her task.
- Try out many possibilities of visual representations and data.
- Close cooperation is key—present your possibilities and let the domain expert explain their thoughts.

13.4   Case Study: Drug Target Prioritization

Marc Streit

This section describes the process, outcome, and lessons learned from a research collaboration between the visualization group at the Johannes Kepler University Linz and a computational biology group at the pharmaceutical company Boehringer Ingelheim. The goal of the collaboration was to develop visual analysis solutions that help researchers to identify new drug targets for cancer therapy. A drug target constitutes the basis for the development of next generation drugs.

13.4.1   Domain Problem

Discovering new drug targets is a challenging process because the domain experts need to take into account a rich spectrum of data sources. The data sources that need to be incorporated in the exploratory analysis include experimental data from patients, animals, and cell lines, but also publicly available knowledge of what we know about biological processes and diseases.

13.4.2   Process and Output

Together with partners from Harvard University, we started to work with a public cancer genomics data set from *The Cancer Genome Atlas (TCGA)* project. TCGA was a large US-based initiative that followed the goal of collecting and analyzing biomolecular data from cancer patients for all major tumor types. Based on continuous feedback from cancer genomics researchers, we developed StratomeX [20], a visual analysis tool for comparing patient subsets in large-scale heterogeneous genomics data. We initially published the work as an application paper at EuroVis' 12. Later on, we extended the tool with guided exploration techniques that support users in picking potentially interesting data subsets during the exploration [26]. In contrast to our earlier work on StratomeX [20] that appeared at a visualization conference, the guided exploration technique was published in Nature Methods.

A core component of the guided exploration workflow was a visualization technique for ranking genes and other entities based on statistical scores and meta-attributes. As ranking problems appear in many different contexts, we generalized the solution and developed the LineUp visualization technique [11]. LineUp allows users to flexibly create and explore multi-attribute rankings. The technique was published at IEEE InfoVis' 13 where it won the Best Paper Award. LineUp was later on also integrated as a component in the Microsoft PowerBI software. LineUp is available as an open-source JavaScript library (https://lineup.js.org) that can be flexibly used as a component in various environments, such as Jupyter Notebooks and R Notebooks. Making the library publicly available not only increases the reproducibility of the visualization research but also increases the potential for adoption of the technique.

Publishing our visual analysis tool for genomics data in Nature Methods helped us to gain interest from pharmaceutical companies, which finally led to a three-year research collaboration with Boehringer Ingelheim. As part of this collaboration, we created the Ordino drug target discovery tool [25] that at its heart also integrates the interactive ranking technique LineUp. To increase the impact in the fast progressing life science community, we uploaded the paper to https://biorxiv.org at the time of the initial paper submission and made the source code available on GitHub.

Making a research prototype ready for productive use goes far beyond what a research collaboration is able to cover. To be able to deploy, maintain, and extend the platform, we founded a spin-off company that goes the extra mile required in terms of software development. Only by being able to demonstrate that we can transform innovative visualization solutions developed as part of a research collaboration into a stable and feature-rich software, we were able to acquire additional funding for the next phase of the collaboration.

As another positive side effect of having a collaborator that actively uses our tools, we had access to a growing provenance graph containing automatically recorded visualizations and user interactions from the visual exploration sessions. To make the provenance information accessible to the users, we developed the KnowledgePearls search and retrieval solution for querying and exploring similar analysis states, which we again published in the visualization community [24].

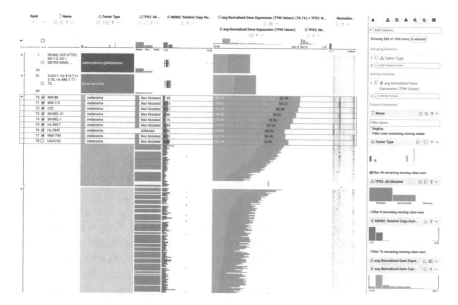

Fig. 13.6 Ordino visual cancer analysis tool for ranking and exploring genes and other entities based on statistical scores and additional meta-attributes

13.4.3 Lessons Learned

One of the most important lessons we have learned is that it is essential to have collaborators who acknowledge visualization as a scientific field that goes beyond creating pretty pictures. Try to find out at the very beginning if your collaborators are convinced that visualization can contribute to solving their domain problems. If this is not the case, convincing them in the course of the project is extremely difficult and frustrating.

Make it clear at the beginning of the collaboration what can be expected as output and—even more importantly—what is out of scope. The typical outputs of a research collaboration are visualization prototypes and publications targeted at the visualization community. Promising more than that will likely fall on your head later on.

The most critical and valuable resource domain experts can contribute is their own time. The more added value they see in the visual analysis solutions, the more time they will contribute. The more time they contribute, the more valuable the outcome will be for their own work as well.

Understanding the domain problem is key to being able to contribute. However, learning about a target domain and understanding the domain-specific language are time-consuming and can take months or even years. One success strategy is to stick with one or few problem domains, if possible. The longer the collaboration lasts, the more productive it becomes.

Having a liaison person on board is highly beneficial. In our project, the actual users of our tools are biologists and other life science experts, while our direct collaborators are bioinformaticians. The value and role of a liaison person are discussed in Chap. 14.

13.5 Case Study: In Situ Simulation Visualization of Parameter Spaces

Thomas Wischgoll

This chapter describes an ongoing collaboration between the Advanced Visual Data Analysis group at Wright State University and researchers at Wright Patterson Air Force Base. This research collaboration resulted in various different research projects, including the simulation and visualization of a dragonfly during takeoff [16]. The following sections will describe a visualization solution that addresses the need for being able to visualize parameter spaces for models from the cognitive science realm.

The ability to do rapid visual assessments of parameter spaces has the potential to change the workflow for both model simulation and model fitting/parameter recovery. It enables the rapid identification of input parameters that result in similar output data or model behaviors. This allows researchers to eliminate redundant input parameters for more efficient use of modeling and simulation computational resources. For example, two parameters should exhibit a strong correlation, one might be held constant while the other varied in order to capture all the unique model behaviors. Further, early visual assessment of the parameter space means that ineffective or incorrect models may be rapidly identified and eliminated from the study. This again results in the effective use of both experimenter and computational time. Finally, parameter space visualizations can reveal unexpected relationships between the parameters and model behavior. If the behavior is incorrect, errors in model design or in model may be more easily found. If the behavior is novel, parameter space visualization will have resulted in new hypotheses or expanded research findings.

This approach [10] is a Web-based solution that is capable of handling larger data sets compared to other commonly available solutions. At the same time, the described solution is directly integrated into the server structure that is used to run the simulations for the models of the cognitive science researchers. As such, it is readily available within the interface for starting and controlling. Hence, the researchers can immediately run the visualization on the simulation data that was calculated so far and make any adjustments to the simulation as necessary.

13.5.1   Domain Problem

Web-based visualizations are of interest in this application area as they can be directly integrated into the high-performance computing (HPC) environment. At the same time, this approach eliminates the need to install additional software on the researchers' computers beyond a browser as security limitations may not allow installation of any software of any kind. The potential for interacting with the data and feeding any resulting visually identified parameter constraints directly into the modeling and simulation process would further improve the modeling workflow.

The data sets typically are larger than many common JavaScript-based tools, such as D3 [5] or Plotly can handle. Such common tool kits are not capable of handling data sets that contain more than half a million data points. At the same time, downloading data sets of that size takes considerable time as well. It is therefore more desirable to generate the visualization on the server directly where the data are computed and stored. Then, only the visual results need to be transferred which is typically a lot less data compared to the entire data set.

13.5.2   Process and Output

This project evolved out of a close relationship between researchers at Wright State University and the 7/11 Human Performance Wing at Wright Patterson Air Force base. We were fortunate enough to have representatives of our collaborators be present at regular research meetings to discuss specific approaches for visualizing high-dimensional parameter space data. Due to the restrictive environment disallowing our collaborators to install software on their own, it was quickly identified that a Web-based solution for our collaborator's visualization needs were a Web-based approach. However, conventional tools or services, such as D3 or Plotly, quickly failed to handle the size of the data sets. With more than half a million data points, the browser typically ran out of memory so that visualization could not be achieved with those tools. In addition, having to download each data set would take too long to be acceptable. Instead, a server-side visualization approach was chosen. This avoids the need for downloading the data and at the same time allows the researchers to visualize their data while it is still being generated for in situ visualization. This server-side approach still utilizes the D3 library. However, it uses node.js to render the results into an image, which is then transferred to the client. Any interactive features, such as axis, are still drawn on the client side to preserve the full interactivity of the visualization approach. The server renders the visualization results in parallel on as many nodes as are available or allocated combined with additional performance enhancements resulting in faster rendering times. In addition, the parallel approach allows us to handle significantly larger data sets compared to the original implementation. Overall, this approach enables our collaborators to visualize their data sets quickly. It is integrated within their Web-based scheduling mechanism and hence readily available to them as they track their simulation progress. The in situ capabilities allow them to adjust parameters on the fly based on our visualization.

13.5.3   Lessons Learned

There are several lessons learned from this project that can be useful in general. The fact that our collaborators actively participated in regular research meetings was a great benefit to the outcomes of the project. It helped make the generated tools better suited for their needs. Due to the fact that the visualization tools were directly integrated into their workflow made those tools directly accessible to the researchers using that high-performance computing platform for their simulations making it as easy as clicking a button on their Web-based scheduling interface. Unfortunately, the project ended as the lead team of researchers at Wright Patterson Air Force Base moved to to a different national laboratory. However, the visualization tools are still accessible for the computing platform.

The fact that this project uses a Web-based platform also has the additional advantage of keeping track of utilization within that server environment. Hence, this provides yet another way of evaluating the visualization tool based on our collaborators voting with their mouse by electing whether to use out tool or not.

Overall, this project resulted in successful implementation of a visualization approach that enabled our collaborators to directly visualize their results. It was very well received among the users of that high-performance computing platform by allowing them to immediately investigate their simulations while they were being computed.

13.6   Case Study: Protein Analysis and Visualization (CAVER)

Barbora Kozlikova

Understanding the structure and behavior of protein molecules is crucial in many biological and biochemical fields, such as drug design and protein engineering. This process requires studying the proteins from many aspects, including their constitution, physicochemical properties, temporal behavior, or interactions with other molecules. These properties and their combination are very hard to perceive and understand using the traditionally used visual representations of molecules and animations of their behavior over time. Therefore, the biochemists require specifically designed visualizations which help them to explore and understand the proteins in more convenient and faster way. This creates very tight connection between the biochemical and visualization fields.

The case described in this chapter captures the interesting aspects of our long-term collaboration with protein engineers from Loschmidt Laboratories at the Masaryk University in Brno, focusing, namely on the exploration of the void space inside proteins and its connection with the protein surface. Such paths, connecting the inner voids with the surface, are denoted as tunnels in literature. There are already several existing algorithms and tools available for tunnel calculation. One of the first tools for tunnel detection was the CAVER tool, whose first version was developed in 2007 [22]. In the same year, the authors of this tool contacted us with the request to improve their original algorithm and to enable them to get insight into the detected tunnels. At that point, our collaboration was established, which lasts until now. Of course, over the years, the research tasks of the biochemists have changed. At the very beginning, they focused on the detection of tunnels in static molecules, which was further extended to molecular dynamics simulations [8]. The possibility to simulate longer and longer trajectories of protein movements resulted in the situation when the biochemists are not capable of observing such simulations frame-by-frame. Therefore, new visual abstractions, enabling the domain experts to drive their focus only on the interesting parts of the simulation, became a necessity [6, 7]. Moreover, recent advances in computational capabilities enabled the biochemists to generate large ensembles of molecular dynamics simulations. This leads to new challenges for visual guidance and comparative visualization, which is our current topic of common interest.

13.6.1   Domain Problem

As already stated, the main focus of our collaborators from the protein engineering group is the detection and analysis of tunnels in proteins. The presence of these void paths significantly influences the reactivity of proteins with small ligands entering the protein inner space and performing a chemical reaction in the protein active site. This specific site is capable of reacting with the ligand and the product of such a reaction can be, for example, a basis of a new drug. On the other hand, the goal of protein engineers is to change the properties of the protein by mutating selected amino acids, i.e., by replacing one amino acid by another. The protein engineers proved that the mutations of amino acids in the close vicinity of tunnels have a large impact on protein properties [17], such as its stability in normal temperature or activity toward ligands and other molecules [21].

With the increasing possibilities to capture large molecular dynamics simulations, currently spanning to hundreds of thousands of timesteps, the domain experts urgently needed help with the exploration of behavior of tunnels in them. They were interested, namely in the development of the shape and properties of the tunnel narrowest site, denoted as the tunnel bottleneck. However, their interest was driven toward the overall behavior of the whole tunnel as well. Here, they were interested, namely in the changes of tunnel shape and its constitution, i.e., the movements of amino acids forming the tunnel boundary. For a better understanding of the conformational changes of these amino acids, they also had to understand their physicochemical properties, such as their hydrophobicity or charges of atoms.

13.6.2   Process and Output

The starting phases of our collaboration were mostly about finding the common language with the protein engineers and understand their needs. In the first stage, we were focusing on the improvement of the original grid-based CAVER algorithm for the detection of tunnels, and we came with the approach utilizing Voronoi diagrams. The next step was to visualize the resulting tunnels and their surrounding amino acids so the biochemists could get a proper insight. The first straightforward solution was to create a plugin for CAVER to the commonly used PyMOL tool for molecular visualization [23]. This enabled us to get our algorithm to the domain experts worldwide. However, as PyMOL was not designed specifically for the visualization of tunnels, it could show the resulting tunnel only in a very basic way which was not sufficient for proper exploration. Therefore, with the protein engineers from the Loschmidt Laboratories, we decided to design and create a new tool for visualization and visual exploration of protein tunnels. This stand-alone tool, called CAVER Analyst, intensified our collaboration even more, as we had to closely discuss not only the functions of the tool, but also the user interface, layout, and interaction. This enabled us to get more insight into the daily workflow of the protein engineers, and on the other hand, the protein engineers also had to look at their research problems from a different viewpoint. The development of the first published version of CAVER Analyst took several years. There were several reasons for that. First, we did not have enough experience with designing such a robust tool which led to several bad design choices which took us significant time to fix. Second, the fluctuation of students at the university made the development complicated, as approximately every three years

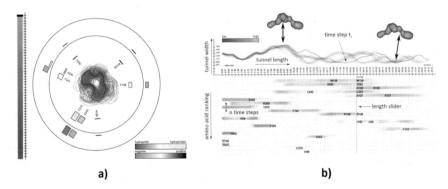

a)   b)

Fig. 13.7   **a** Visualization of the shape of tunnel bottleneck and its changes over time. **b** Visualization of tunnel width along the centerline over time and the surrounding amino acids, with their amount of contribution to the tunnel

the development team had changed. And getting new students to the interdisciplinary topic and knowing the tool always took several months. Before releasing the Beta version, intensive testing by the biochemists had to be performed. Therefore, CAVER Analyst 1.0 was released in 2014, after almost 8 years of its development [18]. The release helped not only with getting the tool designing specifically for tunnel exploration to the community, but created a stable platform for further prototyping of new visualizations. Since 2014, we were working, namely on designing specific visualization methods for visual exploration of large molecular dynamics simulations, which are not anymore observable by traditional animation. We designed very abstracted representations of changes of the shape and surrounding amino acids of the tunnel bottleneck [6] (see Fig. 13.7a) and a method for exploration of tunnel changes along its centerline [7] (see Fig. 13.7b). Both representations helped the biochemists to design proper mutations of amino acids surrounding a given tunnel. The application of these mutations had a significant desired impact on the functions and properties of the corresponding proteins.

In 2018, we released the 2.0 version of CAVER Analyst [15], which contains these techniques for visual exploration of single trajectories of molecular dynamics simulations. Currently, we continue in the successful collaboration in the same manner as it proved to be worthwhile for both partner sides. We collaborate on new techniques for visual exploration of ensembles of trajectories and their comparison, and again, we are using CAVER Analyst as the prototyping environment.

13.6.3 Lessons Learned

This project taught us several important things about successful collaboration which are worth to share.

First, building a trust between two research groups, having the research interest in completely different fields, is a long-term run. Except for speaking the same language, we had to clarify our research goals and expectations from each other. The protein engineers had to understand that our basic research is based on designing new visualization methods and publishing them on visualization venues, and we had to keep in mind the real usability of the designed methods and their benefit in the biochemical research.

Second, designing a tool for prototyping and for public release is completely different. However, creating the tool with keeping in mind the actual users makes also the future prototyping much easier for the developers. In our experience, the prototyping tool is more sustainable if it is paid attention to its usability as well.

The biochemists can also participate on the visualization publications which makes the collaboration even stronger. We decided to include the protein engineers to our publications by helping us with designing, performing, and describing the case studies, demonstrating the usefulness of our newly developed methods.

To conclude, this still ongoing project is already for several years resulting in interesting visualization methods which gave our collaborators the necessary insight into protein tunnels and their behaviors and properties. Moreover, the CAVER tools are well accepted by the community, which was one of the initial goals of our collaboration.

13.7 Small-Scale Visualization Projects

Johannes Waschke and Mario Hlawitschka

Visualization as a field of research aims to support the evaluation and presentation of data delivered from domain experts. Some domains work with highly complex data, which in their native form might be too challenging for the human understanding. This is for example the case with multi-dimensional data like diffusion MRI data, which heavily rely on visualization and therefore attract much research interest. In the shadow of these projects, however, many smaller challenges are waiting for the visualization community. These challenges span time ranges of several working weeks or months, up to a single working year. In contrast to larger projects, which heavily depend on data processing and visualization as a means for further understanding, the purpose of short-term projects might remain in the production of beautiful and informative figures. The focus lies on a better presentation of domain experts' result. Basically, scientific visualization in that case extends the work of a graphic designer, but with difficulty increased for two reasons: First, visualization in scientific context requires for scientific validity. Second, complexity of the utilized data types is usually higher and therefore demands for enhanced knowledge in data processing.

13.7.1 Domain Problem

This chapter explains the *general* concepts of small-scale visualization projects rather than focusing on a single experience. However, we worked on a number of problems arising around trajectory data, which should serve as an example here. The goal for trajectory data is to present a number (up to tens of thousands) of motion paths or connections, which often overlap, twist, or occlude each other. The presentation of such data should emphasize certain data characteristics. Interesting characteristics could be similarities between subsets of the data, as well as additional properties like speed or direction of trajectories. Evaluation of this data cannot be solved with standard image processing software, and thus, we see two options how to proceed.

One option for the experts is to do it by themselves. The methods of evaluation and presentation naturally depend on the interests and abilities of the respective researcher, and they are biased by both official and "unspoken" rules of the concrete research domain. An example is to track the motion of a surgeon's instrument with the purpose to classify the performance of the surgeon. Given the trajectory data, physicians tend to prefer an evaluation performed by a number of experts which should describe or rate qualities of the motion. Examples for such evaluations are verbal phrases like "very direct" or "many attempts" rather than using standardized tests. Other research fields prefer quantified results, which include parameters like speed or distance. Hence, the way of evaluation depends on individual knowledge and the domain's standards. All of these results might be good enough to answer the domain's research question, but they lack visual power for a presentation, and they only contain a limited degree of information.

The second option is a collaboration between domain experts and visualization researchers. Of course, scientific visualization also cannot provide all possible information in one single image, but it can extend the perspective and try to maximize the information level. Furthermore, it should be clear that visualization must consider the abovementioned qualitative and quantitative evaluation steps and therefore should be seen as an extension of traditional evaluation steps. As often stated, the open dialog between both sides, and the interest to understand each other's needs, is crucial.

Besides the knowledge of visualization techniques, as well as experience in possibilities and limitations of visual data presentation, technical factors play a major role. Domains with little relation to computer science are accustomed to work with standard software like Excel and PowerPoint. However, the abilities of these frameworks are limited, and their visualization results often are some kind of generic graph. As soon as we leave common data types, which can quickly be the case for individual experimental setups, data processing is a challenge for domain experts. And how can you visualize data that is even hard for you to simply open, read, and store? For many data types, specialized software frameworks exist. Application of them can be, unfortunately, a complicated endeavor, since they might be hard to find, hard to install (or have to be self-compiled), and hard to use.

13.7.2 Process and Output

The working steps, as we have experienced, are relatively straightforward and similar compared to descriptions from the previous sections. It is unquestionable to have a solid relationship between visualizers and domain experts that builds on mutual interest to understand and help each other. On the one hand, domain experts must be open for new ideas, and they must sacrifice time to formalize the problem and give feedback. On the other hand, the visualizers should carefully avoid ludicrous visual experiments that distract from the actual work of the domain expert. These small-scale projects are meant to benefit the presentation of another domain's research and not to push visualization research to new limits.

In the beginning, the problem should be stated, and the needs of the domain experts should be clarified. It is apparently helpful to consider their ideas and previous work, but we also recommend to keep some distance—to avoid a biased perspective on the problem solution. As an example for trajectory data, we want to bring up a question that concerns the level of abstraction of the visualization. For various domains, it is interesting to (only) have an abstract view on the data, which for example summarizes trajectories to clusters, or which simply presents features derived from the original data. However, in the medical field, absolute positions often have high meaning and abstractions are less popular. Physicians prefer to see unaltered positions in anatomical context and points like these have to be learned during the meetings.

In our experience, further development was performed in numerous iteration steps. After a couple of days or weeks, we presented a prototype and proposed some options how to continue. These prototypes gave the domain experts a quick preview on the realistic outcome. Demonstrations of the prototypes regularly gave the experts new ideas that could be considered for further development of the visualization. This is repeated as long as the resources allow it. Finally, a common result is a set of figures that is planned to be used in the domain expert's publication.

13.7.3 Lessons Learned

First of all, there are technical questions concerning the implementation of the visualization algorithm. While we generally aim for the goal to fabricate reusable visualization tools (which includes a user-friendly interface, documented API, tutorials, and so on), the reality struggles with time pressure and short working periods. In our opinion, a lot of time can be wasted on making the software too "consumer-friendly". Since this is against common software development rules, we want to provide further explanations on that. Small projects often involve very specific data types, and the visualization is highly individual as well. The chances of the visualization algorithm to be ever used again—in that concrete implementation—might be very low. Thus, it simply saves time to handle the technical steps by yourself and provide only the result images to the domain experts. Additional features, like a graphical user interface or a beginner's guide to the software, could cost some months of work but are probably never used—and thus a waste of time.

Some visualization problems are too challenging for standard software (and common knowledge), but not interesting enough to be an active field of visualization research. Nevertheless, solving these problems and proposing creative visualizations help the domain experts to compose better papers with at least improved conveyance of their research results. However, there are negative aspects for the scientific visualizer. Since the quantity of the newly created knowledge is usually small—we are considering projects of several weeks or months—publication as a full paper might be inappropriate. The chances to be successful within the visualization community are higher for long-term projects (with a higher degree of new results) or at least for multiple accumulative projects with similar challenges. The scientific visualizer is here, up to a certain degree, a service provider for the domain experts.

This raises two questions. First, who is responsible for the funding, since low-grade publications do not help to raise money? One answer is the institutes that employ researchers of multiple disciplines. This provides the additional benefit of closely situated working places and therefore good conditions for interdisciplinary collaborations. Outside of these institutes, small-scale visualization challenges might only be worth to keep as a side project.

The second question is about publications and the scientific reward for the work. Since it seems harder to publish in a high-impact visualization medium, the point of co-authorship in a domain-specific journal should be raised early in the meetings with the domain experts. The visualization researcher could be co-author in a high-level journal of the experts' domain and—if the methods are not covered in the paper yet—on a small conference or at a poster session. Another promising function of small-scale projects is to use them for teaching and for thesis projects of students. They form a practical problem and thus motivate the students to grow into the research field of visualization—small-scale projects can be a nice starter for a research career.

Acknowledgements This chapter was supported by NIH grants: NIH P41 GM103545-18 and 1R01EB023947-01.

References

1. Al-Awami, A., Beyer, J., Strobelt, H., Kasthuri, N., Lichtman, J., Pfister, H., Hadwiger, M.: NeuroLines: a subway map metaphor for visualizing nanoscale neuronal connectivity. IEEE Trans. Vis. Comput. Graph. (Proc. IEEE InfoVis '14) **20**(12), 2369–2378 (2014)
2. Al-Awami, A.K., Beyer, J., Haehn, D., Kasthuri, N., Lichtman, J.W., Pfister, H., Hadwiger, M.: Neuroblocks - visual tracking of segmentation and proofreading for large connectomics projects. IEEE Trans. Vis. Comput. Graph. **22**(1), 738–746 (2016). https://doi.org/10.1109/TVCG.2015.2467441
3. Beyer, J., Al-Awami, A., Kasthuri, N., Lichtman, J.W., Pfister, H., Hadwiger, M.: Connectome-Explorer: query-guided visual analysis of large volumetric neuroscience data. IEEE Trans. Vis. Comput. Graph. (Proc. IEEE SciVis '13) **19**(12), 2868–2877 (2013)
4. Beyer, J., Hadwiger, M., Al-Awami, A., Jeong, W.K., Kasthuri, N., Lichtman, J., Pfister, H.: Exploring the connectome - Petascale volume visualization of microscopy data streams. IEEE Comput. Graph. Appl. **33**(4), 50–61 (2013)
5. Bostock, M., Ogievetsky, V., Heer, J.: D3: data-driven documents. IEEE Trans. Vis. Comput. Graph. **17**(12), 2301–2309 (2011). https://doi.org/10.1109/TVCG.2011.185
6. Byska, J., Jurcik, A., Groeller, M.E., Viola, I., Kozlikova, B.: Molecollar and tunnel heat map visualizations for conveying spatio-temporo-chemical properties across and along protein voids. Comput. Graph. Forum **34**(3), 1–10. https://onlinelibrary.wiley.com/doi/abs/10.1111/cgf.12612 (2015). https://doi.org/10.1111/cgf.12612
7. Byska, J., Muzic, M.L., Groeller, M.E., Viola, I., Kozlikova, B.: AnimoAminoMiner: exploration of protein tunnels and their properties in molecular dynamics. IEEE Trans. Vis. Comput. Graph. **22**(1), 747–756 (2016). https://doi.org/10.1109/TVCG.2015.2467434
8. Chovancova, E., Pavelka, A., Benes, P., Strnad, O., Brezovsky, J., Kozlikova, B., Gora, A., Sustr, V., Klvana, M., Medek, P., Biedermannova, L., Sochor, J., Damborsky, J.: CAVER 3.0: a tool for the analysis of transport pathways in dynamic protein structures. PLoS Comput. Biol. **8**(10), e1002,708 (2012)
9. Collins, D.F.: Snapshots of life: muscling in on development. https://directorsblog.nih.gov/2017/07/27/snapshots-of-life-muscling-in-on-development/ (2017).
10. Glendenning, K., Wischgoll, T., Harris, J., Vickery, R., Blaha, L.: Parameter space visualization for large-scale datasets using parallel coordinate plots. J. Imaging Sci. Technol. **60**(1), 10,406–1–10,406–8 (2016)

11. Gratzl, S., Lex, A., Gehlenborg, N., Pfister, H., Streit, M.: LineUp: visual analysis of multi-attribute rankings. IEEE Trans. Vis. Comput. Graph. (InfoVis '13) **19**(12), 2277–2286 (2013). Doi:https://doi.org/10.1109/TVCG.2013.173

12. Hadwiger, M., Beyer, J., Jeong, W.K., Pfister, H.: Interactive volume exploration of petascale microscopy data streams using a visualization-driven virtual memory approach. IEEE Trans. Vis. Comput. Graph. (Proc. IEEE SciVis'12) **18**(12), 2285–2294 (2012)

13. Haehn, D., Knowles-Barley, S., Roberts, M., Beyer, J., Kasthuri, N., Lichtman, J., Pfister, H.: Design and evaluation of interactive proofreading tools for connectomics. IEEE Trans. Vis. Comput. Graph. (Proc. IEEE SciVis 2014) **20**(12), 2466–2475 (2014)

14. Hotz, I., Feng, L., Hagen, H., Hamann, B., Joy, K., Jeremic, B.: Physically based methods for tensor field visualization. In: Proceedings of the conference on Visualization'04, pp. 123–130. IEEE Computer Society (2004)

15. Jurcik, A., Bednar, D., Byska, J., Marques, S.M., Furmanova, K., Daniel, L., Kokkonen, P., Brezovsky, J., Strnad, O., Stourac, J., Pavelka, A., Manak, M., Damborsky, J., Kozlikova, B.: CAVER Analyst 2.0: analysis and visualization of channels and tunnels in protein structures and molecular dynamics trajectories. Bioinformatics (2018)

16. Koehler, C., Wischgoll, T., Dong, H., Gaston, Z.: Vortex visualization in ultra low reynolds number insect flight. IEEE Trans. Vis. Comput. Graph. **17**(3), 2071–2079 (2011)

17. Koudelakova, T., Chaloupkova, R., Brezovsky, J., Prokop, Z., Sebestova, E., Hesseler, M., Khabiri, M., Plevaka, M., Kulik, D., Kuta-Smatanova, I., Rezacova, P., Ettrich, R., Bornscheuer, U.T., Damborsky, J.: Engineering enzyme stability and resistance to an organic cosolvent by modification of residues in the access tunnel. Angew. Chem. Int. Ed. **52**(7), 1959–1963. https://onlinelibrary.wiley.com/doi/abs/10.1002/anie.201206708 (2013). https://doi.org/10.1002/anie.201206708

18. Kozlikova, B., Sebestova, E., Sustr, V., Brezovsky, J., Strnad, O., Daniel, L., Bednar, D., Pavelka, A., Manak, M., Bezdeka, M., Benes, P., Kotry, M., Gora, A., Damborsky, J., Sochor, J.: CAVER Analyst 1.0: graphic tool for interactive visualization and analysis of tunnels and channels in protein structures. Bioinformatics **30**(18), 2684–2685 (2014)

19. Kratz, A., Schoeneich, M., Zobel, V., Burgeth, B., Scheuermann, G., Hotz, I., Stommel, M.: Tensor visualization driven mechanical component design. In: Visualization Symposium (PacificVis), 2014 IEEE Pacific, pp. 145–152. IEEE (2014)

20. Lex, A., Streit, M., Schulz, H.J., Partl, C., Schmalstieg, D., Park, P.J., Gehlenborg, N.: StratomeX: Visual Analysis of Large-Scale Heterogeneous Genomics Data for Cancer Subtype Characterization. Computer Graphics Forum (EuroVis '12) **31**(3), 1175–1184 (2012). https://doi.org/10.1111/j.1467-8659.2012.03110.x

21. Pavlova, M., Klvana, M., Prokop, Z., Chaloupkova, R., Banas, P., Otyepka, M., Wade, R.C., Tsuda, M., Nagata, Y., Damborsky, J.: Redesigning dehalogenase access tunnels as a strategy for degrading an anthropogenic substrate. Nat. Chem. Biol. **5**(10), 727–733 (2009)

22. Petrek, M., Otyepka, M., Banas, P., Kosinova, P., Koca, J., Damborsky, J.: CAVER: a new tool to explore routes from protein clefts, pockets and cavities. BMC Bioinform. **7**, 316 (2006)

23. Schrödinger, LLC: The PyMOL molecular graphics system, version 1.8 (2015)

24. Stitz, H., Gratzl, S., Piringer, H., Zichner, T., Streit, M.: KnowledgePearls: Provenance-Based Visualization Retrieval. IEEE Trans. Vis. Comput. Graph. (VAST '18), 11 (2018)

25. Streit, M., Gratzl, S., Stitz, H., Wernitznig, A., Zichner, T., Haslinger, C.: Ordino: a visual cancer analysis tool for ranking and exploring genes, cell lines and tissue samples. Bioinformatics (2019). https://doi.org/10.1093/bioinformatics/btz009

26. Streit, M., Lex, A., Gratzl, S., Partl, C., Schmalstieg, D., Pfister, H., Park, P.J., Gehlenborg, N.: Guided visual exploration of genomic stratifications in cancer. Nat. Methods **11**(9), 884–885 (2014). https://doi.org/10.1038/nmeth.3088

27. Wan, Y., Lewis, A., Colasanto, M., van Langeveld, M., Kardon, G., Hansen, C.: A practical workflow for making anatomical atlases in biological research. IEEE Comput. Graph. Appl. **32**(5), 70–80. http://www.sci.utah.edu/publications/wan12/Wan_CGA2012.pdf (2012). https://doi.org/10.1109/MCG.2012.64

Chapter 14
Collaborations Between Industry and University

Daniela Oelke and Ariane Sutor

Abstract This chapter describes experiences with collaborations between industry and data analysis experts (incl. visualization experts) at a university from an industrial perspective. The authors are visual analytics and data analysis experts in an industrial research department and are among other things coordinating collaborations between universities and the domain experts of the company. The text summarizes experiences made in many years of collaborations with different universities and institutions. We compare different collaboration models, share our lessons learned and work out success factors for collaborations between universities and industrial partners.

14.1 Why Do We Collaborate with Universities?

In a collaboration, both partners have expectations and goals they want to achieve. For the academic partner, this may be the desire for an interesting research problem, the opportunity to get access to data that otherwise would not be available, or simply the need to raise funding for the own institution.

For enterprises, university collaboration plays a key role to support their strive for sustainability and growth. The most important objective for a university collaboration is to foster innovation; more concretely goals include creating value within the company through alignment with research and to leverage expertise from outside. Besides innovation, another important goal is external branding by getting external recognition towards thought leadership as well as to engage with talents.

The different perspectives of the academic and the industrial partner also become apparent when looking at how success is measured. Our academic partners often tell us that the number and quality of the publications resulting from the collaboration are an important success factors for them. Desired outcomes from university collab-

D. Oelke (✉) · A. Sutor
SIEMENS AG, Munich, Germany
e-mail: daniela.oelke@siemens.com

© Springer Nature Switzerland AG 2020
M. Chen et al. (eds.), *Foundations of Data Visualization*,
https://doi.org/10.1007/978-3-030-34444-3_14

orations for enterprises are manifold. Most obvious is the growth of business or the creation of new business in application projects for the company. Also, the creation of intellectual property is an important asset. Finally, university collaborations are a prime source for establishing contact with new talents through student programs, Ph.D. programs, or hiring of new employees. A less obvious ambition for enterprises is to trigger new lines of research at universities in areas that are of interest.

Ideally, the developed concepts and tools are directly applicable to our own data or a proof-of-concept is made that a new idea can be employed in our application domains. Easy to use demonstrators that also permit to load new data in help us to reuse and communicate new ideas within our company. But also evaluations can be a valuable result of collaboration if they can guide future directions of research and development within the company.

We call projects that meet the needs and expectations of both partners the "sweet spot" research topics. Experience shows that these topics are only a subset of the topics which would be of interest for one of the two partners.

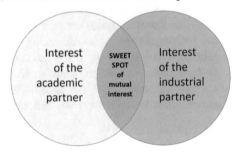

14.2   How Do We Organize Collaboration with Academic Partners? (Collaboration Management)

Being researchers ourselves but at the same time a part of the company, we are able to facilitate the communication between the domain experts and the researchers at the university. We consider it as our job to search actively for suitable tasks and data for such a collaboration and to help the domain experts to define the task in a way that a researcher can work on it. Understanding both worlds and also the technical field of the collaboration partners, we are able to act as liaison persons [1] during the whole process from working out the task description and the contract to the collaboration process and finally, the examination of the results. This has proven to be a successful collaboration management model in the past which we can also recommend to others.

How do we choose suitable academic partners? Of course, the reputation and proven track record of a research group in the concrete field of interest is a key driver for our decision. Furthermore, we aim at establishing a trusted partnership based on personal contact and longer term collaboration. On the other hand, the decision is also influenced by objectives which are set on company level such as focusing on a limited number of strategic partner universities, global presence and a distribution of partners according to the company's needs, or the overall available budget.

14.3 Different Types of Collaboration Models and Their Pros and Cons

Over the years with have tried different collaboration models. In the following, we briefly comment on their advantages and disadvantages from our own point of view.

	Short-term collaboration	Ph.D. student at university	Ph.D. student in company
What is it?	A Ph.D. student at a university is financed by the company for a smaller project, e.g., for 1 year	A Ph.D. student is financed for a period of 3 years by the company but is employed by the university	A Ph.D. student is financed and employed by the company for 3 years
Pros	• Possible to keep the topics close to current needs/interests of the company • Easier to align with 1-year budgets	• Ph.D. student gets input from both sides, the university and the company • Efficiency increases over time (because the partners know each other better and the Ph.D. student has learned about the domain)	• Can also work with confidential data • Learns a lot about the domain and the company and is, therefore, also more likely to create business impact
Cons	• Topic must fit to an existing Ph.D. thesis • Ph.D. student may not have enough time to get acquainted with the domain • Higher overhead costs (financial + time invest when working together)	• Difficult to finance if research budgets are assigned on a year-by-year basis • Difficult to define a "sweet spot" research topic (see Sect. 14.1) for 3 years in advance	• Less contact to academia/the university • Higher time invest for the company in supervision

14.4 Lessons Learned

Looking back at our past collaborations, we recognize that not all of them were equally successful. This raises the question of what fosters or hampers a successful collaboration. In the following, we share our lessons learned.

- **Aiming for "sweet spot" research topics** As explained above (see Sect. 14.1), the industrial and academic partners may have different motivations for working together. Finding tasks that satisfy the needs of both sides is not easy. It requires close collaboration of the two partners (already in the definition phase) and the willingness to accept that not everything that is interesting for one of the partners is a suitable topic for the collaboration. But the resulting win-win-situation is worth the effort. To ensure that collaborations become successful for both sides, it is vital that the academic and industrial partners meet on eye level. This is facilitated if at the side of the industrial partner also researchers are part of the project and an open exchange of ideas is practiced.
- **Longer collaborations are more efficient** This is true for a number of reasons. Most importantly, a longer term collaboration ensures a better mutual understanding and helps to establish "sweet spot" collaboration. Also, the overhead of getting a collaboration started is high. It, therefore, pays off to aim for longer collaborations (which can also be made up of a series of short-term projects though).
- **Mitigating barriers** Barriers may include legal efforts and IP negotiations at the beginning of a project. The concept of working together with trusted strategic partners in multiple long-term projects as well as assigning sufficient budget and management attention helps to mitigate these barriers.
- **Topics of short-term projects should fit a Ph.D. thesis (if performed by a Ph.D. student)** In a short-term project, it is beneficial if the topic of the collaboration project fits the topic of the Ph.D. thesis of the student working on it. This prevents that the student is torn between working for the project and doing work for his/her own Ph.D. thesis and ensures a high motivation of the student.
- **Ph.D. student needs a close by advisor** Especially, junior Ph.D. students need a good supervision if they are to work in challenging research projects. This supervision should be provided by the partner that employs the student (support from afar by the other partner does not work well in our experience).
- **Clearly specifying task and data** In the past, we tried both (a) agreeing on a general research direction but leaving the task and data to use open and (b) clearly specifying the task and the data that the solution should work for. In our experience, in the latter case, chances are much higher that the research will have a business impact at the end. Accepting this higher effort when (collaboratively) defining the project pays off at the end. Another stumbling block is the fact that some data cannot be given to academic partners because of confidentiality restrictions. Therefore, it is vital to ensure early on that there is appropriate data to work on.
- **Regular meetings of both partners** Both partners should meet regularly during the project time—either in person or via conference calls. This ensures that all relevant information is passed on and potential misunderstandings are dispelled quickly.
- **The value of a liaison person** Someone who has never done research by himself may not be aware of what it takes to do good research. Similarly, someone who has always been in academia does not fully understand the goals and constraints of the industrial partner, nor does she/he understand the specific application domain. It has proven useful to have a researcher working in the specific company as a

liaison person coordinating the project. This person is then able to facilitate the communication between both sides and can help to make the collaboration project a success for both partners. (See also [1])

Disclaimer: We are aware that there is a great diversity of needs and premises between different industrial partners when collaborating with universities. This work can only describe our own specific experiences and perspective. Nevertheless, we hope that it helps to raise a common understanding between academic and industrial partners and that our lessons learned will act as a catalyst for successfully implementing new collaboration projects.

Reference

1. Simon, S., Mittelstädt, S., Keim, D.A., Sedlmair, M.: Bridging the gap of domain and visualization experts with a liaison. In: Bertini, E., Kennedy, J., Puppo, E. (eds.) Eurographics Conference on Visualization (EuroVis) - Short Papers. The Eurographics Association (2015). https://doi.org/10.2312/eurovisshort.20151137

Chapter 15
Collaborating Successfully with Domain Experts

Mario Hlawitschka, Gerik Scheuermann, Christian Blecha, Marc Streit and Amitabh Varshney

Abstract The goal of visualization is to provide users with (human) insight into (digital) data, and more than just an action of drawing some pictures based on the data. As most visualization images are not interpreted by visualization experts, but by other users, such as domain experts, "users play a central role in visualization [15]." Practically no one in the visualization community would seriously question this fact. Many researchers in the community stress the relevance of users, including Lorenson [6] who warns of the possibility of the death of visualization without applications, and the members of a more recent IEEE VIS panel [16]. Only users can finally confirm the relevance of visualization because "the overall aim is to achieve the grand vision of enabling data understanding in science, engineering, and society [16]." While visualization has an ambitious goal for serving broader audiences (see Part IV of the book), successful collaboration with domain experts is essential to prevent the possibility warned by Lorenson [6]. In this chapter, we collect experiences and ideas that should help make such collaborations with domain experts a success. It is necessary to note that we take a broad definition of collaboration as the basis of our discussions, i.e., any close cooperation between visualization experts and domain experts. A finer discrimination of different kinds of cooperation can be found in the article by Kirby and Meyer [5]. We present our considerations in three aspects: domain, domain expert, and collaboration methodology. Finally, we discuss how to impact as the main measure of success can be made.

M. Hlawitschka
University of Applied Sciences, Leipzig, Germany

G. Scheuermann (✉) · C. Blecha
Leipzig University, Leipzig, Germany
e-mail: scheuermann@informatik.uni-leipzig.de

M. Streit
Johannes Kepler University, Linz, Austria

A. Varshney
University of Maryland, College Park, MD, USA

© Springer Nature Switzerland AG 2020
M. Chen et al. (eds.), *Foundations of Data Visualization*,
https://doi.org/10.1007/978-3-030-34444-3_15

15.1 Domains Matter

Visualization plays a prominent role in any data-intense science, engineering, social science, or humanities discipline today. Therefore, an individual visualization researcher may choose an application domain among several options. There is always a choice, and due to shortage of time, a single researcher cannot go for all options! Finding answers to the following questions might help in the selection process:

- Why now? What is the technology or methodology catalyst that allows to make a significant contribution in this domain now?
- Why here? What is the advantage of doing this at a specific university? Is there specialized equipment? Are there world-famous scientists?
- Why does a researcher pick this particular domain? Is there a personal interest that drives the potential collaboration? Is there some potential for impact in the visualization domain?

The visualization researcher needs to be aware of the fact that every new application domain comes with some costs. It takes time to learn enough about an application domain. You do not need to become an expert yourself, but you need to understand the language [13], the way of solving problems, and typical problems. Also, the visualization researcher needs to know his/her own field, i.e., visualization [13]. While a specific application domain may utilize some general techniques, existing software, etc., which may be used in many other domains, making the visualization research specially related to this particular domain can yield more meaningful impact. This requires a lot of time. Because a discipline may be at a point where most researchers do not know all aspects of visualization (at least equally well), many visualization researchers are able to deliver successful collaborative research by sticking to one or two domains! In some cases, this has created subfields in visualization, such as flow visualization, biological visualization, or medical visualization—and even more specialized areas like tractography.

15.1.1 The Knowledge of Domain Experts Makes the Difference

Besides the domain, the domain expert is central to the success of the collaboration from the visualization point of view as well. This is also pointed out by Sedlmair et al. [13] paper in their "winnow" phase. Nearly all of the case studies in this part of the book (e.g., Chap. 13) emphasize the expert's role. Without the domain expert(s), these success stories would not have happened. It is as crucial as the first part—the best domain is pointless if the domain expert and visualization researcher do not work as a good team. Clarifying the following questions is important:

- Does the expert know about the most important problems in the field? Does he/she show a taste in elegant problems with impact?
- Is there a common understanding of the goals of the collaboration?
- Are the visualization researcher and domain expert equally knowledgeable in their respective domains?
- Is there an appreciation of career needs of visualization researchers by the domain expert?
- Is there a balance in funding, i. e. are both sides independently funded, or do both go for joint funding, preferably as equal partners? Collaboration does not work well if one partner is funded and the other is not!

If this sounds complicated, keep in mind that a good domain expert needs someone like you. If you do not trust us, look at articles like the work by Jenkins [3] about the relation between computational biologists and classical biologists. Obviously, there may not be an ideal fit in some cases, so a visualization researcher needs to look around, talk briefly to several potential partners, and make an informed decision.

There are some situations that should be avoided. As Pretorius and Wijk [12], Sedlmaier et al. [13], and others have stressed: The expert needs to be able to provide real data. It needs to be in a known format, and it needs to be accessible right at the beginning by the visualization researcher. One author of this chapter has experienced more than once that a highly motivated, talented PhD student waits for more than half a year—or even longer—for data that cannot be accessed because of communication problems, format issues, or problems with non-disclosure agreements. Also, the domain expert needs to reserve sufficient time and interest to discuss the data, its background, the current workflow, and evaluate possible visualization designs. This concerns the funding question, the challenge of having a common interest, and speaking the same language. Authors of this chapter have experienced medical doctors that understand visualization as imaging procedures like MRT or CT instead of using graphics for data analysis. Also, if potential collaboration partners believe that the task of visualization is just about generating nice images for publications, but has no influence on any semantic understanding (which may only be derived by theoretical considerations and quantitative measurements), they are certainly not optimal partners for starting a collaboration. The visualization researcher also has to avoid to be just the software engineer. Doing some engineering as part of the project is fine, but both sides need to be clear about the interest gap [15]: "The visualization researcher aims at publishing in journals and at leading conferences in visualization, and therefore, he/she focuses on developing new and interesting methods and techniques—that is interesting in the eyes of his/her visualization colleagues." Therefore, "the focus is mostly on novelty and not primarily on usability, including down-to-earth issues such as the kind of data sources that can be handled, availability on various platforms, ease of installation, and ease of use. These issues, however, are crucial for the domain expert, who's primarily interested in tools that will help him to work faster and better."

Another problem that is closely related to the interest gap described above is the *"hit-and-run"* mentality that can often be observed in our community. In the *hit-*

phase, visualization researchers develop prototypes that are published in the visualization community as application or design study papers. After the paper is out, PhD students have to move on to the next paper that brings them closer to finishing their PhD. However, they usually do not care too much about the impact of their work in the target domain. This is the beginning of the *run*-phase. Collaborators will quickly realize that the prototypes are nice proof-of-concepts but are often not readily applicable to their real-world analysis problems and needs. Reasons for this are, for example, missing data importers, a lack of scalability of the prototype to larger datasets, or a lack of standard features that were not interesting enough from a research perspective to be implemented. As a consequence of this collaboration model, the funding will not be extended after the initial project period and both sides move on to new collaborations. This is unfortunate because the longer collaborations last, the more productive they usually become.

15.1.2 Methodology Helps

The third ingredient to success is the collaboration approach and methodology. There are few articles on this topic, including Sedlmaier et al. [13], Kirby and Meyer [5], and Wijk [15]. However, many authors derive from their experiences that no single approach works always, see, e.g., [16]. Nevertheless, the literature and our own experience show that successful collaboration is much more often the result of a good plan. As Pretorius and Wijk [12] noted, many visualization researchers without much experience in collaboration would probably start with a basic pipeline for the design of interactive systems like:

(1) Identify user requirements
(2) Develop alternative designs that meet requirements
(3) Implement the designs in interactive prototypes
(4) Evaluate the prototypes.

While the second and third steps are typical for many successful collaborations, the first and last steps require care and are far less straightforward.

Regarding evaluation, visualization defines its goal as providing insight to the domain expert(s). However, as Pretorius and Wijk [12] explain, evaluation is quite difficult based on this goal if one looks at North's characterization of insight [9]:

• Complex data creates understanding that influences further insight.
• Deep insight is sought raising new questions.
• Insight is qualitative and therefore difficult to quantify.
• Unexpected insight makes evaluation difficult, as great discoveries are rare and cannot be guaranteed.
• Insight is based on domain expert's knowledge.

While there are several ideas out there for evaluation, see the respective part of this book, for example, the first part of defining the requirements is even more tricky.

Wijk [15] says: "If the expert is interested in explorative visualization, then he/she is probably aiming at advancing the state of the art in his/her domain. This also means that he/she is not exactly sure and cannot express what he/she is looking for, except that he/she is aiming for new insights." Therefore, it is quite common that the requirements cannot be defined at the beginning, and that collaborators are not able to define what they want to see. It becomes part of the process and a task for the team of domain expert(s) and visualization researcher(s) to define the goals based on their respective knowledge. In the words of Pretorius and Wijk, the domain expert will— based on existing knowledge and new insights— define during the collaboration what he/she wants to see. The visualization researcher will try to define what the data want to be by leveraging his/her knowledge on visualization methods, and new ideas along the way.

Another aspect is quite often overlooked without experiences or good training: The visual analysis process will almost certainly take detours! If users are not able to list typical questions or recurring tasks, and the burden of identifying opportunities is shared between domain experts and visualization researchers, this means that assumptions change during analysis. Consequently, the team needs to be prepared to change them! This may even mean that the whole process becomes a loop rather than a pipeline.

For the second step of developing alternative designs, it is critical to know the visualization literature. Together with phase three, it can be ideal if you have many techniques already available, so different designs can be tested nearly immediately. This was an essential factor in the mechanical engineering case, as described in Sect. 13.3. The group of Ingrid Hotz had nearly all tensor visualization methods implemented in their tool, so we could test them with the engineers within a few weeks. However, this strategy does not work in cases in which the application problem (even application domain) is very specialized and no best practices yet exist. The domain expert may deal with extremely complex phenomena, and only a handful of potential experts will ever look at this type of data. Therefore, only a few or even practically no applicable visualization techniques may exist. The point here is that the visualization researcher knows this, and then he/she may be happy already: If the task cannot be automated and there is real data, it is almost certain that there is a new visualization question involved.

Regarding the implementation phase, it is good to keep in mind that one creates several prototypes in the course of the collaboration. A good advice here is "fail fast." This means fast implementations and early evaluation. It is undesirable if an implementation takes 10 months because your student is very slow, and then your collaborator leaves to another country and a different research direction, so the task is gone! One author experienced this during a collaboration with veterinary medicine experts and got basically nothing out of the project.

However, there is quite a list of very helpful questions that should be asked by the visualization researcher early in the process:

- Is there an application driving the collaboration, but a win-win for the domain and the field of visualization? Visualization wants to be helpful, but it needs to avoid providing technical support only!
- Is the team working on real problems with high potential impact?
- Are both partners participating face to face as equals? Here, the senior visualization researcher must not delegate the cooperation to students or committees!
- Is there an interesting visualization research question?

Regarding the last question, it is typical that it cannot be answered directly. One author experienced this with a recent work on splat detection in fluid flows. The original question was to help understanding heat transport between a wall and a fluid. It becomes clear in the process that due to the high heat diffusion in the liquid metal, the processes directly at the wall are central. Looking at them, the mechanical engineers and the visualization team found out that a so-far overlooked feature called splat is central for the transport; see [10, 11]. A great success story was recently published by Liz Marai's group [7]. In their work on turbulent fluid flows, they realized that the famous mantra from Shneiderman [14] of *overview first, zoom and filter, then details-on-demand* does not fit the needs of their domain experts. They wanted the details first, because they know the overview, i.e., the geometry and the overall flow quite well!

In general, it should be noted that the above remarks on methodology concern mainly collaborations between a small number of research laboratories, typically at universities or research institutes. If the collaboration extends and gets more ambitious to develop software systems that are used by many people for a longer time, usability, sustainability, platforms, funding over longer times come into play. Weber et al. [16] call for such work and indicate the potential and challenges. If the goal is to finally deploy a solution at commercial companies, a recent article by Kasik and Dill [4] is very helpful.

15.2   Creating Impact

While there is a strong personal side to success, there is also the notion of impact on a scientific discipline. One can still see subjective aspects here, but the overall idea is to look at the change in thinking, methodology, or established knowledge in a discipline that is well received by most members of the respective scientific community. This is far less subjective and can often be agreed on, at least by looking back a few years later.

Novel research should find its place in science and important research should ideally have an impact not only in its own field but may reach out to other disciplines. When visualization researchers collaborate with domain experts, typically at least two disciplines are involved and high-quality research may have an immediate impact in both disciplines. In many cases, however, the work is either only visible to researchers from the visualization community or the application community—

depending on where it is published. There are many great application and design study papers published at top visualization venues that have never or rarely been picked up by practitioners from the application domain. So the question is: How can we as visualization community improve this situation?

We postulate that good interdisciplinary visualization research should make an impact in both, the visualization community and the application community. Of course, the impact when working with domain experts directly instead of doing basic visualization research is more direct and immediate. It should be recognized that typically the aspects of the research that creates the impact in these communities may be different depending on the target audience. However, this fact is often not acknowledged by reviewers who see dual publication strategy critically in terms of scientific novelty. In the following two sections, we discuss strategies to increase the impact of interdisciplinary visualization research in both the visualization community and the application domain.

15.2.1 *Impact in Visualization*

A common problem of application and design study papers is that they often focus too much on the domain problem and on presenting a tailored solution that addresses the given domain-specific tasks. While the work can be highly useful for solving the application problem, it is often difficult for the visualization community to learn something from the work. In turn, this leads to the situation that many application and design study papers have no long-term impact in the field of visualization.

Here are a few suggestions to improve this situation:

- Work hard to understand and characterize the domain problem. Explicitly discuss the target users, their tasks, and the data to be analyzed in a structured way. The nested model [8], for instance, provides such a structure.
- Generalize and abstract the problem, data, and tasks by stripping away the domain-specific language. A carefully done abstraction process adds value for the visualization community, because it makes the problem but also the solution applicable to other domain problems. A detailed discussion of abstraction and abstract thinking in the context of visualization is given in Chap. 2.
- The complexity of the target domain can make it difficult for visualization researchers to digest the results and see the potential value in other contexts. Besides abstraction, a successful strategy to alleviate this problem is to pick a toy dataset that is easier to understand for researchers from the visualization community for demonstrating a novel approach. The LineUp ranking technique [2], for instance, was originally developed for ranking of genes in the context of cancer genomics but has been applied to the more general use case of university rankings. This made it significantly easier for visualization researchers to understand the technique and its potential applications.

- If possible, try to apply the solution to problems from two or even more domains. A user interface to interactively query and explore any type of volumetric data is more valuable than a system that just works on neuroscientific data.
- Make your solution available as open source so that other researchers can apply it to different domains and problems. You never know in which context your work might turn out to be useful in the future.

15.2.2 Impact in Application Domains

Application and design study papers often do not make it past the initial software prototype. Long-term adoption is difficult and takes additional effort. A BioVis Dagstuhl seminar group referred to this as the "valley of death (see [1, p. 54])."

In the following, we provide a few suggestions on how to increase the potential impact in domains beyond the field of visualization:

- Make your prototypes available as open source and promote usage of your system. Maintaining the code that is the outcome of academic projects is challenging. However, even unmaintained code can be valuable because someone may pick it up a long time after the end of the project and use it for a different purpose, e.g., for comparing a new algorithm or technique with already published ones. If possible, also make the datasets available that have been used for use cases or case studies. Note that it also makes a difference whether the collaboration partners are from industry or academia. Collaborations from industry often do not care about making the solutions to open domain-specific problems available to the general public. Sometimes this is even actively blocked because making the results public would also make the know-how available to competitors.
- Create Web-based tools and systems that do not rely on specific hardware architectures. This way, the code is easier to adapt and reuse.
- Publish your results in the target domain as well. Otherwise, researchers from other communities will likely not be able to find your work.
- Try to get funding for software developers. Engineers can contribute a lot to the success of a visualization project. For instance, they can take care of implementing features that are not interesting from a scientific point of view but that are extremely valuable for your collaborator. Also, professional software engineers will make your prototypes more stable and scalable, which increase the probability that the collaborators use them for their actual analyses.

References

1. Aerts, J., Gehlenborg, N., Marai, G.E., Nieselt, K.K.: Visualization of Biological Data - Crossroads (Dagstuhl Seminar 18161). Dagstuhl Reports **8**(4), 32–71 (2018). https://doi.org/10.4230/DagRep.8.4.32. http://drops.dagstuhl.de/opus/volltexte/2018/9760

2. Gratzl, S., Lex, A., Gehlenborg, N., Pfister, H., Streit, M.: LineUp: Visual Analysis of Multi-Attribute Rankings. IEEE Transactions on Visualization and Computer Graphics (InfoVis '13) **19**(12), 2277–2286 (2013). https://doi.org/10.1109/TVCG.2013.173
3. Jenkins, J.: What is the key best practice for collaborating with a computational biologist? Cell systems. **3**(1), 7–11 (2016)
4. Kasik, D., Dill, J.: Toward technology transfer evaluation criteria. In: Proceedings of the 52nd Hawaii International Conference on System Sciences, pp. 1590–1596 (2019)
5. Kirby, R.M., Meyer, M.: Visualization collaborations: What works and why. IEEE computer graphics and applications **33**(6), 82–88 (2013)
6. Lorensen, B.: On the death of visualization. In: Position Papers NIH/NSF Proc. Fall 2004 Workshop Visualization Research Challenges, vol. 1 (2004)
7. Luciani, T., Burks, A., Sugiyama, C., Komperda, J., Marai, G.E.: Details-first, show context, overview last: supporting exploration of viscous fingers in large-scale ensemble simulations. IEEE transactions on visualization and computer graphics **25**(1), 1–11 (2018)
8. Munzner, T.: A Nested Process Model for Visualization Design and Validation. IEEE Transactions on Visualization and Computer Graphics (InfoVis '09) **15**(6), 921–928 (2009). https://doi.org/10.1109/TVCG.2009.111
9. North, C.: Toward measuring visualization insight. IEEE computer graphics and applications **26**(3), 6–9 (2006)
10. Nsonga, B., Niemann, M., Fröhlich, J., Staib, J., Gumhold, S., Scheuermann, G.: Detection and visualization of splat and antisplat events in turbulent flows. IEEE transactions on visualization and computer graphics (2019)
11. Nsonga, B., Scheuermann, G., Gumhold, S., Ventosa-Molina, J., Koschichow, D., Fröhlich, J.: Analysis of the near-wall flow in a turbine cascade by splat visualization. IEEE Trans. Vis. Comput. Graph. **26**(1), 719–728 (2019)
12. Pretorius, A.J., Van Wijk, J.J.: What does the user want to see? what do the data want to be? Inf. Vis. **8**(3), 153–166 (2009)
13. Sedlmair, M., Meyer, M., Munzner, T.: Design study methodology: reflections from the trenches and the stacks. IEEE Trans. Vis. Comput. Graph. **18**(12), 2431–2440 (2012)
14. Shneiderman, B.: The eyes have it: a task by data type taxonomy for information visualizations. The Craft of Information Visualization, pp. 364–371. Elsevier, Amsterdam (2003)
15. Van Wijk, J.J.: Bridging the gaps. IEEE Comput. Graph. Appl. **26**(6), 6–9 (2006)
16. Weber, G.H., Carpendale, S., Ebert, D., Fisher, B., Hagen, H., Shneiderman, B., Ynnerman, A.: Apply or die: on the role and assessment of application papers in visualization. IEEE Comput. Graph. Appl. **37**(3), 96–104 (2017)

Part IV
Developing Visualizations for Broad Audiences

The dramatically increasing ubiquity of visualization has greatly expanded the size, breadth, and diversity of audiences who view the visual representations created. No longer are visualizations primarily viewed by an experienced group of scientists who are experts in the data shown in the visualization. Audiences for visualization nowadays commonly include educators, students, decision-makers, self-educated lay experts, museum visitors, website readers, engaged citizens, the casually curious, and random collateral viewers. While these increasingly "broad audiences" are an encouraging sign of the increased importance of visualization, there are also emerging challenges in designing visual representations for, and providing visualization technology to, larger and more diverse audiences.

The chapters of this section explore the challenge of developing visualizations for broad audiences. The discussion begins with definitions and context. It concludes with some challenges and open issues. Between those two bookends are four chapters drawn from the experience of developing visualizations in four different settings: A climate research institute with audiences from the scientific community to policy-makers to the general public, a large government science agency producing visualizations to be broadly distributed online and in public spaces, a science center or museum with interactive installations, and educational settings where students interact with or create visualizations. These chapters discuss challenges of creating engaging and comprehensible displays, abstracting large and complex data, and accommodating diversity of knowledge and goals in the audience.

Chapter 16
Reflections on Visualization for Broad Audiences

Michael Böttinger, Helen-Nicole Kostis, Maria Velez-Rojas, Penny Rheingans and Anders Ynnerman

Abstract Visualizations intended for broad audiences present challenges and potential not typically seen in the more typical situation of creating visualizations through a close collaboration with a domain expert who is already motivated to understand the data through the visualization. In contrast to the explorative character of the process where visualization is used to gain a better understanding of new data within a research workflow, we will focus here on the development and design of fine-tuned visualizations or tools for communication purposes. More specifically, we define visualization for broad audiences as creating visualizations or visualization tools intended for heterogeneous audiences (who may have domain knowledge but differing abilities), distinct audience groups (domain experts in different part of a process who have different goals), large groups of collaborating experts, or the general public (who may have neither domain knowledge nor inherent motivation). Challenges of creating visualizations for broad audiences include defining the characteristics and goals of the audience, engaging those without an inherent motivation to explore the data, and harnessing the techniques of storytelling to create an effective and satisfying communication. This chapter includes some reflections on basic ideas and concepts to address these challenges. More practical examples of successful projects carried

M. Böttinger (✉)
DKRZ, Hamburg, Germany
e-mail: boettinger@dkrz.de

H.-N. Kostis
Universities Space Research Association, NASA Goddard Space Flight Center, Green belt, MD, USA

M. Velez-Rojas
CA Technologies, Santa Clara, CA, USA

P. Rheingans
University of Maine, Orono, ME, USA

A. Ynnerman
Linköping University, Norrköping, Sweden

© Springer Nature Switzerland AG 2020
M. Chen et al. (eds.), *Foundations of Data Visualization*,
https://doi.org/10.1007/978-3-030-34444-3_16

out within different settings are presented in the following chaps. 17, 18, 19 and 20. The final chap. 21 discusses current challenges and open issues.

16.1 Definition of Broad Audiences

In contrast to a focused audience such as a group of domain experts, a broad audience is mostly understood as the exact opposite. As a first approximation, a broad audience can be defined as the general public—i.e., a group that includes virtually everyone, and that has a very wide spectrum in knowledge and interests, varying age groups, varying cultural, geographical, and educational backgrounds.

In this chapter, we aim at a slightly narrower definition: In the context of creating visualizations for broad audiences, we define them as recipients (of visualizations) from which we cannot expect prior expert knowledge of the underlying science or of complex visual data analyses in general, but they could (but not need to) have much more previous and related knowledge in common than the general public. Furthermore, we assume some openness and interest to understand the main messages of the data presented. As an example, semi-focused groups such as groups of students of one specific field of study that at least overlaps with the science presented by the visualizations share a common knowledge base. But we have to expect some spread in their previous knowledge, and this heterogeneity requires us to design our visualizations accordingly.

16.2 The Complexity of Data and the Imperative for Simplicity

The need to store and analyze progressively large and complex scientific data has developed in parallel with the exponential growth in computing power [2]. Today, digital data sources such as numerical computer simulations, remote sensing, or digital imaging produce very large and increasingly complex data. The related paradigm change of the scientific workflow coined the term *data intensive science* [4]. Scientific simulation data, as for example produced by climate models, are three-dimensional, time-dependent, and multivariate. Uncertainty information gained by the use of ensemble simulation techniques adds a further dimension to the data. Similarly, observational data are also associated with noise and uncertainty in the measurement instruments and procedures and, furthermore, gaps in the data due to limits in the techniques used or failures in the systems are also common. Helbig et al. [3] identified the combination of data size, complexity, and heterogeneity as one of the central challenges in Earth system sciences today. Geo-scientific information hidden in data from different sources, at different temporal and spatial scales and using different sampling need to be extracted, combined, and jointly analyzed and visualized to gain insight. Although only limited to the field of meteorological visualization, the survey on visualization in meteorology by Rautenhaus et al. [7] gives a good overview on the data complexity and heterogeneity of data sources and types. Similarly, challenging data are also found in other application areas such as, e.g., biology and medicine [5].

By definition, data visualization reduces the amount of information and hence complexity in order to ideally make relevant parts of the data comprehensible. However, communicating scientific findings to broad audiences on the basis of data visualizations often requires even further information reduction and a cleaner, simplified visualization design compared with that originally used by domain scientists. Too much detail in the display, i.e., concurrent visualization of several quantities (multivariate data), too many and technical annotations might conflict with the aim to communicate a clear message with a visualization. It might also be needed to leave out some of the detail that the domain scientist thinks would be good to include.

Information reduction can be a challenging process. One example: We have visualized the projected development of the sea ice coverage of the northern hemisphere based on global monthly data of future projections for different greenhouse gas scenarios. Since the sea ice area exhibits a strong seasonal cycle, it would be counterproductive to just produce a time animation over all time steps. The high frequency seasonal pattern (largest sea ice extent in March, smallest sea ice extent in September) would dominate the visualization and prevent effectively conveying the intended message that, in the long run, the sea ice would retreat in case of continuing or further increasing emissions. For the pessimistic emission scenario RCP8.5, the model simulated an ice-free Arctic Ocean in late summer from about 2060 on. This is one of the key messages that should be conveyed with the visualization. However, if we would have removed the seasonal pattern and visualized the resulting development of the annual mean sea ice coverage, this would not be conveyed, since new sea ice is formed in the winter season also beyond 2060, and a complete retreat of the sea ice would not be visible. Instead we decided to visualize the sea ice for both seasons (i.e., March and September) at a time together in one joint rendering.

The choice of the colors is one important aspect in the process of information reduction; some specific thoughts on colormap design with respect to visualization for broad audiences can, e.g., be found in Chap. 17. For the visualization shown in Fig. 16.1, we only used two monochromatic colors for the sea ice concentration of the two seasons. To achieve a smooth transition of areas without sea ice to areas with a concentration of 100%, we used transparency mapping with a graded increase of opacity, so that a concentration of 100% is rendered complete opaque. A slight embossing shading was additionally applied to indicate the sea ice thickness.

However, with respect to a planned television documentary we were asked by journalists for a simpler version, i.e., a similar visualization only for the summer sea ice, because showing both seasons together would be too complex and would require further explanations. Accordingly, we produced a video of the sea ice coverage for only the summer seasons, which indeed is easier to convey. However, we asked the journalists to point out within the accompanying narration that this visualization shows the development for the summer seasons only, because otherwise a misleading

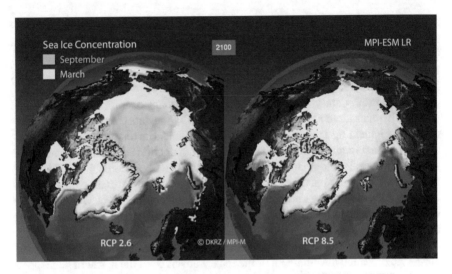

Fig. 16.1 Visualization of the projected sea ice concentration and thickness in 2100 for the pessimistic greenhouse scenario RCP8.5 as simulated with MPI-ESM. The light blue shaded area indicates the projected sea ice extent in September, the white shaded area below that in March. For RCP8.5, the simulation projects a complete loss of the sea ice in September for 2100

message (beyond 2060, no sea ice remains at all) would be communicated. Unfortunately, the resulting documentary broadcasted did not explicitly point out this critical detail. If a visualization of only the summer sea ice is used, but this information is not clearly given, the viewer would think that from 2060 on there wouldn't be any sea ice anymore at all—a wrong message! This example demonstrates that the desire for scientific correctness, completeness, and extensive annotations potentially conflict with a visualization design needed for communication to broad audiences. Visualization for broad audiences is often a compromise between correctness, completeness, and the quest for a simplified visualization design.

We have to ask ourselves which details and aspects in the data are really needed with respect to a specific communication goal and to a specific audience—which information is really needed to tell the story? On the other hand, as, for example, in the case described above, we have to be aware that any information reduction could be interpreted as omitting some detail on purpose. Visualizations produced for broad audiences therefore involve the danger to appear biased. In particular, for policy-relevant subjects, such as climate change, it is recommended to accompany visualizations with precise information about the corresponding data sources and which parts of the data are being visualized.

16.3   Scope, Intent, and Goals

Data visualizations and visualization tools designed to reach broad audiences share a common characteristic. They turn vast quantities of complex data into easily understandable visual information. A clear definition of the purpose of visualization within the context of the target audience is a fundamental step to meet the viewer's information needs. Failure to understand the needs of the audience can lead to misinterpretation of the data and mistrust of the information. However, understanding broad audiences is a challenging process because they are usually composed of diverse groups. Marketing managers successfully address this challenge by grouping their target audiences based on general aspects such as demographic and psychographic information. Demographic information gives designers basic descriptive properties of the audience such as age, gender, occupation, and level of education. These aspects are the basis for the design of a visualization that effectively delivers the desired message. For instance, visualization designers need to consider age when designing the user interface. Psychographic information provides a more in-depth look into sociological factors of the target audience such as hobbies, interests, and financial management aspects. Psychographic information is used to anticipate preferences from the audience, such as a preference between a video presentation and an interactive application.

Having a clear definition of the target audience provides the knowledge necessary to answer critical questions that guide the visualization design process. The first and most important consideration is the intent of the visualization: What are the visualization users looking for? To obtain the correct answer to this question can be difficult; it does not only depend on the information needs of the user, but it also involves the judgment of the visualization designer in the selection of the critical components of the scientific data that are part of the narrative. While experts demand access to the full set of parameters available for exploration of the data, novice users lack the expertise required to operate advanced interaction interfaces. For example, visualization designers can provide predefined sets of parameters that the novice users can select from instead of requiring the adjustment of all individual values. This type of interface redesign contributes to the reduction of cognitive load but still give users the ability to explore the data.

The nature of the target audience also affects the scope, type of visual narrative, dissemination medium, and level of interaction of a data visualization. For example, to define the scope of a scientific visualization in educational settings, it is necessary to determine the level of knowledge in the general topic by the students and the skills needed to interpret the actual visualizations. Once the designer knows that this information is available to teachers before using the visualization, it is possible to ensure that the visualization is rooted in easy-to-understand concepts. It is important to note that designers should not be thinking about finding the lowest common denominator in terms of user knowledge, but the goal is to determine the usefulness of the information for a broad audience. Scientific visualizations need to be designed to support the introduction of more advanced forms of visualization and possibly unknown aspects of the underlying knowledge.

The questions addressed by a scientific visualization and the stories portrayed are closely related to the choice of visual representation as well as the level of interaction. Whether the data visualization takes the form of an infographic, an interactive visualization or a guided visual presentation, it needs to present a narrative structure that captivates the interest of the audience. Compelling data visualizations use small sets of relevant details to create impactful visual stories that resonate with audiences. Depending on the environment, the time, and the interest of the audience in the facts, the information might be better suited to be presented using different techniques. For example, infographics are easy-to-understand overviews designed to guide the audience to a conclusion by following a predetermined storyline. Audiences can access them at any time, and there is no need for specialized technical equipment.

16.4   Manifesto—Why Do We Need to Engage

Scientists are inherently interested in their own research and data. Therefore, they are intrinsically engaged in the exploration and visualization of their research data as well. But if they need to communicate their results to colleagues, they have to be more careful in the visualization design, since these might not have the same specific background knowledge needed to interpret the visualization and comprehend the messages intended to communicate. Accordingly, we have to know the target audience quite well when we design a visualization. Broad audiences, for example, unlike experts for a specific topic, can be expected to be heterogeneous with respect to their individual background knowledge, and their engagement in the context and facts that a visualization intendeds to communicate.

Visualizations intended for communication should inherently be engaging. This means we have to keep the audience interested in the topic, possibly supported by a creative visualization design, by user interaction or by other means. Especially with respect to those, who are not "thrilled" by the underlying science or the visual representation alone, we need to do more to engage the audience.

One technique that could engage audiences is narrative visualization [8], the combination of storytelling and visualization. In principle, each data visualization should tell a story. However, in a presentation or within a TV documentary, the story is being told by the presenter, and the visualization is only used to support the narrative. For interactive visualization-based user experiences, the presenter needs to be replaced by other means, or the respective narration needs to be included. Storytelling techniques and interactivity can be used to explain and communicate complex topics.

16.5 Manifesto—Why It Is Hard to Measure

Measuring the quality or success of visualizations or visualization techniques with respect to specific communication goals is a complex task. Part Two of this book presents a broad range of information around empirical studies in the visualization domain and discusses the underlying theories, the current state of the art, successes, and challenges in great detail. However, in the context of science communication, different design decisions (e.g., choices of colormaps, and exploratory vs. explanatory visualization) and deployed techniques (e.g., animation, storytelling, and virtual environments) need to be evaluated, especially in terms of their effectiveness in engaging broad audiences. At the moment, measuring engagement in this context appears to be a difficult undertaking.

To illustrate the problem, here we would like to briefly mention two exemplary studies. In their study on the use of narrative visualization techniques for engaging users with exploratory information visualization, e.g., Boy et al. [1] conclude that augmenting exploratory visualizations with narrative visualization techniques and storytelling does *not* help engaging users. As this result was unexpected, they conclude that the concept of engagement needs to be better defined in Infovis. In their study, they define engagement as a user's investment in the exploration of a visualization and use timings as a measure. In contrast to Boy et al., McKenna et al. [6] find in their recent work on story reading experiences for data-driven stories that the "visual narrative flow" impacts the readers' preference and engagement. Here, a questionnaire with 14 questions was used to estimate the reader-perceived engagement. They, however, also report that engagement lacks a unified definition in the community.

16.6 Manifesto—A Common Base for Exploration and Explanation

The visual language spans across borders of knowledge, experience, age, gender, and culture, which makes it an effective form of expression in reaching broad audiences. Even though visual representations have been used for communication throughout human history, we are now seeing new opportunities enabled by the rapid pace of digitalization and widespread availability of visual data analysis tools and computing resources. Traditionally, interactive data exploration (exploratory visualization) is a paradigm for supporting domain experts in their discovery and analysis processes, while explanatory visualization is a paradigm for broad audiences who are the recipients of the results emanating from experts' exploration. Since the arrival of the World Wide Web, interactive data exploration has increasingly been available to broad audiences, empowering them to explore and analyze data independently. The general availability of open data and powerful computers is thus enabling a confluence of exploratory and explanatory visualization, and in fact a cross-fertilization

of the two. In a recent viewpoint article, Ynnerman et al. [9] elaborate on this confluence and coin the term *Exploranation* to denote this new paradigm in science
communication and its feedback impact on traditional exploratory paradigms, as it is
noted that also exploratory visualization is increasingly making use of explanatory
methodology.

In this new landscape of technology and methodology for visual communication
to broad audiences, there is a wide range of knowledge areas that need to be mastered
to develop successful visualization beyond the traditional tools for domain experts.
This part of the book elaborates on some of these tools and the design and methodology behind them. Examples we will provide insights into a selection of relevant
topics, which will show that there is a plethora of approaches available to reach
high level of user engagement, not traditionally emphasized in exploratory visualization, such as visual aesthetics and design (look and feel), storytelling and evolving
narratives, annotation and completion (data curation), exploratory gamification, use
of immersive technology, and mediated (guided and collaborative) exploration, and
blending of traditional animations and linear media. Inherent in all approaches, targeting broad audiences is the need for simplification. In Chap. 17, we will address
some approaches shown successful in our selected applications such as tailored visual
abstraction, data reduction and aggregation, model simplification, and constrained
and guided interaction.

16.7   Visualizations in Different Settings

Each visualization project targeting broad audiences is unique: Each one has its own
specific target group, scope, underlying science that needs to be communicated, pur-
sued visual language, domain-specific rules for colormap design, target media, and
type of communication channels, i.e., static (print), dynamic (video), interactive,
immersive, narration, sound, and music. In the next chapters, we present and discuss
several real-world application examples of visualizations that proved successful in
communicating complex scientific data to broad audiences. By including examples
from within different settings, we intend to prepare the ground for identifying mechanisms and principles for a target group-specific visualization design. However, as
we provide examples of the works that have been created in different countries, in
different types of institutes with different obligations, and by differently large teams,
it is not always straightforward to decide if design decisions were taken because of
specific target groups or as a consequence of the team's culture and preferences.

Chapter 17 specifically focuses on visualization work for communication to broad
audiences in the field of climate research done at DKRZ in Germany by a relatively
small team. However, most of the lessons learned also apply to other scientific fields.
The following Chap. 18 discusses several of NASA's activities in the field of data
visualization for outreach. Here, a much larger and transdisciplinary team, the NASA
Scientific Visualization Studio, produces a broad range of sophisticated data-driven
visual products to communicate research findings to the scientific community and the

general public. Chapter 19 presents the likewise advanced visualization work carried out at the Norrköping Visualization Center C in Sweden. Utilizing different display devices such as a dome projection or multi-touch tables, immersive shows and for museum exhibits are created based on visualizations in different thematic fields. This chapter also describes visual design considerations in public spaces and for public audiences. Chapter 20 illustrates the use of visualization in different educational settings. Finally, Chap. 21 discusses a few areas that we think need to be further explored.

References

1. Boy, J., Detienne, F., Fekete, J.D.: Storytelling in information visualizations: does it engage users to explore data? In: Proceedings of the 33rd Annual ACM Conference on Human Factors in Computing Systems, pp. 1449–1458 (2015)
2. Denning, P.J., Lewis, T.G.: Exponential laws of computing growth. Commun. ACM **60**(1), 54–65 (2016)
3. Helbig, C., Dransch, D., Böttinger, M., Devey, C., Haas, A., Hlawitschka, M., Kuenzer, C., Rink, K., Schäfer-Neth, C., Scheuermann, G., Kwasnitschka, T., Unger, A.: Challenges and strategies for the visual exploration of complex environmental data. Int. J. Digit. Earth **10**(10), 1070–1076 (2017)
4. Hey, T., Tansley, S., Tolle, K. (eds.): The Fourth Paradigm: Data-Intensive Scientific Discovery. Microsoft Research, Redmond, Washington (2009). http://research.microsoft.com/en-us/collaboration/fourthparadigm/
5. Kaufman, A.E., Scheuermann, G., Roerdink, J.B.T.M.: Overview of visualization in biology and medicine. In: Hansen, C.D., Chen, M., Johnson, C.R., Kaufman, A.E., Hagen, H. (eds.) Scientific Visualization: Uncertainty, Multifield, Biomedical, and Scalable Visualization, pp. 215–219. Springer London, London (2014). https://doi.org/10.1007/978-1-4471-6497-5_20
6. McKenna, S., Riche, N.H., Lee, B., Boy, J., Meyer, M.: Visual narrative flow: exploring factors shaping data visualization story reading experiences. Comput. Graph. Forum (EuroVis' 17) **36**(3), 377–387 (2017). https://doi.org/10.1111/cgf.13195
7. Rautenhaus, M., Böttinger, M., Siemen, S., Hoffman, R., Kirby, R.M., Mirzargar, M., Röber, N., Westermann, R.: Visualization in meteorology — a survey of techniques and tools for data analysis tasks. IEEE Trans. Vis. Comput. Graph. **PP**(99), 1–1 (2017). https://doi.org/10.1109/TVCG.2017.2779501
8. Segel, E., Heer, J.: Narrative visualization: telling stories with data. IEEE Trans. Vis. Comp. Graph. (Proc. InfoVis) (2010). http://idl.cs.washington.edu/papers/narrative
9. Ynnerman, A., Löwgren, J., Tibell, L.: Exploranation: a new science communication paradigm. IEEE Comput. Graph. Appl. **38**(3), 13–20 (2018)

Chapter 17
Reaching Broad Audiences from a Research Institute Setting

Michael Böttinger

Abstract Data visualization at large can be described as a process that reduces data to mentally comprehensible visual products. Many visualizations are based on very large and complex data, often integrating multiple data sources and complex measures and concepts. Communicating to broad audiences involves drastically simplifying the message, extracting salient concepts and often omitting low-level details. In this chapter we give two examples for data visualizations in the field of climate research that proved to be successful in supporting communication to broad audiences. In a research institute setting, striking a balance between scientific correctness and comprehensibility is key. We describe how a careful design of the visual encoding such as reducing data dimensionality, dealing with data issues (e.g. uncertainty), the number of colors, and choice of visual elements is important to achieve simplicity. Finally, we describe two technical settings that we use for face-to-face communication of climate research results to broad audiences.

17.1   Context, Goals, and Approach

The climate research landscape in Germany is quite heterogeneous. With regard to climate modeling, numerous university institutes and other large research facilities such as the Max Planck Institute for Meteorology or centers funded by the Helmholtz Society jointly form the German scientific community in this field. DKRZ, the German Climate Computing Center, is a domain-specific service facility that provides high-performance computing, data storage, archiving, and associated services to this community.

In this context, scientific visualization of climate research data for both data analysis and communication purposes has been one of the central services of DKRZ for

M. Böttinger (✉)
DKRZ, Hamburg, Germany
e-mail: boettinger@dkrz.de

© Springer Nature Switzerland AG 2020
M. Chen et al. (eds.), *Foundations of Data Visualization*,
https://doi.org/10.1007/978-3-030-34444-3_17

almost 30 years, i.e., since the first IPCC report was published [6]. These visualization services strive to particularly cover the high end part of the visualization work, i.e., projects that go beyond simple 2D techniques used by most domain scientists. Driven by requests of domain scientists and triggered by the growing interest of the media, DKRZ produces data visualizations in the form of stills or animations that are used within the scientific community, but also for communication to the public and to policymakers through various channels. Climate and climate change visualizations of DKRZ have been used for many TV documentaries, newspapers, school books, in science centers, planetariums, climate conferences, and public exhibitions such as several World EXPO exhibitions.

Throughout the last 25 years, DKRZ's visualization and public relations group consisted at most of 2–5 members, some of them part time; however, due to permanent funding, tailored services for climate research visualization and communication could be developed and continuously provided. The repeated interaction with journalists helped the group develop a sense for visualization design suited for communication to broad audiences.

17.2 Climate Research—Climate Change

Some scientific areas receive public attention just due to the nature of the phenomena in focus. Extreme and potentially catastrophic events such as, e.g., tornadoes, hurricanes, volcano eruptions, freak waves or floods have always fascinated people. Such events are severe threats to human life and, at the same time, fascinating manifestations of nature's elemental force, and therefore they are well suited to draw people's attention. And, as Reser et al. [11] concluded, "direct personal experience is a powerful vehicle for acceptance and commitment—and psychological adaptation—in the context of climate change." High-resolution simulations and animated 3D visualizations of such events draw viewer's attention just through their aesthetic appeal as, e.g., the visualization of a tornado within a supercell simulation by Orf et al. ([10] and http://orf.media/). But even simple 2D visualizations of extreme events such as those utilized in weather applications can easily engage people because of their practical relevance. However, climate and Earth system research are a very broad and multidisciplinary research field, and extreme events are only one facet in this context.

Many processes in the climate system are very slow compared to a human's lifespan. As an example, the process of global warming due to man-made greenhouse gas emissions started more than 150 years ago. Today, our world is already about 1 °C warmer than at that time, but due to the long-time horizon, personal experience is not suited to objectively capture the warming process. Climate change can by definition not directly be observed. To capture climate, meteorological observations over a long period of time (e.g., 30 years) and a statistical analysis of the data are needed. To capture climate change, a comparison to a reference climate of another period in time (i.e., the pre-industrial situation) has additionally to be done. Visualization is

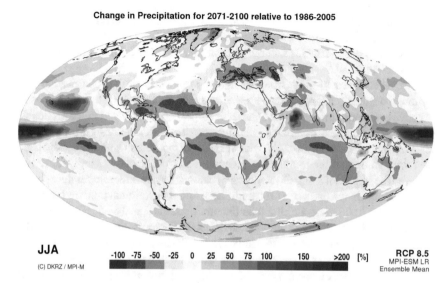

Fig. 17.1 Projected relative change in summer precipitation for 2071–2100 relative to 1986–2005 for the pessimistic greenhouse scenario RCP8.5 as simulated with MPI-ESM. Blue colors indicate an increase, yellow to red colors indicate a decrease, and a bright color is used to show neutral areas with only little changes in precipitation for the summer season

needed in the process of exploring and presenting the underlying data and, finally, the findings of climate scientists [12]. Visualizations of either observations or simulation data are well suited to make the global warming process accessible to both experts and non-experts.

Based on climate scenario simulations with Earth system models, it has become feasible to evaluate possible future climate changes associated with different socio-economic and political storylines. Visualizations of the change in climatological quantities for different future developments indicate potential impacts of human actions on the climate system. However, graphics (such as, e.g., XY-plots representing the temporal evolution of spatially averaged physical quantities, or even animated spatiotemporal displays visualizing the changes in them) are not direct representations of reality; the meaning of the data they represent must be interpreted by the viewer [5], who may need additional assistance in this process. By adding additional information to the visualization, the underlying data can be related to known facts or to the viewer's own experiences. Additional context is mostly added to static visualizations in the form of figure captions, or, in case of video productions, by narrations. These could explain the meaning of the data shown or at least put the information contained into a more general context, but one has to be careful to "remain impartial while attempting to translate data into information and aid users in extracting climate change messages from the data [2]."

Visualizations of climate model results show meteorological and other quantities or changes in them, but potential consequences of such changes to one's life are not communicated along with the pure presentation of the changes. In order to make them fully comprehensible and, ultimately, capture the audience, connections between the change in physical variables and our current and possible future living conditions have to be drawn.

Here, the choice of color schemes and the colormap design play a vital role. Although originally intended for the meteorological domain, the set of guidelines for effective colormaps presented by Stauffer et al. [14] are mostly generally applicable. Furthermore, potential psychological effects of the colormap design used need to be taken into account especially with respect to the actual quantities visualized, the target group and the narrative to be communicated. Schneider and Nocke [13] explored the impact of using different color schemes for the same temperature change visualization. On the basis of a user study, they found the original blue–red–magenta color scheme to be more alerting than the other schemes analyzed; but at the same time, it caused disillusioning associations of powerlessness and fear—feelings that are undesired if engagement of the audience is a desired communication goal.

Figure 17.1 is a typical scientific visualization of climate change data. It shows a geographically mapped visualization of the projected future change in the mean summer precipitation for a strong increase in the greenhouse gas concentration relative to the precipitation simulated for today. A symmetric colormap is used to visualize areas with a projected increase in precipitation in blue and a decrease in precipitation in yellow to red color shades, and a neutral color in areas with no or only little changes. Instead of a continuous colormap, a discrete one was used in order to reduce the information displayed. The continuous data range is mapped to only a few classes of precipitation change to enable easy lookup of the value ranges that correspond to distinct colors.

However, with respect to communication to broad audiences it is not clear if all key messages a climate scientist would infer from the visualization are sufficiently clearly communicated. To relate information of the type "more summer precipitation" or "less summer precipitation" to one's life, consequences of this climatic change need to be discussed in more detail. Of course, a decrease in summer precipitation could mean "a nicer summer" for tourists; but at the same time, agriculture and natural ecosystems could suffer. Depending on the normal local precipitation in an undisturbed climate, a decrease could cause serious economic and social problems. A further problem in the use of this figure for communication to the public is what is not displayed: To correctly assess a change in precipitation given in percent, the recipient has to know the reference, i.e., the geospatial pattern of the mean summer precipitation. This knowledge is required to reason about the potential impact of relative changes. In a relatively dry area, a decrease of 50% can be fatal, while in relatively wet areas a comparable decrease might not be an issue.

Figure 17.2 was produced to underpin this line of thoughts. The bivariate visualization shows, by the height of the bars, the undisturbed mean summer precipitation, and at the same time, by the same colors as used in Fig. 17.1, the projected mean summer precipitation change for a strong greenhouse gas scenario as shown in Fig. 17.1. Since precipitation changes over the ocean are not directly important for human life

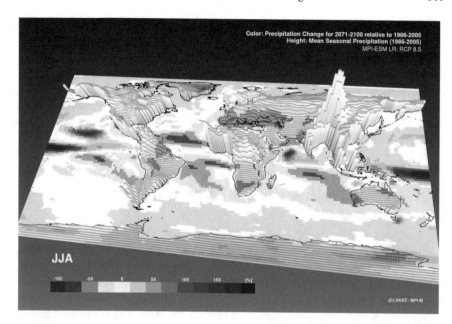

Fig. 17.2 Bivariate visualization of precipitation change: The height of the bars on land indicates the mean summer precipitation for 1986–2005 as simulated with MPI-ESM. The colors show, similar to Fig. 17.1, the projected changes in summer precipitation for 2071–2100 relative to 1986–2005 for the pessimistic greenhouse scenario RCP8.5. Blue colors indicate an increase, yellow to red colors indicate a decrease in precipitation for the summer season, and a bright color is used to show neutral areas with only little changes

conditions, the bars are only rendered on land areas. Furthermore, the visual clutter is reduced in this way and the continents can easily be recognized.

But although Fig. 17.2 includes much more of the information needed to assess the potential impact of the projected precipitation changes on our living conditions, it has a problem: It is much more complex. Viewers may need directions where to look at and how to interpret and assess the combined information displayed. Two different quantities are shown by different techniques. The recipient's view is usually attracted by strong colors; but at the same time, the large bars denoting areas with strong summer precipitation also attract his interest. Without guidance, this might be misleading since the combination of strong decrease in precipitation in the future and relatively little precipitation today is potentially the one where the strongest impact can be expected. In this visualization, this is visible by the relatively small bars turning orange; they denote areas with only little summer precipitation today that might get much dryer in a warmer climate.

At DKRZ, these two visualizations have quite often been shown and explained to visiting school classes or student groups. In this setting, both visualizations were shown consecutively in the form of animations denoting the mean seasonal cycle of the projected precipitation changes. In addition to basic figure captions, guidance on how to assess the visuals was verbally given by the presenter. In this way, the communication was in accordance with guideline 6 of the recommendations to the IPCC and guidance for researchers "Enhancing the accessibility of climate change data visuals" developed by Harold et al. [4] from a cognitive and psychological science evidence-base. Feedback from teachers and students proved the communication achieved with this example to be successful. Furthermore, the aesthetic visualization design of the growing and shrinking bars showing the mean annual cycle in precipitation (c.f. Fig. 17.2) seems to have helped capturing the audiences.

17.3 Climate Projections—The CMIP Multi-model Ensemble

Since 1990, the *Intergovernmental Panel on Climate Change* (IPCC) periodically publishes assessment reports (e.g., [6, 7]) that summarize the current scientific knowledge on climate and climate change on the basis of most recent scientific literature. The readership can be classified as a broad audience, consisting of domain scientists, policymakers, and the general public. To account for the different previous knowledge and needs of different audiences, the reports are comprised of a short *Summary for Policymakers*, a longer *Technical Summary* and the science-oriented thematic chapters of the full reports.

With respect to these IPCC reports and within the *Coupled Model Intercomparison Project* CMIP [3] of the *World Climate Research Programme* (WCRP), all major climate modeling groups worldwide regularly carry out coordinated future scenario simulations using their own climate model systems. The resulting data is stored in a distributed climate database [17] and shared among all research groups. The outcomes of these multi-model ensemble future projections build the basis for estimations of future climatic developments that would result from different economical, technical and societal future developments.

Due to its probabilistic nature, the CMIP multi-model ensembles capture, beyond projected climate changes, internal climate variability, uncertainty in the initial conditions as well as model uncertainty [1, 9, 15]. Analysis and communication of uncertainty or the robustness associated with data play an important role in science in general. Climate simulation results are often visualized in the form of spatial maps showing projected climate changes. For a joint analysis of projection results and the corresponding uncertainty, the latter is often graphically overlaid in form of stippling and/or hatching. Figure 17.3, one of the central figures of the reports' Summary for Policymakers, shows, for two different scenarios, maps of the temperature changes projected for the end of this century compared with the climate of the end of the last century. Note the nonlinear mapping between color and temperature change for small values. The derived robustness of the results is overlaid by stippling/hatching patterns. The corresponding figure caption is taken from the report; however, it is obviously written using a "scientific language" although this summary is directed

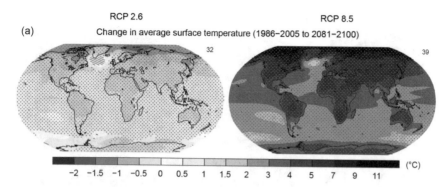

Fig. 17.3 Figure taken from IPCC AR5, Summary for Policymakers [8], Figure SPM.8 (a): Maps of CMIP5 multi-model mean results for the scenarios RCP2.6 and RCP8.5 in 2081–2100 of annual mean surface temperature change. Changes are shown relative to 1986–2005. The number of CMIP5 models used to calculate the multi-model mean is indicated in the upper right corner. Hatching indicates regions where the multi-model mean is small compared to natural internal variability (i.e., less than one standard deviation of natural internal variability in 20-year means). Stippling indicates regions where the multi-model mean is large compared to natural internal variability (i.e., greater than two standard deviations of natural internal variability in 20-year means) and where at least 90% of models agree on the sign of the change (Excerpt from original figure caption of the IPCC's AR5 Summary for Policymakers [8])

to policymakers, media, and the public. Furthermore, by overlaying two classes of robustness, the visualization itself is quite complex.

The figure has two main messages: (a) It visualizes the range of possible future temperature changes: RCP2.6 and RCP8.5 are the most extreme scenarios that had been developed for this project. RCP2.6 is the most optimistic scenario requiring drastic reduction of CO_2-emissions, while in the business-as-usual scenario RCP8.5 the emissions continue to rise throughout the twenty-first century. The mapping between color and physical values is nonlinear and therefore quite complex: For values between -2 and $+2\,°C$, an isovalue interval of $0.5°$ is used for the shading. For larger or smaller temperature changes, an interval of $1°$ is used. (b) The statistical robustness of the multi-model ensemble simulations is shown by the stippling. For both scenarios, the resulting warming is statistically robust for almost the whole planet; therefore, stippling is overlaid almost over the whole world map. This space-filling dot pattern makes it hard to distinguish the filled contours underneath, which counteracts the strive for a clear communication of the results.

With regard to a planned outreach video of the WCRP, we had to develop a simple and clear visualization design for a time-dependent version of the CMIP5 multi-model ensemble data covering the time period from 1981 to 2100. First time animation experiments with the visualization package NCL [16] using a stippling design similar to the one used in Fig. 17.3 clearly demonstrated the general inappropriateness of this method for time animations. Due to the increase of the CO_2 concentrations over time as prescribed by the scenarios, the 2m temperature also increases. In the term of this century, the temperature increase is projected to exceed

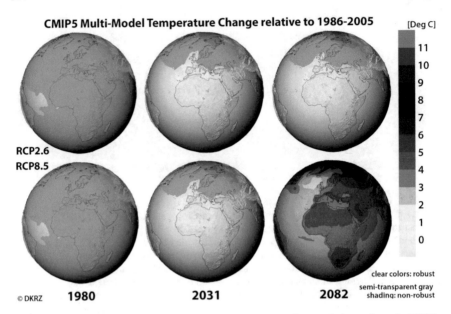

Fig. 17.4 Visualization of the annual mean surface temperature change relative to the period 1986–2005 based on the CMIP5 multi-model mean for the scenarios RCP2.6 and RCP8.5 and the years 1980, 2031 and 2082. Semitransparent gray shading indicates regions where simulated change is small compared to natural internal variability (i.e., less than two standard deviations of natural internal variability). Clear colors indicate regions where the simulated temperature change is statistically robust, i.e., where the multi-model mean is large compared to natural internal variability

the natural variability almost all over the planet. Accordingly, the area denoting robustness gradually grows until it covers almost the full world map. However, due to the implementation of the method, the positions of the single dots were fixed relative to the image space. As a result, the animation showed a growing area with a fixed dot pattern overlaid onto color-filled contours of the actual projected temperature change over time. The fixed position of the dots destroyed the impression of time evolution, and, even worse, made the temperature change displayed underneath nearly unreadable.

Having the communication to broad audiences in mind, we decided to reduce the complexity of the visualization through several measures. First, we used a simple linear colormap for the shading. Second, we only used one level of robustness (two standard deviations) and abstained from additional complex constraints such as the level of agreements between models. Third, we decided to use a semitransparent gray shading of *non-robust* areas overlaid onto the visualization of the temperature change. As a result, robust areas are shown "undisturbed" in clear colors, while the more uncertain (non-robust) areas are presented with a dimmed color scale, guiding the viewer's attention to the trustworthy (robust) part.

The visualizations shown in Fig. 17.4 demonstrate the evolution of temperature change and the robustness with time. The first row shows the warming pattern for the "2-degrees-goal" scenario RCP2.6, while the second one shows the "business-as-usual" scenario RCP8.5. The first column shows the situation for 1980: Relative to the time period 1986–2005, it was slightly colder, but the values are within the range of natural variability almost all over the planet. The second column shows the projected warming in 2031 for both scenarios. In both cases, the warming signal is already larger than the natural variability in a very large area ranging from the tropics to the mid-latitudes. The third row finally shows the situation in 2082. While the warming signal as well as extent of the robust area in RCP2.6 seem to be very similar to those of the visualization of 2031, the RCP8.5 visualization clearly shows a drastic warming, in particular over the continents and at high latitudes. As discussed earlier, the projected warming is larger than the natural variability and therefore almost everywhere statistically robust, denoted by the clear colors.

The corresponding animated version of this visualization shows, for both scenarios, the temporal development of both temperature change and extent of the robust area on two synchronously rotating globes. This projection only allows to see a part of the globe. However, due to the rotation the viewer is able to follow the changes in the global patterns. The initial visualizations were produced with Avizo, a commercial general-purpose 3D visualization system. In addition to rendering the continental outlines, a high-resolution height mapping of the topography is used in order to provide additional guidance. The rendering of the gray semitransparent area was done by rendering a second visualization of the robustness mask followed by post-processing of the resulting two image series. The final composite visualization was used within the short YouTube video "A Short Introduction to Climate Models—CMIP" [18] of the WCRP.

17.4   Displays Utilized for Climate Communication

Apart from the presence of our climate and climate change visualizations in the media throughout the last decades, mostly at the national or regional level, we employ them regularly for face-to-face communications in the context of broad audiences. A straightforward example is the use of visualizations within presentations we regularly give to school classes, students, and other visitor groups. However, for other occasions such as public exhibits, trade shows, conferences or science nights we additionally utilize more or less customized displays that visually attract people and that allow for interactive exploration of the scientific content by the visitors.

In contrast to normal rectangular displays, a spherical display is especially well suited to present geoscientific data as the display shape allows to visualize the data without mapping distortion. With respect to the German pavilion at the World Expo in Shanghai 2010, we developed our "Climate Globe" (Fig. 17.5) based on the commercial globe display solution "Omniglobe." The portable system consists of a spherical display with a diameter of 80 cm (32 inch) and a kiosk with a touch screen that acts as a physical user interface. With respect to this system, we reproduced many animated climate and climate change visualizations specifically for the use with the spherical

Fig. 17.5 Climate globe during the "summer of science," an event organized by the University of Hamburg to present scientific highlights to the public. Visitors could explore the various animated climate and climate change visualizations. This picture shows Michael Böttinger of DKRZ giving detailed explanations to a policymaker

display, complemented by a graphical user interface that allows the viewer to select the content, rotate the displayed content, change the inclination angle of the rotation and start or stop the time animation. However, even without rotation of the projected content, global data can intuitively be explored as it is possible to look at the display from either direction.

Beyond the utilization of the climate globe within our outreach activities, a clone of it, a second one with the same content, has been employed by the Max Planck Society as part of their "Science Tunnel 3.0" exhibition on basic research that has been presented during the last years in several cities around the world.

As the screen resolution of the climate globe is quite limited, it is particularly not suited to present visualizations of very highly resolved simulations. Currently, the highest spatial resolution of climate models run at DKRZ is about 2.5 km on a global 3D grid. Even a 2D slice of the originally unstructured data corresponds to about 16,000 by 8,000 grid cells. The only solution to display the full resolution of the data would currently be a large tiled display. However, since we need a portable and easy-to-use solution, we chose a tilted 55-inch 4k touch screen (Fig. 17.6) for the presentation of such simulation results to experts and specifically also to broad audiences. Here, however, one has to keep in mind that, even with the 4k display,

Fig. 17.6 On the occasion of the public event "Hamburg Science Night 2017," Niklas Röber of DKRZ demonstrates with ParaView the evolution of clouds and rain to young visitors

one pixel of the display represents the information of 16 grid cells; so the simulation is already much more detailed than the display resolution. Interactive visualization with the capability to zoom in to the area of interest could potentially overcome the unbalanced ratio between grid and screen resolution. Equipped with a high-performance GPU, the touch table allows for interactive visualization with tools such as ParaView. However, for very large data fields, the resulting system performance limits the usability for the described purposes, and static animations have to be used instead of communicating such content by means of interactive 3D visualization.

References

1. Cubasch, U., Wuebbles, D., Chen, D., Facchini, M., Frame, D., Mahowald, N., Winther, J.G.: IPCC Climate Change 2013: Introduction, Book section 1, pp. 119–158. Cambridge University Press, Cambridge, UK and New York, NY, USA. www.climatechange2013.org (2013). https://doi.org/10.1017/CBO9781107415324.007
2. Daron, J.D., Lorenz, S., Wolski, P., Blamey, R.C., Jack, C.: Interpreting climate data visualisations to inform adaptation decisions. Clim. Risk Manag. **10**, 17–26. http://www.sciencedirect.com/science/article/pii/S221209631500025X (2015). https://doi.org/10.1016/j.crm.2015.06.007
3. Eyring, V., Bony, S., Meehl, G.A., Senior, C.A., Stevens, B., Stouffer, R.J., Taylor, K.E.: Overview of the coupled model intercomparison project phase 6 (cmip6) experimental design

318

M. Böttinger

and organization. Geosci. Model. Dev. **9**(5), 1937–1958. https://www.geosci-model-dev.net/9/1937/2016/ (2016). https://doi.org/10.5194/gmd-9-1937-2016

4. Harold, J., Lorenzoni, I., Coventry, K.R., Minns, A.: Enhancing the accessibility of climate change data visuals: recommendations to the IPCC and guidance for researchers. http://www.tyndall.ac.uk/sites/default/files/Data_Visuals_Guidance_Full_Report_0.pdf (2017)
5. Harold, J., Lorenzoni, I., Shipley, T.F., Coventry, K.R.: Cognitive and psychological science insights to improve climate change data visualization. Nat. Clim. Chang. **6**, 1080–1089 (2016). https://doi.org/10.1038/nclimate3162
6. IPCC: Climate Change 1990 The Science of Climate Change. The Intergovernmental Panel on Climate Change (1990)
7. IPCC: Climate Change 2013: The Physical Science Basis. Contribution of Working Group I to the Fifth Assessment Report of the Intergovernmental Panel on Climate Change. Cambridge University Press, Cambridge, UK and New York, NY, USA. www.climatechange2013.org (2013). https://doi.org/10.1017/CBO9781107415324
8. IPCC: Summary for Policymakers, Book section SPM, pp. 1–30. Cambridge University Press, Cambridge, UK and New York, NY, USA. www.climatechange2013.org (2013). https://doi.org/10.1017/CBO9781107415324.004
9. Knutti, R., Sedláček, J.: Robustness and uncertainties in the new cmip5 climate model projections. Nat. Clim. Chang. **3**, 369–373 (2012). https://doi.org/10.1038/nclimate1716
10. Orf, L., Wilhelmson, R., Lee, B., Finley, C., Houston, A.: Evolution of a long-track violent tornado within a simulated supercell. Bull. Am. Meteorol. Soc. **98**(1), 45–68 (2017). https://doi.org/10.1175/BAMS-D-15-00073.1
11. Reser, J.P., Bradley, G.L., Ellul, M.C.: Encountering climate change: 'seeing' is more than 'believing'. Wiley Interdiscip. Rev.: Clim. Chang. **5**(4), 521–537 (2014). https://doi.org/10.1002/wcc.286
12. Schneider, B.: Climate model simulation visualization from a visual studies perspective. WIREs Clim Chang. **3**(2), 185–193 (2012)
13. Schneider, B., Nocke, T.: The feeling of red and blue - a constructive critique of color mapping in visual climate change communication. In: Leal Filho, W., Manolas, E., Azul, A.M., Azeiteiro, U.M., McGhie, H. (eds.) Handbook of Climate Change Communication: vol. 2. Climate Change Management, pp. 289–303. Springer, Cham (2018). https://doi.org/10.1007/978-3-319-70066-3_19
14. Stauffer, R., Mayr, G.J., Dabernig, M., Zeileis, A.: Somewhere over the rainbow: how to make effective use of colors in meteorological visualizations. Bull. Am. Meteorol. Soc. **96**, 203–216 (2015). https://doi.org/10.1175/BAMS-D-13-00155.1
15. Tebaldi, C., Knutti, R.: The use of the multi-model ensemble in probabilistic climate projections. Philos. Trans. R. Soc. Lond. A Math. Phys. Eng. Sci. **365**(1857), 2053–2075 (2007). https://doi.org/10.1098/rsta.2007.2076. URL http://rsta.royalsocietypublishing.org/content/365/1857/2053
16. UCAR/NCAR/CISL/TDD: The NCAR Command Language (Version 6.4.0) [software]. https://www.ncl.ucar.edu/ (2018). https://doi.org/10.5065/D6WD3XH5. Accessed 20 Sept 2018
17. Williams, D.N., Balaji, V., Cinquini, L., Denvil, S., Duffy, D., Evans, B., Ferraro, R., Hansen, R., Lautenschlager, M., Trenham, C.: A global repository for planet-sized experiments and observations. Bull. Am. Meteorol. Soc. **97**(5), 803–816 (2016). https://doi.org/10.1175/BAMS-D-15-00132.1
18. World Climate Research Programme: A Short Introduction to Climate Models - CMIP. https://www.youtube.com/watch?v=wTBkq9nWNEE (2017). Accessed 04 Oct 2018

Chapter 18
Reaching Broad Audiences from a Large Agency Setting

Helen-Nicole Kostis, Miguel O. Román, Virginia Kalb, Eleanor C. Stokes,
Ranjay M. Shrestha, Zhuosen Wang, Lori Schultz, Qingsong Sun,
Jordan Bell, Andrew Molthan, Ryan Boller, and Assaf Anyamba

Abstract NASA's missions, engineering accomplishments, and scientific findings
have inspired generations and advanced our understanding of the world we live in.
NASA's Earth science data are acquired by various sources, including satellites,
aircraft, and field measurements. Captured data, their by-products, and their visual
representations developed by research teams become available within few hours
after satellite overpass or processing through a variety of NASA's imaging, mapping
services, and portals. Such online services as the Global Imagery Browse Services
(GIBS) [10], Worldview [13], LANCE [3], and LAADS DAAC [35] are freely and
openly available thanks to NASA's Earth-Observing Satellite Data and Information
Systems (EOSDIS) [9]. These services provide access to products created over the last
30 years, support a broad range of users from the scientific community to the general
public, and cover a multitude of applications such as basic and applied scientific
research, natural hazard and disaster monitoring, and social and educational outreach.

H.-N. Kostis (✉) · A. Anyamba
Universities Space Research Association and NASA Goddard Space Flight Center,
Greenbelt, MD, USA
e-mail: helen-nicole.kostis@nasa.gov

M. O. Román
Universities Space Research Association (USRA), Columbia, MD, USA

V. Kalb · R. Boller
NASA Goddard Space Flight Center, Greenbelt, MD, USA

E. C. Stokes · Z. Wang
University of Maryland, College Park, MD, USA

R. M. Shrestha · Q. Sun
Science Systems and Applications, Inc., Greenbelt, MD, USA

L. Schultz · J. Bell
University of Alabama in Huntsville, Huntsville, AL, USA

A. Molthan
NASA Marshall Space Flight Center, Huntsville, AL, USA

© Springer Nature Switzerland AG 2020   319
M. Chen et al. (eds.), *Foundations of Data Visualization*,
https://doi.org/10.1007/978-3-030-34444-3_18

In order to illustrate the significance of the overall work, the visualization products, and the broad range of users, we present three case studies: NASA's Black Marble Product Suite, the Global Imagery Browse Services (GIBS) and Worldview, and the scientific visualization production process to communicate results to the scientific community and the general public.

18.1   NASA's Black Marble Product Suite

Helen-Nicole Kostis, Miguel O. Román, Virginia Kalb, Eleanor C. Stokes, Ranjay M. Shrestha, Zhuosen Wang, Lori Schultz, Qingsong Sun, Jordan Bell, and Andrew Molthan

Mesmerizing **night light** views of our planet from space fascinate the public (Fig. 18.1). Beyond stirring curiosity, these *pretty images* serve as global daily measurements of nocturnal visible and near-infrared (NIR) light that drive earth system science and applications, including social, environmental, and economic research. NASA's Black Marble nighttime lights (NTL) product suite is available at 500 m resolution daily since January 2012 with data from the Visible Infrared Imaging Radiometer Suite (VIIRS) Day/Night Band (DNB) onboard the civilian Suomi National Polar-orbiting Partnership (SNPP) satellite [32]. Suomi NPP observes nearly every location on Earth at roughly 1:30 p.m. and 1:30 a.m. (local time) each day, as it images the planet in vertical 3,000-kilometer (2,000 mile) strips from pole to pole. The VIIRS instrument detects photons of light reflected from Earth's surface and atmosphere in 22 different wavelengths and makes quantitative measurements of light emissions and reflections,
which allows researchers to distinguish the intensity, types, and the sources of night lights and observe how they change over several years [16].

The Black Marble product suite provides high-quality cloud-free daily Nighttime Light (NTL) imagery that has been corrected for atmospheric, terrain, snow, and other effects [34]. The corrected nighttime radiances, resulting in a superior retrieval of nighttime lights, enable the first-ever quantitative analyses of daily, seasonal, and annual variations at the native DNB pixel scale. The Black Marble product suite has two parts—Daily At-sensor Top of Atmosphere (TOA) and the Daily Moonlight-adjusted Nighttime Lights (Final level 3 product). Both products use the standard suite of VIIRS land products as input and are integrated as part of NASA's Black Marble processing chain, which contains both daytime and nighttime branches. Each processing branch produces a unique set of ancillary and quality assurance (QA) flags [31]. The nighttime branch is the path that actually generates the final Black Marble products. The process begins with the at-sensor TOA nighttime radiance, along with the corresponding nighttime cloud mask, multiple solar/viewing/lunar geometry values (including moon-illuminated fraction and phase angles), and the daily snow and aerosol status flags. These Science Data Sets (SDS) allow open access to the primary inputs used to generate the NASA's Black Marble NTL time series record, thus ensuring reproducibility of the final outputs. A series of temporal and spatial gap-filling techniques are also employed to improve the coverage of the product. The product generation pipeline is described in [32]. The product offers scientific quality daily data that are global and meets the needs of all major disciplines in nocturnal studies such as artificial light, cloud/aerosols, ocean at night, as well as various applications, such as energy access, disaster risk reduction, conflict, and migration.

To monitor nighttime patterns at finer scales, a new suite of Black Marble products is also developed by the research team [4]. The Black Marble HD product is generated through the synergistic use of the daily NASA's Black Marble standard product with data from other Earth-observing satellites (e.g., Landsat-8, Sentinel-2) and ancillary data sources (street, building, and other GIS layers) [30]. Black Marble products are used and visualized to monitor seasonal and long-term changes in cities, while also enabling the assessment of specific sectors that are impacted by disturbances in power delivery, such as those resulting from conflict, earthquakes, and hurricanes [24], including Hurricane Maria that tore across Puerto Rico in September 2017—the longest power outage in the US history [23] (Fig. 18.2).

Fig. 18.1 Earth at night as seen from space (2016). These full-hemisphere composite views of Earth at night utilize NASA's Black Marble Nighttime Light (NTL) products and clouds, and sun glint from MODIS Blue Marble Next Generation imagery. Credit: NASA Black Marble Team

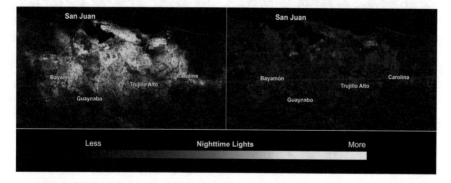

Fig. 18.2 Image pair above shows the extent of electric lighting across the city of San Juan in Puerto Rico before the storm (left) and two months after (right) Hurricane Maria passed over Puerto Rico. We see a massive drop in night-light intensity due to loss of power. The color bar ranges from purple (less) to yellow (more) and aims to illustrate the comparative reduction of night light times when comparing pre- and post-event images. Credit: NASA/Scientific Visualization Studio and NASA's Black Marble Team

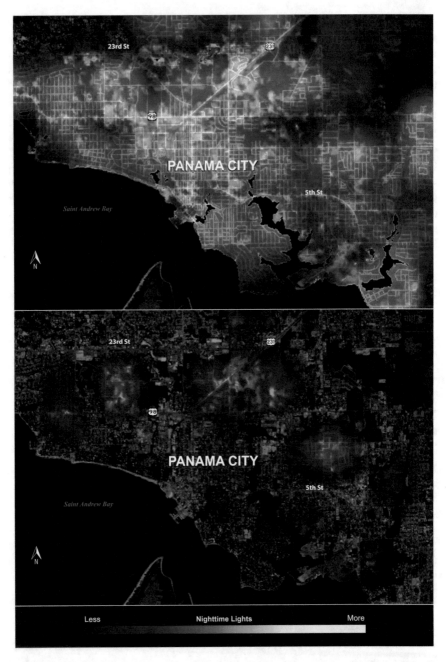

Fig. 18.3 Data visualization of nighttime lights before (top) and after (bottom) Hurricane Michael's effect in Panama City, Florida. The image pair shows conditions before the storm on October 6, 2018, (top) and after on October 12, 2018 (below). The color bar ranges from purple (less) to yellow (more) and aims to illustrate the comparative reduction of night light times when comparing pre- and post-event images. Credit: NASA/Earth Observatory and NASA's Black Marble team

Immediately after the hurricane devastated Puerto Rico, it rapidly became clear that the destruction would pose daunting challenges for first responders. Quickly knowing where the power is out—and how long it has been out—allows better deployment of rescue and repair crews. To support recovery efforts the research team [4] combined night light, Landsat [33] and OpenStreetMap [7] data to monitor where and when electric power was restored. In addition, the research team analyzed demographics and physical attributes of neighborhoods affected by the power outages [30].

On October 10, 2018, Hurricane Michael, a category 4 storm, made landfall in the Southeastern USA and knocked out the power of at least 2.5 million customers. To aid first responders, the research team quickly scaled the observations onto a base map that emphasized the location of streets and neighborhoods. The base map incorporated data from Landsat and Sentinel-2 satellites, as well from OpenStreetMap to show locations for streets and neighborhoods [15]. Figure 18.3 shows data visualizations of nighttime lights of Panama City in Florida before and after Hurricane Michael.

The absence of nighttime lights, as seen in Figs. 18.2 and 18.3, offers a new way to visualize storm impacts. It is a visible indicator from space that critical infrastructure beyond power may be damaged, including access to fuel and other necessary supplies. The power outage maps generated using the Black Marble product helped disaster response efforts in the short-term as well as long-term monitoring during the crucial stages of disaster recovery. This generally is the case when monitoring the impact of disaster (or any event) through power outages. The initial phase of disaster recovery typically is to identify the areas that suffer significant impact in power infrastructure. By visualizing and tracking these outages over time, we can monitor the recovery effort from the vantage point of space, and better understand the state of basic service provision in local communities, and how vulnerable populations with poor access to resources are being affected. NASA's Black Marble visualized products aim to inform disaster response efforts in the short term, but also improve our understanding of how communities can become more resilient to disasters in the long term. For more information and future developments of NASA's Black Marble products, please visit: https://blackmarble.gsfc.nasa.gov/.

18.2   Visualization of Earth Observations with NASA's Worldview and the Global Imagery Browse Services (GIBS)

Ryan Boller on behalf of the Worldview and GIBS team and Helen-Nicole Kostis

Since the 1960s, NASA has built and operated a fleet of Earth-observing satellites to better understand how our home planet is changing and to conduct scientific research for societal benefit. This fleet is continually updated and collects a large volume and wide variety of data in support of a diverse range of disciplines including atmospheric composition, weather, and climate variability. The Earth-Observing System Data and

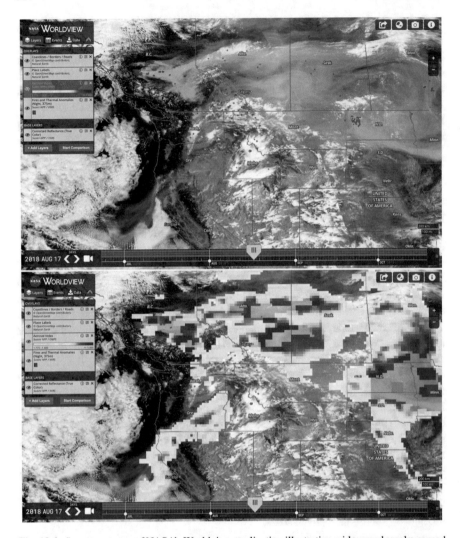

Fig. 18.4 Screen captures of NASA's Worldview application illustrating widespread smoke caused by fires in Western North America and captured by the NASA-NOAA Suomi National Polar-orbiting Partnership satellite on August 17, 2018. The top image shows the corrected reflectance (true color) and fires and thermal anomalies products from the visible infrared imaging radiometer suite instrument. The bottom image adds the aerosol index product from the Ozone Mapping and Profiler Suite instrument which helps to illustrate the extent and intensity of the smoke. Credit: NASA Worldview Team, NASA Goddard Space Flight Center

Information System (EOSDIS) [8] collects, archives, and distributes these data to a diverse user community. Given the diverse nature of these scientific measurements and the users they support, the resulting data products are also diverse: varying data formats, processing levels, spatial and temporal resolutions, coordinate systems, etc.

Despite these challenges, the unique global, detailed, and precisely calibrated nature of the measurements contains tremendous potential for this data set to be used for wide societal benefit. Since many off-the-shelf software tools are not designed to handle this diversity in addition to the time-varying nature and scale of these products, EOSDIS began building an open-source, open access visualization system in 2011 in an effort to improve the data's usability. This system was separated into its backend, the Global Imagery Browse Services (GIBS) [10], and its frontend, the Worldview Web application [11], as shown in Fig. 18.4.

GIBS collaborates with scientists and engineers from each NASA Earth-observing satellite instrument team to identify which data sets should be turned into *visualized data products* and then works to develop them to match users' expectations. These products include mapping the reflectance of light from the Earth's surface as well as quantifying scientific properties (e.g., land surface temperature) into appropriate color ranges (Fig. 18.4). GIBS continuously ingests data as new satellite observations are made, stores them in an optimized manner for distribution, and serves the visualized data products to the public through standards-based Web services [10] as imagery. This imagery and other information are stored and served at the native resolution of the instrument to preserve as much detail as possible and are often made available within three hours of observation to support near real-time decision making. The imagery is also standardized into four map projections to allow inter-comparison between products and to minimize distortion, especially near the poles.

While providing open services via GIBS may be sufficient for certain technically inclined users, the Worldview Web application was designed to be a general-purpose frontend to all of the imagery provided by GIBS. The core of the application allows users to select multiple imagery products and explore them with a pan and zoom interface. Examples of features designed to encourage users to explore changes over time include a prominent time widget to interactively select dates, an animation widget to loop through a series of dates, and a comparison tool to compare imagery from different dates or instruments.

The imagery products from GIBS are thematically organized by scientific discipline and hazard/disaster type to guide users to the most relevant products for their interests. The application also allows users to download imagery from within the app as well as the underlying data if they would like to perform a deeper analysis.

As a whole, GIBS and Worldview have seen a broad uptake in new users across a diverse set of disciplines. Science examples include studying Saharan dust transport across the Atlantic and investigating processes involved with high intensity wildfires. In applied sciences, users monitor sea ice for nautical navigation, river ice breakup to predict flooding near remote settlements, and agricultural fires to understand their impact on air quality. The system is used in academia for research and to train the next generation of scientists. It is used in planetaria to show an accurate representation of our planet in the context of the cosmos. Computer scientists train artificial intelligence applications to automatically identify natural phenomena contained in the imagery. News and weather organizations use the imagery to illustrate current natural events, and the public shares imagery relevant to their interests on social media.

In summary, by providing a visual starting point to using satellite data, Worldview and GIBS have broadened usage of NASA's data across a diverse set of disciplines, enabling new use cases and streamlining existing ones. In some cases, these previously unavailable visualizations are sufficient for a user to complete their task rather than needing to download and learn to use the underlying data. Users have also found value in having a single interface to explore visualizations across different satellite platforms in a homogeneous manner. For more information and to use NASA's Worldview, please visit: https://worldview.earthdata.nasa.gov/.

18.3 Visualization Production Untangling Scientific Complexity

Helen-Nicole Kostis and Assaf Anyamba

The role of science is not complete until its results have been communicated to the public [6, 19]. NASA's Science Storytelling Team aims to fulfill that critical role by engaging and informing the general public about the latest research findings of the agency's missions and results of its engineering and scientific endeavors. Based at NASA's Goddard Space Flight Center, in Greenbelt, Maryland [20], the team comprises five groups: the Scientific Visualization Studio (SVS) [29], the Conceptual Image Laboratory (CiLab) [1], the Goddard Media Studio, science writers, and social media experts. Team members work collaboratively between the different groups and closely with scientists for the creation of visualizations, animations, and media engagements (using social media and live interviews) in order "to promote a greater understanding of Earth and Space Science research activities at NASA, and within the academic research community supported by NASA" [29]. While the majority of the team is based at NASA's Goddard Space Flight Center, it supports the entire agency and works across the agency to distill messages and deliver science news stories to the general public.

In order to translate these research findings from the private realm of the science teams to the realm of the general public, data-driven storytelling is employed. The Scientific Visualization Studio (SVS) brings in key expertise by developing production-quality data-driven computer graphics animations (movies) and still images that cover all of NASA's science themes and by producing a wide range of products in different formats (e.g., HD, UltraHD for tiled displays and hyperwalls, Dome shows, mobile, 360 videos). The visualizations are developed closely with scientists, producers, and science writers and strive for scientific accuracy and data integrity. The narrative impact of these data-driven animations or movies stems most often from creating visual representations of scientific visualization (sci vis) data–spatial data with an inherent structure that is invisible to the naked eye. Such movies allow us to perceive data, and therefore science, in otherwise impossible scales and perspectives. For example, with such movies we can *see* satellite orbits, solar flares, hurricane structures, climate phenomena, and planetary discoveries. Upon approval from the scientists, the movies are released to the public through the SVS website [29]. Since the productions (movies and animation frames) are freely and openly accessible to the public, they may then take a life of their own thanks to the broader scientific community, museums, filmmakers, documentarians, educators, news outlets, etc.

Creating visualization productions of this kind is definitely a distinct and niche practice within the communities of computer graphics, production, and visualization. Such productions are also called cinematic scientific visualizations, as they are created with movie making tools with camera direction, good composition and artistic aesthetics [14]. The content creation tools used include packages that are utilized by science teams (e.g., IDL) and production studios (e.g., Maya, Renderman). IDL [2] is commonly used in many NASA-related scientific fields, like satellite imaging, remote sensing, geosciences, astronomy, etc. Inherently, SVS uses IDL to process data and perform data analysis, and over the years has even become part of customized processes that are called on render time. It is important to mention that this type of visualization productions has additional levels of requirements that the animation production packages (e.g., Maya and Renderman) cannot cover, such as extreme time precision requirements and ever changing data types and formats that need to be ingested. For this reason, these packages have been extended internally (within SVS) over the years to support its scientific data and storytelling needs. When post-production and compositing are needed, SVS members typically use Adobe's After Effects package, while productions that include conceptual animations and/or video shots might utilize additional compositing packages.

It should be noted that even though this type of visualization production is a niche field, it has a rich history of almost 50 years in the making. Exemplary historical efforts and contributions include NASA's JPL Computer Graphics Laboratory, where computer graphics pioneer Jim Blinn developed between 1977–1987 the groundbreaking films of the planetary flybys by the Voyager spacecraft, when almost nobody had seen a computer generated image [22]. In the 1980s, Donna Cox based at the National Center for Supercomputing Applications (NCSA) formed the *Renaissance Team* [18] and over the years produced a series of groundbreaking data-driven works for films and documentaries (e.g., *Cosmic Voyage*, *A Beautiful Planet*, *Hubble*) that defined the field of cinematic scientific visualization.

Among the few and niche teams that engage in visualization production that aims to explain science results, NASA's Science Storytelling Team presents unique aspects that are worth mentioning. For example, a big part of the broader storytelling team (visualizers, animators, producers, writers) works out of the same building; the team is located at the same center with the biggest pool of NASA scientists within the agency, which facilitates dialog and fosters collaboration; the majority of the visualization projects are normally assigned to a single visualizer, who serves as a lead for the specific project. The lead visualizer carries the project through from the very beginning to its release to the public. In that sense, visualizers serve as generalists and typically wear multiple hats in comparison with their colleagues in animation and production studios where specialized experts are required for separate aspects of the production (e.g., lighting, modeling, compositing, etc). It is common for a visualizer to serve within a project also as a producer, to contribute to the script and narration of the story, and even handle post-production and compositing as well. It should be emphasized that the majority of the team members (especially the visualizers) become deeply involved with the science and data behind the stories they work on; therefore, a keen interest or background in science is helpful.

Case Study: Tracking Diseases from Space
How the 2015–2016 El Niño Triggered Dengue Outbreaks in Southeast Asia

El Niño is an irregularly recurring climate pattern characterized by warmer than usual ocean temperatures in the equatorial Eastern Pacific Ocean, which creates a ripple effect of anticipated weather changes in far-spread regions of Earth. Over the last 20 years, Dr. Assaf Anyamba, a research scientist at NASA Goddard Space Flight Center, using remote sensing data has been monitoring how these changes in precipitation, land surface temperatures, and vegetation create conditions for transmission of diseases in humans and animals around the world; in simpler words, Dr. Anyamba has been tracking diseases from space. As part of his research efforts, Dr. Anyamba led a study whose methods and results were published in 2019 in the Journal Nature Scientific Reports [12]. This study was the first of its kind to comprehensively assess the public health impacts of the major climate event of El Niño on a global scale. This task was possible by analyzing satellite data and modeling to track climate anomalies, along with field measurements, public health records, and disease outbreak reports.

The Scientific Study

NASA's remote sensing data are captured continuously by various Earth-orbiting satellites and become available to the scientific community and the public within a few hours from a satellite overpass. This continuous collection of data is referred to as a time series and from the vantage point of space provides unique capabilities to the scientific community, as scientists can monitor changes on Earth and its atmosphere and oceans on local, regional, and subsequently global scale. In their research practice, scientists who utilize remote sensing data strive for the collection of field (i.e., ground) measurements as well, which serve as reference points for the validation of remote sensing data at a single location. Dr. Assaf Anyamba for this study [12] orchestrated and participated along with partner distributed teams in fieldwork in multiple locations in the Free State Region of South Africa. For this endeavor, field work included activities such as setting up weather stations to collect a variety of data measurements (e.g., temperature, rainfall, wind direction, and speed) and even collecting samples of mosquito vectors from the same location (Fig. 18.5). For research studies, field measurements of that kind are project-oriented and require advance planning since the location has to be coincident with the satellite measurements that will be monitored over a predefined period of time. Looking at ground/field measurements is paramount, as they eventually give specific meaning to the remote sensing data. In this study, as part of the field measurements, mosquito vectors were collected from the region of Free State in South Africa, in order to determine how various climate measurements relate to local disease vectors over time. Therefore, the field team collected mosquito vector samples during the rainfall season and then these samples were analyzed to identify their type and the viruses they were carrying.

Fig. 18.5 (Top) Field measurements by Dr. Assaf Anyamba with field partners in Free State, South Africa. On the top left photo, Dr. Assaf Anyamba geolocates a weather station with Dr. Claudia Cordel (Field Team Lead) and Hermann Zwiegers (Field Technician). On the top right photo, Dr. Assaf Anyamba and Lara Van Staden (Field Technician) are collecting mosquito vectors from the field. (Bottom) Visualized remote sensing data from the same region where field measurements are performed, subregion of Free State, South Africa. Credit: Assaf Anyamba, Universities Space Association and NASA Goddard Space Flight Center

In addition, the scientific team cross-referenced remote sensing data and field measurements with disease reports on ProMed—the Program for Monitoring Emerging Diseases. ProMed is an open-source Internet-reporting system dedicated to the rapid global dissemination of information on outbreaks of infectious diseases and acute exposures to pathogens that affect human health, including those in animals and in plants grown for food or animal feed [5]. The synergistic approach and synthesis of fieldwork measurements, remote sensing data, (Fig. 18.5) and disease reports for specific regions over time are what led to confluence and the determination of when, where, and how a subset of diseases occur. The study indicated during El Niño periods a worldwide uptick in reported cases of diseases such as plague and hantavirus in Colorado and New Mexico within the USA, cholera in Tanzania, and dengue fever in Brazil and Southeast Asia, among others (Fig. 18.6).

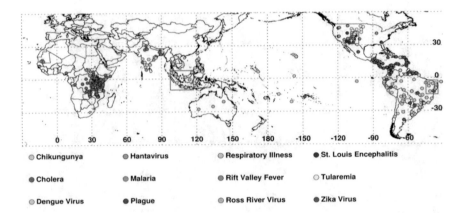

Fig. 18.6 Geographic distribution of various disease outbreaks (indicated with color) on a global scale, between the period of April 2015–March 2016. Regions (in boxes) that are historically susceptible to disease outbreaks can be found within the USA, Brazil, Tanzania, and Southeast Asia. This figure was created using Interactive Data Language (IDL) software (version 8.6.0) and was included in the journal publication [12]. Credit: Anyamba, A., Chretien, J.-P., Britch, S. C., Soebiyanto, R. P., Small, J. L., Jepsen, P., Forshey, B. M., Sanchez, J. L., Smith, R. D., Harris, R., Tucker, C. J., Karesh, W. B., Linthicum, K. J

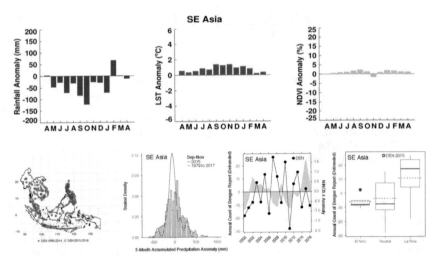

Fig. 18.7 Plots to demonstrate the regional climate conditions and disease outbreaks for dengue (DEN) in Southeast Asia. Plots of this figure were created using R software (version 3.4.1) in support of the journal publication [12]. Credit: Anyamba, A., Chretien, J-P., Britch, S. C., Soebiyanto, R. P., Small, J. L., Jepsen, P., Forshey, B. M., Sanchez, J. L., Smith, R. D., Harris, R., Tucker, C. J., Karesh, W. B., Linthicum, K. J

Science Results in the Private and Public Realms

Exploratory data analysis and visualization are inherent and critical processes for science teams and are performed on a constant basis. And while the publication of the journal paper [12] unveils the methodology and data analysis behind the research study, it has been written for a specific type of audience—subject matter experts (SMEs) and the scientific community. Visualizations in the format of plots (Figs. 18.6, 18.7) included in the published paper are created to validate results and support the scientific process and follow the highly technical language of the paper. In order to communicate results to broad audiences using visualization, it is worthwhile to take a look at the different types of audiences and also the practices of visualization involved. In 1990, DiBiase penned the article *Visualization for Earth Sciences* [19], and he articulated the range of functions of visual methods in an idealized research setting, within the private and public realms and laid out the challenges for the field. In the paragraph below, we explain these realms and their differences and augment DiBiase's suggested range of functions.

A scientist or a science team explores data to reveal pertinent questions and test hypotheses based on data. They employ visualization to validate data, and the resulting visualizations may stay within the science team—the private realm of the science team. As part of their research practice, scientists share their results and findings with the scientific community by incorporating visualization techniques to showcase their data analysis in scholarly publications and through presentations in professional conferences. Even though they are traversing from the private to the public realm, their visualizations often require subject matter expertise, and they are developed mostly to communicate results within the scientific community. Bringing the findings to the attention of a wider audience and the general public is a different practice, as it [...] "is quite a different matter to compel attention and understanding in a diverse, hurried, skeptical population of readers than to communicate with an eager, familiar group of associates," [19] and this requires a shift from the practice of visual communication of results toward the craft of visual storytelling with data. This [...] transposition of a tentative personal investigation to a public one is a synthetic process. Synthesis in this sense entails summarizing and generalizing the results of exploratory and confirmatory analyses and articulating a new, integrated conception of how the components of the research problem interrelate. It is a bridge from the private to the public realms [19]. NASA's Scientific Visualization Studio (SVS) specializes in creating explanatory visualization productions to serve both the scientific community and the general public, while respecting the needs of each audience—something that presents its own set of challenges.

Cinematic Scientific Visualization Production: Process

Cinematic scientific visualization makes data and science results more accessible. By combining the powers of art and the craft of storytelling with movies, it brings science to the attention of a broader audience and makes scientific data interesting to the eyes of the general public. As Scott McCloud says "The creation of media can serve as windows into our world" and, in that sense, cinematic scientific visualizations serve as windows of re-entering our world, "and when you do that it allows people to triangulate the world they live in and see its shape" [17].

In the paragraphs below, we present the team and give an overview of the storytelling process of creating visualization productions to communicate the results of Dr. Anyamba's study [12]. The overall process described can serve as a guiding template to similar productions. It is important to mention that each project typically presents its own set of requirements and limitations, therefore the process will need to be adjusted accordingly. In addition, providing a detailed guide of cinematic scientific visualization production is outside the scope of this article. The course publication [14] offers a deep dive into this topic and sheds light even into technical aspects of such productions, including workflows and pipelines for various formats.

As soon as a project comes into SVS, a meeting is scheduled with participating members: lead visualizer, science team, SVS Director (Horace Mitchell), and SVS Team Lead (Greg Shirah). During that initial meeting, the science team describes the research work, including the methodology used, the data involved, and unique aspects of the findings. Typically, a producer and a science writer are assigned to support the visualization production, contribute to the story script, coordinate press releases, and create narrated videos if needed. For this specific project, the visualization production team consisted of:

Science Visualizers: Helen-Nicole Kostis (Lead), Greg Shirah (SVS Team Lead)

Producer: Matthew Radclif (Lead), Helen-Nicole Kostis

Scientists: Assaf Anyamba (Lead), Radina Soebiyanto

Science/Data Support: Jennifer Small

Most visualizations are incremental work, and they reflect a deep background that includes infrastructure, processes, and institutional knowledge. Therefore, the first author sees all team members of the SVS as contributors to each visualization. It should be also mentioned that in this production, Team Lead Greg Shirah provided oversight to the entire project and refined the process that mapped the disease locations as particles during render time.

As it typically is in such productions, after the initial meeting, the immediate next step is to receive data from the science team, process them, and ingest them into the production infrastructure and pipeline (e.g., Maya, Pixar's Renderman) to produce drafts as soon as possible with the goal to *see* what the data look like.

A Practical Guide for Storytelling in Scivis Production

Foundational Principles
Data integrity and Scientific Accuracy

- **Purpose:**
  - Why am I creating the visualization?
  - Who is the audience?
  - What are the key messages?
  - How will it be consumed? (Medium, Format, Resolution(s))

- **Content:**
  - What data are needed to tell the story?

- What are the types of data? (e.g., information, scientific, or both)
- What is the dimensionality of the data? (e.g., 2D, 3D, Volumetric)
- What is the source of the data?
- Who funded the data collection?
- What are the data relationships?

• **Structure:**

- What is the relationship with our world? (scale, time, and space)
- What is the best visual encoding and/or visual metaphor for the type(s) of data? Encoding is typically dependent on types of data (e.g., information, scientific, or both)

• **Abstraction:**

- What is relevant to the story? (keep it and remove the rest)
- What relationships are critical to illustrate and include? (remove the rest, what is not included is equally important to what is included)
- Less is more? Yes, except when it is not. Science presents complex phenomena, with many relationships and lots of data. Sometimes, it pays off to keep parts where the visualization is visually complex to carry that message to your audience.

• **Composition:**

- What are the visual variables?
- What textual elements are needed?
- What are the dynamic variables?
- Framing: What is in the frame and why? What is the camera choreography?
- Lighting: What is the setup?
- Rendering: What are the inputs and outputs ?
- Post-Production: What is done in post-production and why?

Please note that the above guide is an adaptation from multiple sources, by reading articles, blog posts, and attending storytelling and visualization presentations. It is a work in progress and is continually refined. Whether one uses this guide or not, it is important to uphold the two foundational principles of scientific storytelling—respect the integrity of the data and the accuracy of the science.

Early on, the visualization production team identified two challenges based on previous experience with El Niño-associated productions that would be critical in defining the storytelling approach and subsequently the visualization design:

1. How to visualize the concept of teleconnections for the given study?
2. How to convey the broader story given the multiple regions, types of diseases, varying patterns, and timescales?

Through a series of iterations and close collaboration with the science team, the storyboard was refined (Fig. 18.8), and the production team decided to address challenges 1 and 2 above by breaking down the story into four separate visualization productions and by combining on-screen visualization of scientific and information

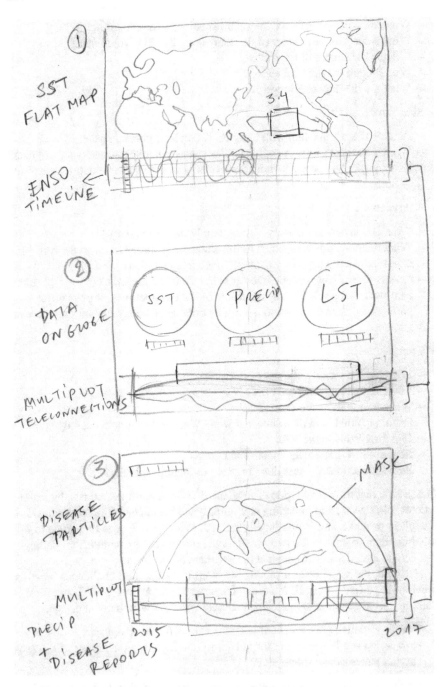

Fig. 18.8 Storyboard sketches of visualization production, Credit: Helen-Nicole Kostis, University Space Research Association and NASA Goddard Space Flight Center

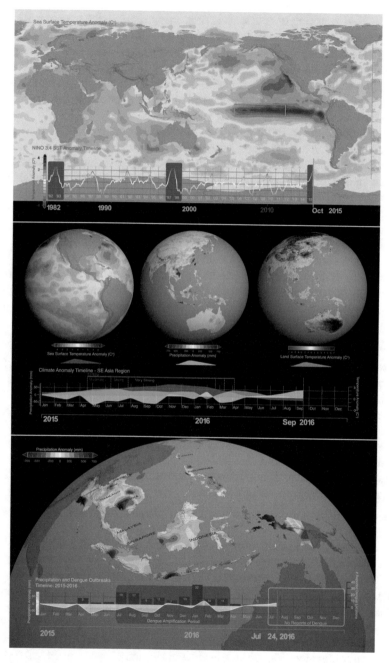

Fig. 18.9 Representative frames of produced visualizations. Credit: Scientific Visualization Studio, NASA Goddard Space Flight Center

data. This led to pairing remote sensing data (mapped onto a flat map or a 3D globe) with plots and multiplots within the same frame in all visualizations to support continuity from one visualization production to the next one, when shown in sequence. In addition, to tell the broader story of the research given the available resources the production team decided to select a case study and illuminate the relationships of a single disease (in this case dengue outbreaks) within a specific region (in this case South East Asia) for the period of 2015–2016.

Dengue fever is a painful, debilitating disease, transmitted between people by mosquito vectors. It is a predominantly tropical disease affecting approximately 400 million people annually in many areas of the global tropics including Southeast Asia. Dengue epidemics worldwide occur in urban areas where there is a coincidence of large numbers of vectors and people with no immunity to one of the virus types. During the 2015–2016 El Niño event, the South East Asia region received below than normal precipitation resulting in drier and warner than normal conditions, which increased the populations of mosquito vectors in urban areas, where there are open water storage containers providing ideal habitats for mosquito production. In addition, the higher than normal temperature on land shortens the maturation time of larvae to adult mosquitos and induces frequent blood feeding/biting of humans by mosquito vectors resulting in the amplification of dengue disease outbreaks over the South East Asia region.

The four separate visualization productions were:

- *Visualization 1*: Showcase the relationship between sea surface temperature (SST) anomalies and Niño 3.4 index for the period 1982–2017 and highlight major El Niño years.
- *Visualization 2*: Demonstrate the concept of El Niño-Southern Oscillation (ENSO) teleconnections in South East Asia for the period 2015–2016.
- *Visualization 3*: Reveal the relationship between precipitation in South East Asia and dengue disease outbreaks.
- *Visualization 4*: Reveal the relationship between land surface temperature in South East Asia and dengue disease outbreaks.

For a complete description of the productions, please see *Appendix A* at the end of this chapter.

These four visualizations were created in close collaboration with the science team. The visualizations were incorporated in a narrated video production to accompany an official NASA press release [21] and social media posts announcing the findings of the article. Upon approval from the science team, visualizations were released to the public through the SVS website [25–28]. Customized 4 K versions of these visualizations were created for hyperwall (tiled displays) that exist within NASA Goddard facilities and for traveling systems that are exhibited in scientific events and conferences. Dr. Assaf Anyamba and the team have been using the visualizations in presentations. Currently, the science and storytelling team is building on this work as they are focusing on creating visualizations to demonstrate differential teleconnection response patterns between El Niño and La Niña events and Rift Valley fever outbreak patterns in the region of South Africa.

Appendix A

Detailed descriptions of visualization productions.

Visualization 1:   Data-driven visualization of the El Niño phenomenon for the period 1982–2017. This production visualizes monthly sea surface temperature anomalies (SST) around the world from 1982 to 2017 and the Niño 3.4 index on a corresponding timeplot graph. The Niño 3.4 index represents average equatorial sea surface temperatures in the Pacific Ocean from about the International Date Line to the coast of South America. In the timeline, the major El Niño event years are highlighted. In these years, SST anomalies peaked for example during 1982–1983, 1997–1998, and 2015–2016. This visualization production creates visual associations between increases in sea surface temperature anomalies displayed on a flat map, Niño 3.4 indices in the timeplot, and the actual El Niño events. To learn more, please visit: https://svs.gsfc.nasa.gov/4695.

Visualization 2:   ENSO teleconnections in South East Asia Region during the El Niño event. The production starts in 2014 showing sea surface temperature (SST) anomaly data on a 3D globe. As time passes, in 2015 sea surface temperature anomalies in the equatorial Pacific Ocean (left) give rise to precipitation (center) and land surface temperatures (right) anomalies in Southeast Asia during the period 2015–2016. A multiplot on the bottom illustrates the interplay of Niño 3.4 index values and regional values for the South East Asia region of land surface temperature anomaly and precipitation anomaly. In the multiplot, the dengue outbreaks period is highlighted. Higher than normal land surface temperatures and therefore drier habitats drew mosquitoes into populated, urban areas containing the open water needed for laying eggs. As the temperature increased, mosquitoes had the urgency to bite more frequently but also reproduce and mature faster, resulting in an overall increase in population and mosquito bites, therefore the dengue outbreaks. To learn more please visit: https://svs.gsfc.nasa.gov/4697.

Visualization 3:   Visualize the correlations of precipitation, disease reports, and dengue outbreak period for the South East Asia region. The corresponding time-plot reveals the relationship between precipitation anomaly in Southeast Asia and dengue outbreaks. Drier than normal habitats draw mosquitoes into populated, urban areas containing the open water needed for laying eggs. Drier conditions induce higher than the normal temperature which have similar impacts as above. As time unfolds, dengue reports are mapped on to the region and the periods of dengue amplification are highlighted. To learn more, please visit: https://svs.gsfc.nasa.gov/4693.

Visualization 4:   Visualize the correlations of land surface temperature, disease reports, and dengue outbreaks period for the South East Asia region, using similar techniques with visualization 3 above to link the aftermaths of the drought persistence (lack of precipitation) to increased temperature on land, which increased mosquito vectors in the region. As time unfolds, dengue reports are mapped onto the region and the periods of dengue amplification are highlighted. To learn more, please visit: https://svs.gsfc.nasa.gov/4696.

Appendix B

Data sets used:

- **Remote Sensing Data**:

  - Sea Surface Temperature (SST) Anomaly, time series: 1982–2017. Global monthly SST data known as optimum interpolation (OI) SST version 2 data set produced by NOAA can be accessed from: https://www.ncdc.noaa.gov/oisst.
  - Land Surface Temperature (LST) Anomaly, time series: 2002–2017. Global monthly 0.05" LST MOD11C3 data set available at https://lpdaac.usgs.gov/dataset_discovery/modis/modis_products_table/mod11c.
  - Precipitation Anomaly, time series: 2002–2017. Global Precipitation Climatology Project (GPCP) Global 1° Monitoring Product, available at: ftp://ftp-anon.dwd.de/-pub/data/gpcc/html/-monitoring_download.html
    .

- **Information Data**:

  - Disease reports for South East Asia region. All global disease occurrences are georeferenced as sourced from https://www.promedmail.org/.
  - Niño 3.4 SST index. It can be obtained from the National Oceanic and Atmospheric Administration (NOAA)'s National Center for Climate Prediction online archives at: http://www.cpc.ncep.noaa.gov/data/indices/sstoi.indices.
  - Land Surface Temperature (LST) average for SE Asia Region (monthly time series: 2015–2016). Processed and provided by the Science Team for the visualization production.
  - Precipitation average for SE Asia Region (monthly time series: 2015–2016) Processed and provided by the Science Team for the visualization production.
  - Numbers of disease reports for SE Asia Region (monthly 2015–2016). Processed by Science Visualizer (Lead) from global disease occurrences provided from https://www.promedmail.org and approved by the Science Team.

- **Cartographic Data:** (developed internally at SVS)

  - Country outlines
  - Water mask (global)
  - South East Asia region mask
  - Latitude coordinates
  - Niño 3.4 region subset

References

1. Conceptual Image Lab. https://svs.gsfc.nasa.gov/cilab/index.html
2. IDL: Interactive Data Language. https://www.harrisgeospatial.com/Software-Technology/IDL

3. LANCE: NASA near real-time data and imagery | earthdata. https://earthdata.nasa.gov/earth-observation-data/near-real-time
4. NASA's Black Marble research team: https://blackmarble.gsfc.nasa.gov/#people
5. ProMED international society for infectious diseases (2019). https://www.promedmail.org/
6. 'Science is not finished until it's communicated' - UK chief scientist (2013). https://www.climatechangenews.com/2013/10/03/science-is-not-finished-until-its-communicated-uk-chief-scientist/
7. OpenStreetMap: https://www.openstreetmap.org/
8. Earth Observing System Data and Information System (EOSDIS) | earthdata (2019). https://earthdata.nasa.gov/eosdis
9. Earth Science Data Systems (ESDS) Program | earthdata (2019). https://earthdata.nasa.gov/esds
10. Global Imagery Browse Services (GIBS) | earthdata (2019). https://earthdata.nasa.gov/eosdis/science-system-description/eosdis-components/gibs
11. Worldview | earthdata (2019). https://earthdata.nasa.gov/worldview
12. Anyamba, A., Chretien, J.P., Britch, S.C., Soebiyanto, R.P., Small, J.L., Jepsen, R., Forshey, B.M., Sanchez, J.L., Smith, R.D., Harris, R., Tucker, C., Karesh, W., Linthicum, K.: Global disease outbreaks associated with the 2015–2016 El Niño event. Sci. Rep. **9**(1), 1930 (2019). https://doi.org/10.1038/s41598-018-38034-z
13. Boller, R.: Worldview: explore your dynamic planet (2019). https://worldview.earthdata.nasa.gov/
14. Borkiewicz, K., Christensen, A., Kostis, H.N., Shirah, G., Wyatt, R.: Cinematic scientific visualization: the art of communicating science. In: ACM SIGGRAPH 2019 Courses, Article No.: 5, p. 1–273.. ACM (2019). https://doi.org/10.1145/3305366.3328056
15. Carlowicz, M.: Lights out in Michael's wake. https://earthobservatory.nasa.gov/images/92887/lights-out-in-michaels-wake
16. Carlowicz, M.: Night light maps open up new applications. https://earthobservatory.nasa.gov/images/90008/night-light-maps-open-up-new-applications
17. Cisneros, M.: You are an Artist + Viz and Tell (2018). https://www.meetup.com/Data-Visualization-DC/events/256745546
18. Cox, D.: Renaissance teams and scientific visualization: a convergence of art and science. In: Collaboration in Computer Graphics Education, SIGGRAPH'88 Educator's Workshop Proceedings, pp. 81–104 (1988)
19. DiBiase, D.: Visualization in the earth sciences. Earth Miner. Sci. **59**(2), 13–18 (1990)
20. Garner, R.: NASA's Goddard Space Flight Center (2019). https://www.nasa.gov/goddard
21. Garner, R.: 2015-2016 El Niño triggered disease outbreaks across globe (2019). https://www.nasa.gov/feature/goddard/2019/2015-2016-el-nino-triggered-disease-outbreaks-across-globe
22. Gómez, J., Blinn, J., Em, D., Rueff, S.: SIGGRAPH 2017 history of the JPL computer graphics lab (2017). https://www.youtube.com/watch?v=ctvLmjonqyE
23. Gray, E.: Night lights show slow recovery from Maria. https://earthobservatory.nasa.gov/images/144371/night-lights-show-slow-recovery-from-maria
24. Kalb, V.: NASA Black Marble Product (2019). https://viirsland.gsfc.nasa.gov/Products/NASA/BlackMarble.html
25. Kostis, H.N.: Land surface temperature anomaly and dengue outbreaks in South East Asia region: 2015–2016 (2019). https://svs.gsfc.nasa.gov/4696
26. Kostis, H.N.: Niño 3.4 index and sea surface temperature anomaly timeline: 1982–2017 (2019). https://svs.gsfc.nasa.gov/4695
27. Kostis, H.N.: Precipitation anomaly and dengue outbreaks in South East Asia: 2015–2016 (2019). https://svs.gsfc.nasa.gov/4693
28. Kostis, H.N.: ENSO teleconnections in South East Asia for the period of 2015–2016 (2019). https://svs.gsfc.nasa.gov/4697
29. Mitchell, H.G.: Scientific visualization studio. https://svs.gsfc.nasa.gov/

30. Román, M.O., Stokes, E.C., Shrestha, R., Wang, Z., Schultz, L., Carlo, E.A.S., Sun, Q., Bell, J., Molthan, A., Kalb, V., et al.: Satellite-based assessment of electricity restoration efforts in Puerto Rico after Hurricane Maria. PloS One **14**(6), e0218883 (2019). https://doi.org/10.1371/journal.pone.0218883
31. Román, M.O., Wang, Z., Shrestha, R., Yao, T., Kalb, V.: Black Marble User Guide version 1.0 (2019). https://ladsweb.modaps.eosdis.nasa.gov/missions-and-measurements/viirs/VIIRS_Black_Marble_UG_V1.0_March_2019.pdf
32. Román, M.O., Wang, Z., Sun, Q., Kalb, V., Miller, S.D., Molthan, A., Schultz, L., Bell, J., Stokes, E.C., Pandey, B., et al.: NASA's Black Marble nighttime lights product suite. Remote Sens. Environ. **210**, 113–143 (2018). https://doi.org/10.1016/j.rse.2018.03.017
33. Roy, D.P., Ju, J., Kline, K., Scaramuzza, P.L., Kovalskyy, V., Hansen, M., Loveland, T.R., Vermote, E., Zhang, C.: Web-enabled Landsat Data (WELD): Landsat etm+ composited mosaics of the conterminous United States. Remote Sens. Environ. **114**(1), 35–49 (2010)
34. Wang, Z., Shrestha, R., Román, M.: NASA's Black Marble Nighttime Lights Product Suite Algorithm Theoretical Basis Document (ATBD), version 1.0, April 2018 (2018). https://viirsland.gsfc.nasa.gov/PDF/VIIRS_BlackMarble_ATBD_V1.0.pdf
35. Wolfe, R.E.: LAADS DAAC. https://ladsweb.modaps.eosdis.nasa.gov/

Chapter 19
Reaching Broad Audiences from a Science Center or Museum Setting

Anders Ynnerman, Patric Ljung and Alexander Bock

Abstract Research has shown that learning outcomes can be improved by interactive visualization and exploration. This has led to the appearance of interactive installations on a range of platforms from handheld devices to large immersive dome theaters. One of the underlying principles of this data-driven visualization for broad audiences is the notion of the confluence of exploratory and explanatory visualization into the concept of "Exploranation," meaning that explanation and exploration converge in the same application. However, it is necessary to apply specific visualization and interaction design principles to enable engaging storytelling and user-driven discovery in interactive installations targeting a general audience. The design principles are unique for different platforms and uses. We here present an account for some results, challenges and areas in need for further research. We also describe a set of different cases in which visualization has been used to reach broad audiences. Based on the examples, lessons learned are described and general principles and recommendations are provided.

19.1 The Need for Visualization in Science Centers and Museums

Visual representations have traditionally played a fundamental role in the quest to convey knowledge about the content of museum collections containing items such as rare historical artifacts or valuable pieces of art. They also serve a purpose in framing events such as the narratives of pivotal moments in human history. In science centers, visualization, and particularly interactive visualization, is one of the core technologies in the illustration of scientific knowledge, major findings and method-

A. Ynnerman (✉) · P. Ljung · A. Bock
Linköping University, Norrköping, Sweden
e-mail: anders.ynnerman@liu.se

© Springer Nature Switzerland AG 2020
M. Chen et al. (eds.), *Foundations of Data Visualization*,
https://doi.org/10.1007/978-3-030-34444-3_19

©at Magnus Johansson

Fig. 19.1 Leading museums and science centers in the world are to an increasing degree using interactive visualization. The image shows the interactive mummy installation at the Mediterranean Museum in Stockholm. Photograph courtesy of Ove Kaneberg/Medelhavsmuseet

ologies across all disciplines, and to describe how everyday technology works and the ongoing innovation process behind them (Fig. 19.1).

The list of museums in the world contains the most well-known such as the British Museum in London and the Louvre in Paris, as well as smaller regional museums of limited size and scope. There is also a clear distinction between science museums, such as the American Museum of Natural History in New York and the Natural History Museum in London, and science centers, such as the Exploratorium in San Francisco. The latter prioritizes engagement of visitors through interactive installations and hands-on experimentation. These large museums and science centers have large-scale in-house research units in the areas of focus and are also developers and curators of both static and traveling exhibits. In the production of exhibits, external partners are often contracted or partnered with, and visualization research institutes and companies are frequently consulted. It should be noted that the vast majority of museums and science centers are, however, of smaller scale with smaller or no research and development capacity and are depending on external partners to produce exhibits and individual installations.

As it will be argued in this chapter, visualization has an increasingly important role to fill in all of these public venues. This is happening in the wake of visualization technology development, data availability, visualization methodology and the increasing role of visual media in our daily lives, as well as the continuous strive to increase visitor engagement in the collections and to increase amount and quality of knowledge provided.

19.2 Exploranation—Explanation Through Exploration

The increased gathering and processing of scientific data, may it be medical, climate, geophysical or various other types, ultimately needs to be conveyed into contexts suitable for different scenarios and for wide and diverse audiences. Intelligent functionality and guidance on how the scientific information should be presented is of increasing importance, and thus, an essential part of explanatory visualization efforts, such that the correct understanding of the resulting conclusion can be made by decision makers, single individuals and/or the general masses. The use of visualization in public spaces, such as science centers and museums, has traditionally been based on static representations of phenomena, often in the context of illustrations produced by exhibition designers and curators, or linear media such as videos shown on screens embedded in installations. In recent years, we are, however, seeing a rapid increase of the use of interactive visualization enabling visitors to explore scientific data from experiment or simulations. An important example of this is Earth observation and climate simulation data as described in Chap. 17. It should be noted that the use of interactive models is also becoming popular in learning situations such as understanding of the organ functions of the human body, but in this chapter, we will focus on how interactive visual exploration of scientific data can provide engaging and rewarding experiences for visitors with vastly different backgrounds in terms of age, language, culture, etc.

On a general note, it can be argued that the science of learning and communication and its practical application is facing a paradigm shift through the introduction of interactive visual data exploration into a field that has traditionally been dominated by explanatory approaches. In this on-going confluence of exploratory and explanatory approaches, we see many challenges, opportunities and synergies between the two visualization paradigms. To denote this convergence between exploratory and explanatory visualization, the term "Exploranation" was coined by Ynnerman et al. [30]. It can be noted that this was done in the context of research on science communication, but the convergence can be seen in many areas and uses of visualization. The underpinning trends that drive the introduction of exploranation in public spaces are based on several parallel areas of development: the availability of data which is one of the cornerstones in this change, the rapid development of computer hardware, in particular GPUs as an enabling factor, and finally, the maturing visualization methodology plays a fundamental role in providing the framework for meaningful and informative visual representations.

In our work, we have identified a set of core challenges in bringing interactive visualization to public spaces, and we will in this chapter elaborate and reflect on them in some detail. The challenges and demands, which in comparison to other uses of visualization, can be extreme, and solutions can from a visualization perspective be grouped into four different areas:

- **Rendering quality and performance:** Visitors to public spaces have varying levels of understanding of computer graphics and visualization. In particular, visitors of the "gaming generation" have high expectations on visual quality and performance. It is a challenge to convey the notion of connection to data, its sizes and complexity, which may affect the rendering quality and frame rates.

- **Interaction and navigation:** Complexity in interaction and navigation is one of the obstacles in both effective exploration and explanation using visualization. With a starting point in the understanding of human attention and awareness, user interfaces have to be tailored to specific audiences without hampering the freedom to explore. In doing this, apart from user and application parameters, aspects to consider are: time available for exploratory visualization, regions of interest in the data and specific communication goals.
- **Robustness and reliability:** Installations in public spaces or use of software for demonstrations with large, paying, audiences put extreme demands on reliability and robustness both from a software and hardware perspective. Elimination of potential point of failures in mechanical interaction is key as visitors will interact with devices in unforeseen and forceful manners and in public shows a software issue can cause both audience and presenter to lose immersion and engagement in the science presented.
- **Storytelling and performativity:** One of the key challenges in bridging the gap between the data and the visitor is to provide the user with guidance and explanations that do not interfere with the notion of exploration. This inevitably entails approaches to support mediated experiences through physical and/or virtual guides using various embodiments and storytelling approaches. The transfer of performativity between individuals, different presenters, in facilitated experiences is one major obstacle for widespread use of interactive live presentations in immersive environments such as planetariums.

We are still in the early days of exploranation in public spaces, and much research needs to be conducted to support this area, which has the potential to become one of the largest uses of visualization with impact reaching far beyond the traditional research and development domains targeted by visualization research. In the remainder of this chapter, we will, with a starting point in the work conducted at the Norrköping Visualization Center C, in Sweden, describe identified areas of challenges as well as provide example scenarios from science centers and museums.

19.2.1 The Norrköping Visualization Center C

Much of the work, challenges, approaches and results, come from the long-term build-up and 10 years of operations of the Norrköping Visualization Center C [26] and its connected research groups at Linköping University, in Sweden. The Center is unique in the sense that it combines several facets of research, development, public dissemination, commercialization and collaboration in one organization with close interplay between its units. The public part of the center has more than 150 000 visitors per year, and the infrastructure at the center is encompassing both public spaces, galleries and research laboratories. The primary facility is a fully immersive dome theater seating 100 visitors. At the time of writing, the dome is the only

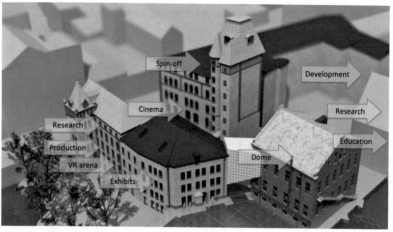

©at Kristoffer Jansson & Jonathan Klittmark

Fig. 19.2 Norrköping Visualization Center C. The mission of the center contains a combination of research, science communication, commercialization and spin-off and collaborations with society, as well as teaching university-level programs

dome in the world with full 6P laser projection capability, and with stereoscopic viewing on a 15 m diameter dome with 8 K resolution. The research unit of the center consists of research groups in Computer Graphics and Image Processing, Immersive Visualization, Information Visualization, Scientific Visualization, Graphical Design and Image Reproduction and Visual Learning and Communication. In total, 120 staff are employed in the environment at large.

With a starting point in the visualization research and its applications, the center has, since its inception in 2010, systematically worked on the introduction of interactive visualization in the public spaces at the center including both stand-alone exhibits and visualization environments targeting facilitated experiences. The comprehensive experience of science communication at the center, and the state-of-the-art visualization and computer graphics research conducted, the close proximity between research, application and public spaces, has made it possible for the center to spearhead many efforts in the introduction of interactive visualization in science communication and indeed leading to the notion of exploranation. The success stories, and failures, in this work form the basis for this chapter (Fig. 19.2).

19.3   Reflections and Challenges

In this section, we present perspectives on development and production of visualization in public spaces and highlight some of the challenges we have encountered in our work. These challenges are intended to provide concise examples of areas where research efforts are needed to promote the field and support producers of content for science centers and museums.

19.3.1 A Visual Learning and User Perspective on Visualization in Public Spaces

In developing visualization for public spaces, it is imperative to closely connect with state-of-the-art research in science education. It is also apparent that learning research has largely in the past been dealing with cognitive and sociocultural aspects of visual representation and multimedia learning [11, 19, 23]. Only recently, studies gathering and evaluating digital platforms, games, simulations, animations and traditional visualizations in teaching and learning have been conducted [7, 8, 22].

From this literature, we learn that the content itself, how it is presented, people's prior knowledge, interest and engagement, also affect the result. To be successful in developing new visualization approaches, the learning effect of interactive visualization has to be evaluated. In addition, design principles for different content have to be established. To understand how these new types of visualizations should be designed, we have to investigate how they stimulate exploration, engagement, interest and learning. With the emerging use of interactive data visualization in public spaces, there are unique possibilities to do so. In this multidisciplinary approach, involving both visualization and learning researchers will be necessary and will enable perspective studies on developed technology and methodology as well as ensuring sound implementation of visualization. In our work, we have identified several challenges that call for attention from the visual learning perspective such as (1) exploration of how storytelling, using explorative visualizations, affects human cognitive structure and experience, (2) investigation of the limits of the massive information that are necessary to integrate to understand complex data and (3) investigation of the learning effect of immersive virtual reality on students and other audience.

19.3.2 Provenance and Managing Heterogeneous Data to Bridge the Gap Between Experts and Laymen

To deliver an authentic experience of interactive exploration where the users are, in fact, interacting with the real scientific data poses several challenges. In communicating science, it is important to be able to answer where data is originating from, who produced it and with what methods. Provenance is broadly the record of activities performed by agents that produce (digital) artifacts, including presentations. The concept can be applied at different levels of granularity as well as with different perspectives [6]. Provenance is also an essential piece in reproducible research, as shown in the visualization community [10]. It can be seen that visualization in exploranation draws a similarity with reproducible research but also aims at providing a higher-level of interpretation by including provenance in the application software for user interaction as well, methods to aggregate interaction steps into high-level components for a visual scripting language, enabling nonlinear storytelling and integration of automated interaction, camera navigation and view selections. Preservation of the gathered data with provenance provides not only the final result but also a record of how science was conducted and tells a story in itself, provenance and preservation, how curators are incorporating annotations and interpretations and how exhibition designers provide stories and context. At the core of this lies the traceability and preservation of all related information and embedded knowledge that is made available to the public, and in this case, the museum or science center visitors. To support provenance in exploranation in public spaces, several requirements can be identified leading to areas that need further exploration and research: (1) underlying architectures to handle heterogeneous digital assets (scientific data, interpretations, narratives, storytelling scripts), (2) provenance tracking that enables the user to explore the presented data in ways previously not supported, such as going behind and beyond the data presented, potentially re-run a computation with different context or parameters and (3) the exploranation visualization process itself and the users' interactions also constitute a process where provenance enables further research into science communication and learning.

19.3.3   Tailored Visualization and Interaction Techniques

Scientific visualization methods and representations are as clearly demonstrated throughout this book typically developed for domain expert users that carry years of training and expertise. The visualizations are thus information-rich and are aiming to provide as much, and condensed, information as possible, at a rate the domain expert can handle.

In a science communication context, these assumptions no longer hold, the user is untrained on both the visual representations as well as operating advanced user interfaces. Experts often demand a large set of parameters to dynamically adjust, and advanced interaction schemes may further require frequent use to stay fluent. Experts often demand full exploration possibilities, resulting in interfaces with large sets of parameters and advanced interaction schemes, which may cause steep learning curves and require frequent use to stay fluent. In our prior work, such as the visualization table [16], see Sect. 19.5, we have gained valuable experience in tailoring advanced scientific visualization methods to be meaningful and easy to use. In this work, adaptation of visual representations and user interfaces is a key part in enabling a walk-up-and-use experience. It can be argued that in communicating science it is helpful to favor qualitative information over quantitative data. The purpose is to reduce the cognitive load on users without losing the contact with the original data. This entails tailoring and adaptation of established scientific visualization techniques and development of new approaches. An example of adaptation of the user interface for volume rendering is shown in Fig. 19.3, where the visual representation has been changed and the parameter space was condensed.

Challenges in tailoring visualization and interaction are found in a wide range of visualization areas. Scalable visualization (from smart phones to domes and large displays), how to efficiently use display real-estate at varying scales, has been a focus for our research [4]. Another rather unexplored area is visual representations of provenance data, see Sect. 19.3.2, and interaction techniques to analyze and modify data to create derived provenance data. In short, a visual language to explain and describe scientific processes and data is needed. Some examples of core requirements for tailoring of visualization and interaction that we have addressed can be summarized as: (1) de-cluttering of display surfaces for clarity of meta-level instructions and deep exploration, (2) use of established design principles to produce aesthetically pleasing rendering to promote engagement (illuminations, color schemes, etc., (3) support for non-invasive and embedded storytelling tools and (4) use of illustrative techniques for emphasis of salient features.

19.3.4 Support for Collaborative and Performative Interaction

In an imminent future, the need for intuitive and autonomous approaches to support the presentation of scientific knowledge will be of crucial importance. The scenarios and target applications at hand are affecting which and what interaction that can be used and the kind and quantity of information that can be conveyed in a facilitated presentation. To develop intuitive interaction and view support, reducing the cognitive load on users, and presenters, in various science communication scenarios is essential.

©at Daniel Jönsson

Fig. 19.3 Simplified interface and visual representation using dynamic image galleries. The user can interactively explore volumetric data without the non-intuitive transfer function concept and select color and opacity settings for the full range of data values

©at Thor Balkhed

Fig. 19.4 Presenter using interactive visualization in a dome theater need high-level interfaces that must not interfere with the narrative of the presentation. The system should automatically produce and conduct camera moves to enhance salient features matching the narrative. Photograph: Thor Balkhed, Linköping University

An example of a guided scenario is shown in Fig. 19.4. In this, understanding of how exploratory immersion can be engineered and orchestrated is pivotal, and knowledge can be drawn from the abundance of literature on insights to human attention and awareness.

Essential parameters to consider in facilitated experiences are estimated available time with the target user and which parts of the scientific data that are of key importance for the exploratory presentation. This in turn calls for the use of semi-automated control of the interaction and viewing parameters during a facilitated live presentation. In the context of the OpenSpace project [4], we have investigated support for such interaction, resulting in a high-level scripting for interactive nonlinear storytelling with integrated support for automated interaction, camera navigation and view selections. It is also of high interest, in addition to direct interaction, to explore capturing of speech and gestures as input for automation of interaction and navigation. Topics that are of highest interest for future exploration in this domain are: (1) dynamic and contextual visual salience detection for optimization of camera views and visual layout in immersive environments, (2) multi-modal and multi-user interaction such as context-dependent gesture recognition and speech recognition to support facilitated interactive visualization and (3) constrained camera path optimization and even learning-based approaches to view selection.

19.3.5 Natural User Interfaces and Semantic Level Interaction

An important aspect of the learning experience of a visitor to a museum or a science center is the ability to directly interact with and discover notable aspects of the artifact or datasets in a self-directed manner. These self-guided discoveries often result in follow-up questions that a traditional exhibition piece is ill-equipped to deal with, as it has to serve audiences with a large range of prior knowledge for the topic and thus a large range of potential questions. Traditionally, this challenge is addressed by the presence of a knowledgeable visitor guide who is able to answer detailed questions and thus customize the experience to the knowledge level of the particular visitor. However, particularly for a small science center, it is not possible to provide enough knowledgeable guides to satisfy this need for all incoming visitors at all times.

An exciting new development in computer graphics and data analysis that can help alleviate this constraint is the embodiment of the visualization/narration system itself as a virtual avatar, which could be manifested as a photo-realistic digital human [15]. The availability of these advanced query systems provide visitors with readily available access to heterogeneous, high-dimensional data that can further augment an exhibition piece or an installation. One of the most natural human-friendly approaches to interfacing with the system is to simply be able to ask informed questions about the application domain and the exhibition piece, and to expect contextualized, insightful answers. Regardless of whether the modality used in the interfaces is an avatar or visualization of complex data, one still wants to ask questions as if one is having a dialogue with a literate expert. As an example, new levels of interaction with human representations are rapidly opening up through recent advances in visual representation of human avatars imaging using light field displays supporting virtual human style interaction [14, 24].

In recent years, the development of these expert query systems have rapidly accelerated such and our prediction is that soon they will be feasible for smaller museums and science centers. IBM introduced Watson, a technology platform based on natural language processing and machine learning intended to analyze large amounts of unstructured data, and to query the results to provide deep insight about a particular application domain. The combination of natural interaction with human representations using technologies such as light field displays, where avatar representations act as conveyors of information and ultimately, with such a Q&A systems could lead to a virtual museum guide that would further increase the possibilities to provide engaging and explainable visualization for general audiences. For this particular example, significant challenges need to be met, before the vision can be realized, in areas such as (1) light field capture and synthesis, (2) light field compression and on-screen rendering and (3) the development of a Q&A system designed for contextualization and presentation of scientific data.

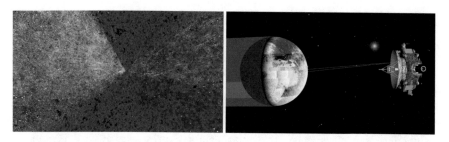

Fig. 19.5 Important aspect of the communication of scientific discoveries is explaining the vast scale of the cosmos. (left) Showing the available deep sky surveys illustrating the large-scale structure of galaxy distribution, the earlier existence of quasars, and the first light of the universe, the cosmic microwave background radiation, behind. (right) Smaller scale exploration of the New Horizons spacecraft taking humanity's first pictures of Pluto in 2015

19.4 Facilitated Experiences in Large-Scale Immersive Environments

Astronomy and the exploration of space is perhaps the primary example of a research area where science communication has played a central role. Historically, the general interest in human understanding of the cosmos has been one of the driving forces behind scientific exploration. It is therefore no surprise that the concept of a planetarium showing a full night sky with projections of the stars and planets has been the model for facilitated immersive science communication. With the digitalization of planetariums, replacing analog high-precision instruments with projection systems, the planetarium has changed into a general-purpose immersive dome theater in which engaging stories of science can be told. The digital dome theaters are also used as high-end playback systems for rendered productions with varying themes. We will not consider prerendered material in this chapter, but focus on interactive use with a knowledgeable facilitator interacting with the audience and dynamically adapting the content displayed (Fig. 19.5).

Despite the possibility to tell interactive stories in dome theaters using any scientific content, the use is primarily in the astronomy domain and vendors of planetarium solutions are providing software that supports interactive exploration [9, 18, 28, 29]. The software Uniview is one of these exceptions and contains interactive visualization of neuroimaging data in dome theaters. A recent addition is also the molecular dome project which will enable interactive browsing of large-scale molecular data based on animations by Berry [1], and the work on interactive molecular visualization software by Muzic et al. [25]. However, given the demands put on the facilitator still today few places in the world have live facilitated presentations in the regular public program, and most experiences are still in the astronomical domain. We thus here provide description of such a case and extract experiences and recommendations.

19.4.1 Interactive Astrovisualization in Digital Dome Theaters

Visual representations of data have been employed in the field of astronomy since the very beginning. Early human's attempts of creating structure in the night's sky by connecting stars to create familiar images, the constellations, show this as much as Galileo Galilei using drawings of the positions of the moons of Jupiter to gain insight into their movements. The abstract nature of many astronomical phenomena, such as orbits, lend themselves especially for using visual abstractions for knowledge discovery and for use in explanations to broad, public audiences. The need for visual representations has dramatically increased in the last decades with the increased availability of open data and systematized information flow. For example, the Apollo astronauts did not only use their cameras to take accurate photographs used for scientific discoveries, but they were also trained in photography to make these images as appealing as possible without compromising their scientific value. This has continued into the present with, for example, the Juno mission orbiting Jupiter, carrying a camera whose sole purpose is to be controlled by citizens and thus support and inspire the next generation of scientists and to stimulate scientific curiosity in its own right. Missions, such as the Apollo endeavor, are one important pillar in bridging the gap between astronomy experts and the interested public, and the endorsed scientific explorations are indirectly influenced by the level of public engagement in the missions.

The second important pillar of this bridging is the engaging communication of scientific findings from the scientists back to the public. While this is an ongoing effort in many fields, astronomical research might be the most advanced field to spearhead this movement as there is an innate interest in many people to space-related research. One of the last remaining obstacles for the expert knowledge dissemination is the segregation of software tools used for public presentations and scientific software aimed at domain experts. This leads to the challenge that scientists are required to invest a large effort to prepare and conduct presentation of their scientific findings and supporting data, thus preventing many of these presentations from occurring. From our experience, most researchers are highly interested in disseminating their scientific findings to the public if there is an easy channel for doing so that does not significantly distract them from their main scientific work. Conversely, the general public is interested in hearing and learning about scientific discoveries from the scientists themselves as it provides a closer relationship to the acquired scientific data.

To support domain experts in astronomy and space science, OpenSpace [3], an open-source astrovisualization software that serves different visualization environments, such as desktop environments, virtual reality systems, as well as planetarium venues, was developed. The software package provides the general user with tools to visualize and contextualize astronomical research accurately and, at the same time, enables scientists to perform their analysis to generate these discoveries. Given that the software and the majority of the data is freely available, it can then be used for public presentations either by the scientists themselves or other interested parties in addition to being provided to the general population for self-directed learning on their own home computer [4]. In addition, as both discovery and dissemination use the same software, the burden of converting data is much lower, providing more opportunities for the scientists to dedicate time to conduct presentations. One additional approach to increase the reach of these public presentations is through the use of *Astrocasting*, in which multiple remote instances of OpenSpace are linked. Linking the visualizations and streaming of video and audio of the presenter, this further increases the reach of these presentations to include other planetariums and users at home on their own connected computers. Furthermore, this enables new presentation modalities as multiple geographically separated presenters can now present the same topic to a broad audience. Alternatively, this technique also allows a team of scientists that is not in the same physical location to share a virtual space around their data to discuss findings and thus improve team communication. This is yet another area in which both public presentations as well as scientific discoveries can utilize the same techniques.

OpenSpace has been used in a number of live presentations as an interactive visualization tool and continues to find increasing use for scientific discoveries. An example of a remote presentation is the fly-by of the New Horizons spacecraft at Pluto, which was simultaneously visualized at 13 planetariums with an accompanying online video stream [5]. During the event, participating planetariums around the world submitted their audiences' questions to the mission scientists who were commenting on the actions of the spacecraft in real-time. This was particularly of interest as the event coincided with the closest approach of the spacecraft at Pluto. This way, this remote presentation was able to reach more than 2000 people around the world simultaneously that could actively participate in this scientific discovery. At the same time, for many of the mission scientists, it was the first time they saw the mission plan which was devised years ahead of time executed in a 3D environment.

Another example was a presentation given by an Apollo Lunar Module Pilot, who provided a detailed explanation of the Apollo 16 mission using data from the Lunar Reconnaissance Orbiter, see Fig. 19.6. During the presentation, Charlie Duke revisited the mission he was part of and elaborated on the interesting stops made during their EVAs during their 71 h stay on the Moon, while OpenSpace was providing the visual context to these explanations, including the visualization of the distance travelled from the lunar module. Integrating the ability to render high-fidelity terrain models of planetary bodies [2], such as the Moon, that include high-resolution imagery as well as 3D terrain models, is of vital importance for the presentations as they provide the necessary context for the explanation that is unable to be provided by static pictures alone. Being able to record the explanations from these pioneers of space exploration will be increasingly important as sadly fewer of these explorers are able to tell their stories directly as the years go by (Fig. 19.7).

Another use case for the interactive use of such a software is for team communication and hypothesis generation. An example for this case were discussions surrounding the release of the Gaia DR2 dataset, which contains detailed information for more than a billion near-by stars in the Milky Way. Algorithms had to be

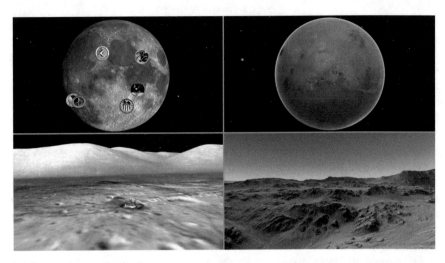

Fig. 19.6 (top left) Visualization of all six Apollo landing sites rendered using OpenSpace. High-resolution images for all landing sites are publicly available. (bottom left) The Apollo 17 landing site rendering using OpenSpace and data from the Lunar Reconnaissance Orbiter. The tracks from the Lunar Rover are clearly visible. (top right) Humanity's next target for human exploration, Mars, as so far only been visited by 11 landers that provided data back from this distant world of which a large amount of surface data is available (bottom right)

Fig. 19.7 Particularly in the case of the Moon, it is important to faithfully recreate the observed data to match the available observed imagery. (left) The Earthrise picture taken on December 24, 1968 by Apollo 8. (right) Recreation of the same moment in time in OpenSpace, showing the same lunar features and the Earth in the background

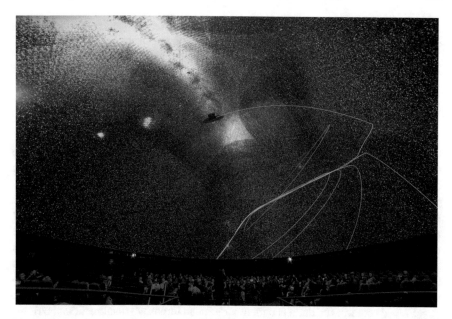

Fig. 19.8 Gaia Science Sprint at the Hayden Planetarium in New York. OpenSpace was used to interactively visualize the Gaia DR2 dataset for a large group of domain experts. The image shows a rendering of 900 million stars on the planetarium dome and the Gaia spacecraft and a few iterations of its orbit. The event is an example of how software and data can be used for both public dissemination of science as well as team communication and scientific exploration by domain experts. (c) 2019 IEEE. Reprinted with permission

developed to efficiently render this large number of stars simultaneously in 3D inside OpenSpace. These capabilities were presented to a Gaia science sprint in New York in 2018 to a large group of astrophysicists whose research is focused on the results of the Gaia mission and who are thus intimately familiar with the dataset. One of the sessions involved the scientists inspecting the data in a large planetarium and being able to dissect the dataset interactively, see Fig. 19.8. In addition to generating new hypotheses, this session was also used to confirm prior hypotheses of the movements of subsets of stars for which velocities are available. This makes it possible to animate the position of the stars over a range of a hundred of thousands of stars and enables the expert to visually detect co-moving stars, or clusters with low relative velocities. The enthusiasm and engagement of the scientists in adopting new tools is a very promising result of this collaboration.

A future class of use cases is going to be enabled through the use of recorded or procedurally generated flight paths. This will enable the use of OpenSpace inside other learning environments, such as school classrooms, by facilitating semi-interactive presentations. This kind of presentations are mostly scripted, but contain episodes of self-driven discovery, which can increase the learners appreciation for the dataset. Semi-interactive techniques can also be utilized in planetarium environments that provide regularly scheduled shows by providing a mix-and-match approach of storytelling, which can adapt based on the needs, or questions, of the audience by providing an array of options to facilitate a nonlinear visual storytelling approach.

19.4.2 Experiences and Design Recommendations

Creating prerendered and interactive visualization content for use in large-scale immersive environments require the consideration of a number of unique design challenges. Planetarium dome surfaces encompass a significant, if not the entire, part of the user's field-of-view and thus require great care when designing content for these display environments, some of which are shared with challenges for virtual reality headsets. For example, great care must be taken to visually guide the audience members attention to a location of interest as the use of traditional pointing devices are limited in a dome environment and require the utilization of other methods to the same effect. One fundamental design principle for immersive environments is to restrict the focus location to a small section of the available surface area and keep the content that requires the audience member's focus limited to that area. Planetariums are particularly suited for this as there is a natural center direction prescribed by the geometry of the dome surface, either being the structure's apex in a traditional dome, or a forward direction in a tilted dome configuration. Showing the content of interest in this small focus area while using the rest of the dome surface to provide ambient context allows for a more immersive experience of the audience in our experience.

The ability to provide a focus for the audience can be enhanced through the use of sound. In addition to using an instrumental music track to provide a constant background ambience, most immersive display environments are equipped with at least 5.1 speaker setups which make it possible to utilize directed audio to guide the user to a specific location on the planetarium surface. This technique can enhance already employed focus techniques such as compositional flow to guide the audience to a particular area of interest on the display surface.

Particularly, for interactive visualizations, smooth camera movements are important due to the immersiveness of the display, as rapid movements are liable to create nausea in audience members who are not accustomed to immersive environments. A general rule of thumb is to slow down the camera movements by about a factor of 3 compared to desktop-sized small screen display setups. If the entirety of the dome is used to present information, rather than just to provide context and thus negating the effect of a focus area, a reduction of $5\times$ is advisable. While motions with these speeds will seem slow when viewed on a small screen, the speed of movement of objects in the immersive display will be greatly enhanced. Audience members are also sensitive to different types of camera movements. For instance, when zooming directly toward or away from an area of interest, only a minority of the population will be nauseated by it, regardless of the speed of camera movements. However, lateral pitch movements are more challenging and have to be executed more slowly as the audience member loses the fixed point of reference that a central object provides.

However, rolling the camera around the axis between the center of a planetarium and the focus area is to be avoided at all costs, as it induces nausea in the greatest amount of people. There are, however, techniques to mitigate these unwanted effects. For instance, it is possible to combine a slow rotation with a lateral pitch movement and thus "hide" the roll such that most audience members will not notice the roll movement and thus not be affected by it.

Another major issue in most immersive environments is the available brightness of a projector system, particularly when displaying stereoscopic content. An often overlooked side effect of the low brightness is the transition of the audience members into scotopic vision in which the ability to discriminate between colors becomes impaired. This effect requires the visualization design to overemphasize color contrast and increase the color saturation beyond limits that are usually applied on traditional display systems. As ongoing technological progress enables an increase brightness, this problem is unlikely to disappear as the maximum acceptable brightness in a closed environment is limited due to the possibility of interreflection between the different parts of the planetarium.

Lastly, we want to emphasize the utilize of examining content by creating content using fisheye projection techniques. It is feasible to gaining familiarity with planetarium display devices by creating content in fisheye and transferring these to planetarium display surfaces repeatedly. This leads to a new mantra when desiring to effectively create content for immersive environments: "Think Fisheye, Work Fisheye, Live Fisheye"!

19.5 Interactive Multi-touch Tables at Science Centers

Museums and science centers visitors come in several different constellations, it may be an individual, a family, a group of students, or other groups ranging from industry to government agencies. There are furthermore different scenarios in terms of whether visitors are exploring the exhibit on their own, or if they are given any assistance. Guidance may be spontaneous, open-ended or more tightly scripted storytelling, depending on the skills and abilities of the facilitator and the specific agenda or plan for the visit. From our own experiences, and many others, multi-touch tables have proven to be a facilitating technology that can serve most visitor constellations and scenarios [12, 31].

Unguided individuals and smaller groups often employ the concept of walk-up-and-use; the touch table is intuitive and easy to use for the average visitor and requires no prior information or training. With multiple users in a visitor group, it is possible to have multi-touch interaction with the table as they are gathered around it, and thus forming a social context. They can easily interleave their interactions and undergo a discussion or dialogue about the topic and related questions that may arise.

Facilitated, or guided, sessions are often adapted to the specific visitor constellation and depend on the current scenario. Again, the ease of interleaving interaction between the guide and the visitors make the touch table an effective tool for story-

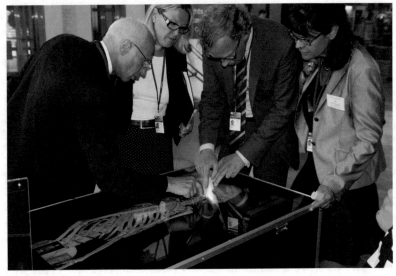

©at Amanda Sundberg

Fig. 19.9 Virtual Autopsy Table on tour. Anders Persson, director of CMIV at Linköping University, is showing a case. Photo courtesy of David Karlsson/RISE Interactive

telling and communicating science to a broader audience and the general public. In this chapter, we provide a few examples of installations based on touch tables and conclude with a set of recommendations regarding design choices.

19.5.1 The Virtual Autopsy Table

The use of multi-touch tables at the Visualization Center C in Norrköping originates from the research on techniques for direct volume rendering (DVR) of large medical datasets acquired through full-body CT scans. The challenge faced was volumetric data larger than what fits into the GPU's memory, and adaptive decompression technique was employed to optimize resolution in different parts of the volume based on the predicted final appearance in the rendered image [21]. A few years later, a multi-touch user interface was added on top of the research software that had been developed [20]. Much effort was given to provide a touch-based user interface tailored for public spaces, and the final result was an intuitive selection of the data to visualize. Based on the selected dataset, a set of possible *transfer function* settings were shown and the user could swipe different thumbnail representations of these onto the data. This triggered the software to optimize the resolution for the new settings, and the user could rotate or cut through the volume using a clipping plane. A transfer function setting in the context of DVR defines how data values in the volume are interpreted as different materials, or organic tissues, and are subsequently associated with appearance properties in the rendering process, such as colors and opacities. In professional applications, it employs a complicated user interface, in most cases based on visualization of a binned histogram of data values. This system was showcased in 2009, a year before the Apple iPad was introduced, but since then, touch interfaces have become truly ubiquitous. Today, the memory of typical GPUs are able to hold the full resolution of these datasets and the original software have been replaced and adapted to the feature sets of modern GPUs. Nevertheless, the user interface design principles have remained similar and simple to adhere to the walk-up-and-use principle (Fig. 19.9).

Originally, the touch table-based visualization system was coined the Virtual Autopsy Table as the first application and datasets were based on a project where cadavers from forensic cases underwent a full-body CT scan with a subsequent reading and examination before a regular autopsy. The content has since been evolving and grown into a database of animal scans, other museum artifacts, and more. The *Inside Explorer* table and associated content database is now being developed by the spin-off company Interspectral [13]. Another branch of development from this original Virtual Autopsy Table is the medical visualization table by Sectra [27] that is targeting medical professionals and thus is advanced in terms of user interface features and functionality. In this case, however, the end user is not the general public but radiologists, doctors and nurses, where an emphasis is placed on the use in educational settings.

19.5.2   The Gebelein Man at the British Museum

A particular interesting application of interactive multi-touch visualization is found at the British Museum in Gallery 64. Not only can visitors explore the Gebelein Man, a more than 5,500-year-old naturally mummified cadaver, through volume rendering [31] but, here lies also the physical mummy, albeit within a controlled environment and behind glass. Visitors can explore the mummy and learn about his story through interactive exploration of the CT-scanned digital representation. The story itself turns out to be spectacular; by using virtual clip planes, the explorer can see behind the left shoulder from the inside and discover a fracture on the fourth rib. By touching the info-icon next to it the user can read about the discovery that the Gebelein Man was murdered by a stab in the back (Fig. 19.10). There is an interesting and relevant observation that can be made here: the technology the visitors can experience is the very same techniques that scientists used themselves, with the very same data the scientists explored. For the visitors, the data is rendered using the same underlying volume rendering techniques as for the scientists and they both can explore the data by using different transfer function settings. This fact also becomes part of the story to tell about the installation and adds yet another dimension of science communication to the exploranation concept. Worth to note is that the introduction

of the digital exploration installation in Gallery 64, the fraction of visitors viewing
the physical mummy increased from 59% to 83%.

19.5.3 Discovering and Learning About the Microcosmos

Another example where multi-touch tables have been used for science communication to the public is the Microcosmos table, described by Höst et al. [12]. In this
application, users can select a topic and browse among a collection of image and
video cards floating around on the display, each presenting a different aspect of the
topic. The topics are Proteins, Viruses, Cells, Molecules, Genes, Life processes, and
Diseases, where the image cards may be present in one or more of these topics. The
concept is to create a free exploration session in unguided scenarios, or a storytelling
session in a guided situation, about the biological structures at the microscopic and
submicroscopic level. In a study to learn what users preferred and what they were
exploring during sessions interaction data was captured and logged [12]. Some of
the findings from this study indicate that the touch interface provides an intuitive,
effective and primarily engaging form of interaction with the content (Fig. 19.11).

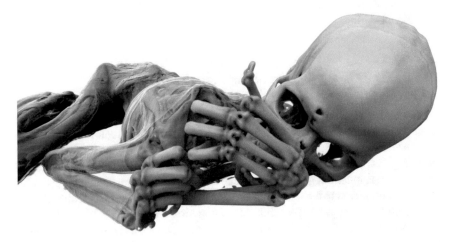

©at Daniel Jönsson

Fig. 19.10 Gebelein man mummy from the British Museum rendered using photon mapping by
Jönsson et al. [17]

19.5.4 Design Recommendations: Artifacts at the Tip of Your Fingers

Research on the benefits in learning and public dissemination from interactive visualizations using multi-touch tables is still in early stages, and much more work is needed to define and verify rigorous design principles. Nevertheless, some conclusions that have been observed and reported [31]. The recommendations are based on observations of a large number of installations world-wide and also in-depth studies of our own solutions.

- Object focus: The artifact on display should always be the main focus so the user is experiencing a notion of direct interaction with the rendered object. This in turn yields serious constraints on the interaction paradigms and robustness of the implementation.
- Judicious interaction constraints and micro-level interaction freedom: The user experience needs to be perceived as free and explorative without the system interfering. However, as the user explores the artifact and specific points of interest

©at Gunnar Höst

Fig. 19.11 Multi-touch table being using in a public exhibition space used to educate visitors to science centers about the microcosm. At the same time, the usage of the table was anonymously studied and found that the table attracted visitors attention, which also made extensive use of the available gestures to explore the datasets. Copyright 2018, Gunnar Höst et al. [12]

are reached they should be revealed non-invasively and guide the user to explore them.

- Minimalistic icons: To keep the focus on the explored artifact, icons need to be sparse and blend into the scene. Each icon need a clear relation to the artifact and the story being told.
- Limiting multi-user interaction: Making a robust experience without dramatic scene changes and rapid changes requires extra design constraints efforts and testing as multiple eager and uninitiated users may concurrently interact with the table.

In Fig. 19.12, we show an example of how these design criteria have been implemented in interface of the Inside Explorer software by Interspectral AB. The software is a portal to a range of scanned subjects and objects ranging from humans with interesting medical conditions to exotic animals and rocks such as meteorites with interesting interior.

Based on our experiences, we have found that multi-touch interfaces together with interactive visualization provides one of the most effective paths to successful installations in public spaces and is a primary example of the realization of the exploranation concept.

©at Thomas Rydell

Fig. 19.12 Interactive exploration of a volumetric representation of a housefly. The user can select among a number of predefined transfer function settings, use cut planes and freely rotate the object. There are also embedded information hotspots that lead the user through the narrative. Copyright 2019, Interspectral AB

References

1. Berry, D.: Molecular animations (2017). http://www.molecularmovies.com/movies-/viewanimatorstudio/drew%20berry/
2. Bladin, K., Axelsson, E., Broberg, E., Emmart, C., Ljung, P., Bock, A., Ynnerman, A.: Globe Browsing: Contextualized spatio-temporal planetary surface visualization. IEEE Trans. Vis. Comput. Graph. **24**(1), 802–811 (2017)
3. Bock, A., Axelsson, E., Costa, J., Payne, G., Acinapura, M., Trakinski, V., Emmart, C., Silva, C., Hansen, C., Ynnerman, A.: OpenSpace: A system for astrographics. IEEE Trans. Vis. Comput, Graph (2019)
4. Bock, A., Axelsson, E., Emmart, C., Kuznetsova, M., Hansen, C., Ynnerman, A.: OpenSpace: Changing the narrative of public dissemination in astronomical visualization from what to how. IEEE Comput. Graph. Appl. **38**(3), 44–57 (2018)
5. Bock, A., Marcinkowski, M., Kilby, J., Emmart, C., Ynnerman, A.: OpenSpace: Public dissemination of space mission profiles. In: 2015 IEEE Scientific Visualization Conference (SciVis), pp. 141–142 (2015)
6. Buneman, P.: The providence of provenance. In: Proceedings of the 29th British National Conference on Big Data (2013)
7. Clark, D.B., Tanner-Smith, E.E., Killingsworth, S.S.: Digital games, design, and learning: a systematic review and meta-analysis. Rev. Educ. Res. **86**(1) (2016)
8. D'Angelo, C., Rutstein, D., Harris, C., Bernard, R., Borokhovski, E., Haertel, G.: Simulations for stem learning: systematic review and meta-analysis. In: SRI International, Menlo Park (2014)
9. Evans, Sutherland: Digistar Planetarium Software (2018). https://www.es.com/Digistar/
10. Freire, J., Silva, C.T.: Making computations and publications reproducible with VisTrails. Comput. Sci. Eng. **14**(4) (2012)
11. Gilbert, J.K., Reiner, M., Nakhleh, M.: Visualization: Theory and Practice in Science Education, vol. 3. Springer Science & Business Media, Berlin (2007)
12. Höst, G., Schönborn, K., Tibell, L., Fröcklin, H.: What biological visualizations do science centervisitors prefer in an interactive touch table? Educ. Sci. **8**(4), 166 (2018)
13. Interspectral: Inside explorer visualization table (2019). https://interspectral.com
14. Jones, A., Unger, J., Nagano, K., Busch, J., Yu, X., Peng, H.Y., Alexander, O., Bolas, M., Debevec, P.: An automultiscopic projector array for interactive digital humans. In: ACM SIGGRAPH 2015 Emerging Technologies, p. 6. ACM (2015)
15. Jones, A., Unger, J., Nagano, K., Busch, J., Yu, X., Peng, H.Y., Barreto, J., Alexander, O., Bolas, M., Debevec, P.: Time-offset conversations on a life-sized automultiscopic projector array. In: Proceedings of the IEEE Conference on Computer Vision and Pattern Recognition Workshops, pp. 18–26 (2016)
16. Jönsson, D., Falk, M., Ynnerman, A.: Intuitive exploration of volumetric data using dynamic galleries. IEEE Trans. Vis. Comput. Graph. **22**(1) (2016)
17. Jönsson, D., Kronander, J., Ropinski, T., Ynnerman, A.: Historygrams: enabling interactive global illumination in direct volume rendering using photon mapping. IEEE Trans. Vis. Comput. Graph. **18**(12), 2364–2371 (2012)
18. Klashed, S., Hemingsson, P., Emmart, C., Cooper, M., Ynnerman, A.: Uniview - visualizing the universe. In: Eurographics - Areas Papers. Eurographics Association (2010)
19. Kress, G.: Multimodality: A Social Semiotic Approach to Contemporary Communication. Routledge, Abingdon (2009)
20. Ljung, P., Lundström, C., Rydell, T., Persson, A., Frishert, W., Ynnerman, A.: The virtual autopsy table (2009). https://www.youtu.be/bws6vWM1v6g
21. Ljung, P., Winskog, C., Persson, A., Lundström, C., Ynnerman, A.: Full body virtual autopsies using a state-of-the-art volume rendering pipeline. IEEE Trans. Vis. Comput. Graph. **12**(5), 869–876 (2006)
22. Manches, A., Bligh, B., Luckin, R.: Decoding learning: the proof, promise and potential of digital education (2012)

23. Mayer, R.E.: Incorporating motivation into multimedia learning. Learn. Instr. **29**, 171–173 (2014)
24. Miandji, E., Kronander, J., Unger, J.: Compressive image reconstruction in reduced union of subspaces. In: Computer Graphics Forum, vol. 34 (2015)
25. Muzic, M.L., Autin, L., Parulek, J., Viola, I.: cellVIEW: a tool for illustrative and multi-scale rendering of large biomolecular datasets. In: Bühler, K., Linsen, L., John, N.W. (eds.) Eurographics Workshop on Visual Computing for Biology and Medicine. The Eurographics Association (2015)
26. Norrköping Visualiserings AB: Norrköping Visualization Center C (2019). http://visualiseringscenter.se/
27. Sectra: Visualization table (2019). https://sectra.com/medical/product/sectra-terminals/
28. Skyskan: DigitalSky 2 Software for Definiti Theaters (2018). https://www.skyskan.com/products/ds
29. Society, A.A.: World Wide Telescope (2018). http://www.worldwidetelescope.org/home
30. Ynnerman, A., Löwgren, J., Tibell, L.: Exploranation: a new science communication paradigm. IEEE Comput. Graph. Appl. **38**(3), 13–20 (2018)
31. Ynnerman, A., Rydell, T., Antoine, D., Hughes, D., Persson, A., Ljung, P.: Interactive visualization of 3D scanned mummies at public venues. Commun. ACM **59**(12), 72–81 (2016)

Chapter 20
Reaching Broad Audiences in an Educational Setting

Penny Rheingans, Helen-Nicole Kostis, Paulo A. Oemig, Geraldine B. Robbins
and Anders Ynnerman

Abstract Visualization can be a powerful tool to enhance learning and to better support the learning process. Tailoring visualization to the specific audiences and goals of these situations can increase the likelihood of effective communication. In many cases, visualizations which require limited prior specific domain knowledge are helpful. Similarly, when crafting visualization development experiences for students in visualization courses, selecting easy-to-understand application domains for projects helps students to leverage their existing intuition about the domain in order to be able to focus on developing skills in visualization. Visualizations offer the potential to enrich educational settings, making concepts more engaging, concrete, and accessible. Visualizations for student audiences present both challenges and opportunities. The challenges are grounded in the potentially limited background knowledge of students, the need for examples to be engaging and accessible, and the need to support explicit curricular goals. The opportunities stem from a freedom to choose methods, data, and even application domains to address curriculum focus and learning objectives. This chapter discusses issues in adapting visualizations for educational settings, tuning visualizations to support curriculum goals, scaling visualization to fit on student devices, and crafting visualization development projects in course set-

P. Rheingans (✉)
University of Maine, Orono, USA
e-mail: penny.rheingans@maine.edu

H.-N. Kostis
Universities Space Research Association & NASA Goddard Space Flight Center,
Greenbelt, MD, USA
e-mail: helen-nicole.kostis@nasa.gov

P. A. Oemig
New Mexico State University, Las Cruces, NM, USA

G. B. Robbins
ASRC/AFSS NASA Goddard Space Flight Center, Greenbelt, MD, USA

A. Ynnerman
Linköping University, Norrköping, Sweden

© Springer Nature Switzerland AG 2020
M. Chen et al. (eds.), *Foundations of Data Visualization*,
https://doi.org/10.1007/978-3-030-34444-3_20

tings. The observations discussed in this chapter result from our experiences that include adapting visualizations produced by a science agency for the general public for use in the K-12 classroom, scaling and adapting interactive visualizations developed for a museum setting for classroom use, and designing visualization exercises and projects in a college-level computer science course.

20.1   NASA Visualization Explorer: Adopting the Newspaper Model in the Classroom

Helen-Nicole Kostis, Paulo A. Oemig, Geraldine B. Robbins

The digitization age and instant transmission of information via data platforms have enabled new practices in communication. The coming of age of the digital world offers tremendous benefits, but it also presents challenges for consumers. Making sense of the world involves developing critical skills, evaluating claims against evidence, asking questions. This article discusses the history of and opportunities stemming from the NASA Visualization Explorer (NASA Viz)—a science storytelling project aimed to inform the general public about NASA's research. The project created and released visualization-based stories to highlight research findings from NASA's exploration of the Earth, sun, moon, planets and universe [1] (Fig. 20.1). Although the intended target audience was the general public, upon release, the project was utilized by teachers in the classroom. In the paragraphs below, we provide an overview of the project, describe its objectives and present a *newspaper* model to bridge the delivery of information from print to a *multiliteracy* understanding of graphic/visual representation of data. Furthermore, we present how the project was used in the classroom.

Overview and Project Objectives
NASA Viz stories highlight findings and research efforts from all four NASA science themes—Earth science, Heliophysics, Planetary science, and Astrophysics—and include technology and science mission accomplishments. The project comprises an iOS universal application, an Android application, and a mobile-friendly website featuring all the released science stories including links to the source material and related content. This effort officially went live in July 2011 and was actively developed until March 2017, when the Android beta version application was released to the public. During that period, the project released two visualization-based stories every week and since then it released a story every other week. As of August 2019, the project had released 577 stories.

In 2010, the iPad was released and its screen size lent itself to the data-driven and visually rich media content developed and produced by NASA's science storytelling team (for more information, see Sect. 18.3 of this book). As the device hit the market,

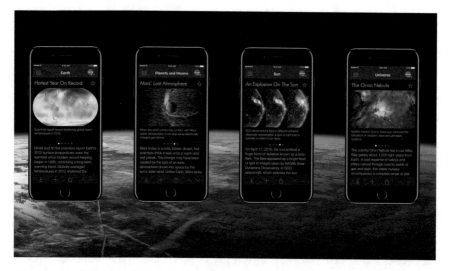

Fig. 20.1 NASA Visualization Explorer—a science storytelling project to inform the general public about NASA's research. The project released visualization-based stories about NASA's exploration of the Earth, sun, moon, planets and universe. Credit: NASA Visualization Explorer Team

Fig. 20.2 Team member Helen-Nicole Kostis demonstrates the iPad app and shares visualization-based science stories with kids at the NASA Visualization Explorer pavilion during Earth Day events taking place at Union Station in Washington, DC. Credit: NASA

a small group[1] from the storytelling team embarked on a three-month pilot project to experiment with the new medium in order to develop the visual and editorial language for the project and define the product. Through a series of iterations, mockups and support from NASA's Inclusive Innovation (I2) Award administered by Goddard Space Flight Center, the team brought the project to life and scaled into full production by releasing two visualization-based stories per week. All aspects of the project were developed in-house, ranging from mobile and server-side database development, to operations, content production, and editorial. The project objectives were to: (a) serve as a conduit to engage and inform the public about NASA's research and science efforts through visualization and multimedia content; (b) repackage already produced data-driven visualized content into bite-sized science stories/features; (c) engage new audiences by leveraging iOS devices; and (d) become an early adopter of the rapidly evolving touch screen and mobile technology.

There is value in creating visually compelling and accessible content for the public, and in developing new means of digital distribution to bring that content to the public. Even for those who are already engaged in science, either as amateur practitioners or just as curious consumers of science, we need to continue nurturing their engagement to help them reach the next level and maybe even contribute back to science: for example, they might pursue careers in STEM fields, advocate for funding scientific research, and develop educational programs for the next generation. For those who are not yet engaged we need to captivate them with innovative materials that are scientifically accurate [13]. Over the last decade, Helen-Nicole Kostis, NASA Viz project manager, had the opportunity to interface with students and the general public during outreach events, such as the World Science Festival, Earth Day (see Figs. 20.2 and 20.4) and school visits in support of NASA Viz. These experiences made her realize that when we take a moment to explain to audiences of any age that the content they are seeing is developed based on scientific observations and data, that same content takes a whole new meaning in their eyes. And that very moment can become a pivotal one that could define their future engagement on science and STEM fields, because there is an immediate increase in interest and appreciation for the visual content they are seeing. Therefore, developing data-driven scientific visualizations and the conduits that distribute them coherently, while leveraging storytelling techniques, can serve as a catalyst in conveying complex research results and in engaging new audiences in the sciences.

As the project manager and science visualizer, the first author's goals were to meet and exceed the objectives described above. Kostis, inspired by a deep appreciation for data-driven content, aspired to develop with the project team[2] a first-of-its-kind storytelling conduit that framed data-driven visualizations as the core elements of the story and placed them front and center. Up until 2010, news features in typical new media outlets were mostly text-based and in some cases included one or two

---

[1] Team members that equally contributed to the initial concept of the project: Helen-Nicole Kostis, Horace Mitchell, Neema Mostafavi, Wade Sisler, Christopher Smith, Michael Starobin. With database expertise provided by Joycelyn Thomson Jones.

[2] NASA Visualization Explorer Team: https://svs.gsfc.nasa.gov/nasaviz/credits.html.

visuals (images or videos). In that news model, text served as the core element of the feature story and visuals only had a secondary role and mostly were there to support, or sometimes to *break* the text. Alternatively, Kostis was interested in exploring the landscape where data-driven visualizations serve as the core element of a science story and text is provided to accompany the visuals and add context, when needed. She saw this project as an exploration that reversed the traditional text-laden news media approaches and beamed stories to share NASA's scientific knowledge and visual wonder developed by the science storytelling team.

NASA Viz went live on July 26, 2011 and it was initially available only on the iPad.[3] The project attracted a niche and highly engaged audience. One of the unexpected outcomes was an overwhelming interest from the education community. The team started receiving emails from teachers who were using the application in the classroom, inquiring about stories on a specific topic, phenomenon, or grade level, and even requesting educational features within the app. Due to the overwhelming amount of emails and need for guidance from subject matter experts, Kostis reached out to NASA Goddard's Director of Education Dr. Robert E. Gabrys. This led to a series of educational efforts and a collaboration that continued over the next four years that significantly defined the direction of the product development. A summary of these education-targeted efforts follows next.

Engaging the Educators

In order to address the enthusiastic interest from teachers and under Dr. Gabry's guid-ance the NASA Viz team immediately formed a working group with Albert Einstein Distinguished Educator Fellows (2011-2012 Cohort), who were using iOS devices in their classrooms. Einstein Fellows are accomplished K-12 educators in the fields of science, technology, engineering, and mathematics (STEM), who are selected yearly to serve in the national education arena. Fellows spend eleven months working in Federal agencies or in US Congressional offices, applying their extensive knowledge and classroom experiences to national education programs and/or education policy efforts [3].

This working group was led by 2011–2012 Einstein Fellow Geraldine Robbins, who was based at NASA Goddard during this period working under the supervision of Dr. Robert Gabrys and Carmel Conaty (Information Education Lead, Office of Education). The NASA Viz team shared with the working group the feedback and inquiries sent by teachers and requested from the Fellows to examine the NASA Viz application, its released content, and the editorial plan. Within two weeks, the working group held a meeting at NASA Headquarters, in Washington, DC, where the Fellows explained:

- How devices are administered in public versus private schools. For example, they explained the difference between classroom or school-shared devices and personal devices, how shared devices are administered and the intricacies of shared versus personal devices that can be taken home.

---

[3]https://www.nasa.gov/centers/goddard/news/releases/2011/11-044.html

- The limitations that public schools encounter in various US districts, including access to technology, Internet and network capabilities, and number of available devices (if any) that are typically administered for classroom use.
- Teacher workflows, schedules, and limitations.

In addition, keeping in mind that the project was designed and developed to serve the general public, Dr. Gabrys suggested to follow the ***newspaper model*** for the NASA Visualization Explorer project, meaning that the project could be used in the classroom similar to how a teacher brings a newspaper article to spark interest and/or explain a relevant topic typically at the beginning of the class. Since the project was not designed initially as an educational application, it would require a separate version and the development of educational content in order to turn the application into a fully educational tool serving teachers and students. The input provided by the Einstein Fellows was instrumental, as it informed the software development of the project for the next four years, within the constraints of available resources that were allocated for this endeavor. With the newspaper model in mind, it was agreed to proceed with the following principles:

- Aim to develop features that would serve the general public, teachers, and students.
- Strive to grow the project and make its content accessible to as many as people as possible, including users who do not have access to iOS devices.
- Aspire to serve educators and students from public schools, while keeping in mind the lowest common denominator of the technological and infrastructure challenges they face.

For example, when the Einstein Fellows suggested that the application and its content should be accessible to as many educators and students possible, we developed and released iPhone/iPod touch versions of the application and later on a universal iOS application (see Fig. 20.3). In parallel, in order to serve the population that did not have access to iOS devices we designed and developed an interactive and mobile-friendly website and the Android beta version. In addition, after careful consideration the NASA Viz team identified and developed features within the application that would help teachers and students, while at the same time they would still be useful for the general public. Such features, to name a few, include the ability to classify stories in NASA's four science themes (Earth, Planets & Moons, Sun, and Universe); search capabilities; ability to save stories for offline use, since accessibility to wireless networks was not guaranteed; and ability to save NASA Viz stories in custom story playlists. These playlists could be used by teachers or students in classroom projects, but were also used by the general public and NASA personnel in outbound communication efforts. These playlists could be also viewed on the web, if the user had no iOS device available.

The project continued with the above framework in place. In parallel, it was critical to check if our efforts were fulfilling the needs of the educational community. For this reason, we developed a follow-up pilot project, under the guidance of Dr. Gabrys and Paulo Oemig (Einstein Fellow at NASA Goddard, cohort 2012–2013)

Fig. 20.3 NASA Visualization Explorer iOS Universal application. Credit: NASA Visualization Explorer Team

targeting the academic calendar of 2012–2013. For this project we collaborated with the Maryvale Preparatory School [2] a private girls' school serving grades 6–12, that was incorporating iOS technology in the classroom for all subjects and all grade levels.

Six teachers (grades 6–12) participated in the pilot project, and at its conclusion, they provided the following information for each NASA Viz story they used in the classroom: (1) title of the story, (2) course name the story was used in, (3) grade level, (4) the date they used the story, and (5) how the story was used. In addition, teachers were asked to provide general comments and feedback on features of the application (e.g., search, themes, saving stories, and custom playlists), the content of the stories and how the application was used in the classroom. The pilot demonstrated that the application features were actually utilized in the classroom and the content from NASA Viz was most often introduced using the *newspaper model* as predicted by Dr. Gabrys or within projects created by the students. Furthermore, teachers commented on its versatility of use, reported that NASA Viz facilitated differentiated instruction and that the accompanying text enabled greater understanding of the visual/audio components, in comparison to using them standalone. As expected, not all stories were adaptable to all grade levels; for example, a subset of stories were suitable for Advanced Placement (AP) Environmental students, but were too advanced for tenth graders. In additional comments, the teachers expressed interest in classifying material per grade level and course and mapping content to the Next Generation Science

Standards (NGSS) [5]. Regarding the content, teachers shared that they would like to see simplified colorbars based on grade level; concepts introduced in stories; units utilized within colorbars to be explained in detail; keywords should be defined and possibly develop a glossary within the application; labels to be included in videos explaining geographic areas (cities, countries) and indicators (like pointers) highlighting the phenomena explained in data visualizations; content categories based on whether visuals are data-driven or concept-driven; and expressed preference on fully produced stories accompanied with narration and labels. After careful consideration, the requests from the teachers made it clear that a completely different approach was required to retrofit NASA Viz for the classroom. That would require integral collaboration with teachers and districts across the country, so that we would be able to map content based on Next Generation Science Standards, the curriculum and grade levels. This type of effort would require a team and resources that were outside the scope of NASA Viz. Nevertheless, the project continued to release stories and engage the public according to its original vision.

Multiliteracies in Today's World

Literacy in its simple form refers to the ability to read and write. Today's world demands a broader and more nuanced understanding of the term. The practice of generating and interpreting information requires three components: technology, knowledge, and skills [22]. Learning is contextual and social. Learning is contextual because the context in which a technology/tool (e.g., a tablet device) is used and its purpose for using that technology are situationally driven, and it is social because it requires skills learned through socialization. The tools of literacy are various (e.g., texts, pen, digital media), and at the most basic level, the process of producing literacy combines decoding and encoding for reading and writing. This process today extends to multimodal forms of data representation and expression. More appropriately understood, literacy can be conceived as multiliteracies [11]. The spectrum of literacies across societies, institutions, and disciplinary literacies is diverse. The general public and students alike are confronted with a "multiplicity of communication channels and media [and] increasing salience of cultural and linguistic diversity [11]." The multimodal forms of communicating ideas require particular skills to discern information. It is important to recognize that different texts (expository/informational, descriptive, persuasive, and narrative) make use of different text structures (description, sequence, cause and effect, comparison, and problem-solving), and features (captions, glossary, hyperlinks, index, keys, graphs)—[8, 23] all of which also play a role in the general use of the NASA Viz application, but particularly for its use in the classroom.

NASA Viz in the Classroom: A Case Study

Upon completion of his Einstein Fellowship, the second author returned to his teaching assignment in New Mexico. Paulo Oemig incorporated the use of the NASA Viz application in his eighth-grade science classes from 2013–2014 school year. Roadrunner Middle School (pseudonym) is a public suburban school located in New

Fig. 20.4 Students explore the NASA Viz application during Earth Day events taking place at Union Station in Washington, DC., Credit: NASA

Mexico and at the time of the study, it had a student population of approximately 830 students. Strong interests in science and bilingual education, his experience at NASA and doctoral studies led him to assess his own instruction and elements of his curriculum. With the understanding that learning is enhanced when it is contextualized, the approach in his classroom was very active. At the core of his physical science classes were laboratory experiences (laboratories performed once per week, ranging from fully structured to open-ended during the school year), demonstrations (at least three times per week), daily warm-ups (science and math prompts/problems related to the topic to be addressed that day), public service announcements (research performed in teams of up to four and presenting to peers/public), discussions (seminar style), animated lectures (one per week, up to thirty minutes), reading–writing and note-taking (extracting relevant information to explain to peers), and writing of trade books [14].

The total number of students in Paulo's five science classes were 154 (81 female, 73 male). Out of the five classes, one was composed of students who were either English Learners (ELs), have been in a bilingual program in elementary, or parents had requested their students to be in the class. The class had 27 students (14 female, 13 male), 9 students were born in Mexico, one had only been in the USA for three months, and all but three students qualified for free/reduced lunch.

The NASA Viz application became an instructional element in the curriculum of all five science classes. The same stories were used with all classes, and a new story was selected approximately every two weeks. During the second half of the school year students had an opportunity to pick a story and share what they learned with their peers. The readability of the stories according to Degrees of Reading Power [4] varies between 6–8 grade bands of the Common Core State Standards [7]. The NASA Viz stories were used as warm-ups, as reading comprehension prompts leading to practice text structures and as problem-based learning challenges, which encourage writing to practice different types of texts. As warm-ups, the second author connected the iPad to the projector and asked students to note an interesting fact, record a question, and make a connection to something they already knew or was covered in class. For use as prompts to practice text structures, the text or script of the story was introduced first and students identified signaling words for different text structures; for instance, some signaling words for cause and effect text structure are: *because, if/then, since, consequently, as a result.* Following the reading of the story and identification of signaling words, students wrote the main idea using at least two signaling words. After students shared in small groups the main idea from the story, the visualization was played for the class. Some stories have narrative accompanying the visualization production and some only have music as background. The visualization always added details to the text/narrative and expanded students' understanding of the main idea. The story *Mapping Earth's Gravity* does not have a narration; this story was used to illustrate the relationship between mass and gravity. It was introduced in the middle of a unit on Newtonian physics. As problem-based learning challenges, the second author used two stories, *America on Fire* and *Fishbone Forest* toward the end of the 2012–2013 school year; the instructional lessons [6] developed for these two stories brought together an opportunity for students to elaborate on writing different types of text (e.g., persuasive, descriptive, expository).

At the beginning of the school year, all students were given a science attitude protocol consisting of a Likert scale (1 strongly agree, 2 agree, 3 undecided, 4 disagree, 5 strongly disagree) and five items: 1. Learning science is important to me, 2. I feel comfortable reading about science, 3. I enjoy the laboratory period (hands-on) in science classes, 4. I feel comfortable writing about science, 5. I am confident I can understand science. At the end of the school year, students received the same protocol with five additional items: 6. Writing trade books (children's book) about a science topic helped me understand science, 7. Writing laboratory reports helped me understand science, 8. The NASA Viz stories helped understand science, 9. Presenting public service announcement posters/multimedia to peers helped understand science, 10. The demonstrations in the classroom made me more curious about science. A teacher evaluation was also administered at the end of the school year. The aggregate feedback and data received from the protocol focus group interviews, one-on-one *chats*, and teacher evaluation indicate that students overall enjoyed doing the laboratories, although did not care so much for writing laboratory reports; enjoyed working in groups to put together public service announcements; in general, demonstrations made students more curious and engaged in the learning. In regard to writing trade books about different science topics, all five classes wrote three—the EL class wrote four, students enjoyed the experience, particularly those who had younger siblings. Overall, students reported that the NASA Viz stories contributed to their understanding of science. However, most of the students in the non-EL class reported not liking identifying signaling words and writing the main idea. In contrast, for the most part students in the EL class enjoyed learning text structure via the stories. It increased their confidence level in reading and writing about science. This cannot be attributed solely to the stories or the author's class; students were also taking language arts, and some were taking a READ 180 class, which is a reading intervention program and other classes contributing to a broader exposure of literacy. Most students, regardless of class enrollment, reported appreciating and enjoying choosing their own stories to share in small groups and/or to the whole class.

Conclusion

The NASA Viz stories were conceived with the general public in mind and aimed to relate information on recent and current NASA scientific missions. As described, the use of the stories can be adapted with careful preparation and pedagogical knowledge to the classroom. The positive impact on informing and engaging the public with NASA's science and work is amplified by its potential to inspire students in classrooms as well. Further study is recommended to determine the utility of the NASA Viz in the classroom. For instance, a pilot study, with a control class or school, at a school district level could examine the effectiveness of NASA Viz stories, which have been pedagogically developed. In addition, students could research a project within a NASA mission or theme and create their own stories with a multimedia program, write a storyboard, and present them to peers and the school board.

20.2 Scaling Visualizations to Fit in Curriculum and on Student Devices

Anders Ynnermann

The process of taking advanced visualization into the educational setting often entails downscaling of large-scale visualization projects in terms of hardware platforms, software complexity, and content scope. In doing this at least three interdependent factors need to be taken into account when integrating and adapting any visualization in an educational setting [20]

1. The state of students' conceptual knowledge of relevance to the visualization in question,
2. The suite of cognitive skills that may be required for students to interpret and interact with the visualization,
3. The external nature of the visualization itself—the actual graphical conventions, visual symbols, and interactive features that constitute the visualization.

One urgent issue that remains for adapting visualizations to educational settings is to meet Chandler's concern [10] raised almost a decade ago: moving beyond the "wow" factor. The mere presence of a multimodal or engaging visualization does not automatically equate to beneficial learning. Instead, it is crucial to empirically identify what, and how, specific affordances of newly emerging visualization environments actually influence processes of knowledge acquisition and learning. In view of this, it is interesting to see not that scalable exploratory visualizations are emerging as important educational tools for providing students with necessary cognitive skills in various disciplines [21]. It is also apparent that successful attempts in this direction using static or interactive visualizations is tightly integrated into the curriculum matching their intended learning outcomes. Although this is far from a trivial task, educational research from the last decade or so offers some research-based guidelines for doing so [15, 18] and include the following: Reflect on the fundamental factors that may influence students' ability to interpret and interact with the visualization; take cognizance of current theory on how individuals process and learn from visualizations; make the intended conceptual knowledge communicated by the visualization explicit; ensure students have the knowledge of the visual language and interactive features incorporated by the visualization that depicts the content in question; and make students aware of the representational and interactive limitations of the visualization.

Another aspect that is emphasized in recent literature is that visualization development experiences in educational contexts must take cognizance of students' "visual literacy [19]." In the sciences, visual literacy is fundamental to the development of intended conceptual understanding and includes specific cognitive skills such as decoding the visual/graphical language composing a visualization; using a visualization to find a solution to a particular problem; connecting and switching between different multiple visualizations that represent the same concept or principle; and discerning the power and limitations of a visualization. In the digital age, scaffolding students' expert visual literacy skills in course settings calls for identifying and explicitly teaching visual literacy as part of carefully designed visualization assessments that specifically probe students' representational competencies as part of authentic tasks [15].

In an effort to gain widespread use of interactive visualization in the class room and following the guidelines above, an implementation of a platform for digital learning was produced by the Norrköping Visualization center (NTA-Digital) [9, 17]; see Fig. 20.5. Initially two themes, the human body and space, were developed. The interactive environment consists of frame stories creating a mental setting for the topic, tasks matching the learning goals in the curriculum, and a 3D model that can be interactively explored. The graphics were implemented in WebGL to enable use on a range of lightweight platforms. Currently, the platform has 10000 registered users.

©at Anders Ynnerman

Fig. 20.5 Solving the "human body puzzle" task in the NTA—digital learning platform. The platform uses gamification approaches and interactive storytelling to frame the learning goals

20.3 Crafting Visualization Development Experiences in Course Settings

Penny Rheingans

Data visualization courses have become common electives in many degree programs, both undergraduate and graduate, in disciplines including computer science, data science, information systems, and data-intensive application domains. Typical objectives of such courses are for students to develop an understanding of the stages of the visualization pipeline, become familiar with the rich vocabulary of visualization components and approaches, explore the capabilities and limitations of different visualization methods, become familiar with implementation approaches to key algorithms, practice creating visualizations for example situations, and gain experience working with domain experts who serve as project clients. Skills developed in data visualization courses can be used by students later in a variety of settings, becoming a powerful tool for use in future endeavors.

I have taught a university data visualization course many times to a mixed class of graduate students and advanced undergraduates, from either computer science or an information-rich application domain. Most students in this course have at least two years experience in programming and data structures; a few students bring only a little computing experience but deep expertise in some application domain where they wish to use visualization. This course takes an active-learning approach that draws inspiration from team-based learning [12]. It combines interactive quizzes, small group discussions, small group exercises, lectures on requested topics in visualization, individual assignments on visualization design and algorithm implementation, and a multi-phase group project for a client with real data and goals [16]. The examination, analysis, and development of visualizations are a theme throughout, but it is particularly important in the in-class exercises and term project.

Students in visualization courses represent a particular kind of broad audience with distinct goals. In this setting, students are the creators of the visualizations, rather than the consumers; while the visualizations are a vehicle for learning about the visualization process, rather than a mechanism for learning about an application domain. Appropriate visualizations for this setting share some constraints with those intended for other types of broad audiences, but also have distinct requirements. Some requirements will vary with the particular student characteristics and course goals, but several recommendations are useful across a range of settings.

When I began teaching this course in the mid-1990s, the examples and term projects were overwhelmingly from the scientific domains, reflecting both my own background and the dominant topics of the time. Examples, data, and clients were easy to find, but students frequently struggled to understand enough of the application domains to be informed consumers or developers. When I taught a version of this course intended for first-year undergraduates in 2007, I found myself thinking much harder about how to choose examples, domains, and assignments that were accessible to beginning students. Through addressing that challenge, as well as course revisions of the more regularly taught advanced course, I have identified some ways in which intentionally chosen visualization examples and application domains can enrich learning in visualization courses.

1. *Students can learn as much from bad examples as good ones.* At the beginning of the course, students tend to believe that all visualizations are good visualizations. Even talking about design does not necessarily break that mindset. An assignment where students bring in examples of bad visualizations to present to the class is a lot of fun and helps them think more critically about how to improve an initial visualization to be more effective.

2. *Working with an application domain client with actual information discovery goals creates deeper learning experiences than exercises with generic data.* Through such experiences, students have the chance to explore the impact of design choices in the context of discovery goals, giving them a framework for measuring the impact of different design choices. Additionally, students find working on a project that someone actually wants to use to be really motivating. The real-world aspect of the project serves to make the other aspects of the course more relevant.

3. *The domain context behind examples, assignments, and projects should be easy to understand and explain, both for experts and students.* Ease of understanding is particularly important for activities in which students will be developing their

own visualization examples and prototypes, since students will need to make design and implementation choices informed by their understanding of the data and goals. When the application is sufficiently accessible, students can rapidly become sufficiently close to being domain experts to use their visualizations to make discoveries. Understanding of the application data and goals makes it easier for students to judge whether the visualizations they have developed are successful.

4. *Learning how to address the ambiguity in domain goals is an important component of learning about visualization development.* It is important for students to practice responding to ambiguity in data semantics, domain goals, domain expert expectations, and the characteristics of audiences. In many cases, domain experts themselves may not have clear ideas of what they are interested in beyond the traditional and even less idea what visualization approaches might enable. Real-world problems with real clients and real messy elements provide a valuable opportunity to explore an underdefined design space.

5. *Students can learn the most from learning opportunities where they are given the opportunity to fail safely.* In order to be able to have the freedom to explore possibilities, students need to have the freedom to fail (i.e., to produce results that are less than optimal) without catastrophic consequences. A course setting is a much more appropriate opportunity than in a production setting in a first job. Providing that there is an important course element. Course projects are a learning exercise first, sometimes with results that are useful to the client. With this in mind, it is important that potential clients understand that student teams are not guaranteed to produce usable products.

20.4   Common Themes and Final Thoughts

The educational settings, instructional goals, and learning roles of visualizations can vary greatly between the integration of visualization into a course setting. Across those differences, some common themes emerge:

1. Visualizations and visualization systems developed for other settings must often be adapted for classroom use.
2. Limitations of classroom and student equipment, constraints on available time, and lack of deep data domain understanding all impact how visualization is most effectively integrated into a classroom setting.
3. Visualization designers and developers should engage educators in conversations about their goals and constraints.

Addition investigation and discussion would be valuable to increasing the ease and effectiveness of visualization in educational settings. Interest topics include further examination of target characteristics for effective visualizations for educational settings, a methodology for adapting visualizations for educational settings, rubrics for assessing the suitability of a visualization for educational use, and formal evaluation of the adoption of visualization in specific educational settings.

References

1. NASA Visualization Explorer. https://nasaviz.gsfc.nasa.gov/. Accessed 2019
2. Maryvale Preparatory School. https://www.maryvale.com/
3. Albert Einstein Distinguished Educator Fellowship (AEF) Program. https://science.energy.gov/wdts/einstein/
4. Degrees of reading power. http://www.questarai.com/assessments/district-literacy-assessments/degrees-of-reading-power/. Accessed 2019
5. Next science generation standards. https://www.nextgenscience.org/. Accessed 2019
6. Earth system science education alliance courses, NASA Viz app. http://essea.strategies.org/vizapp.php. Accessed 2019
7. Common core state standards initiative. http://www.corestandards.org/. Accessed 2019
8. Altieri, J.L.: Reading science: Practical strategies for integrating instruction. The Electronic Journal for English as a Second Language **20**(4), (2017)
9. Bohlin, G., Göransson, A.C., Gericke, N., Tibell, L.A.E.: NTA-Digital – Tema Kroppen. In: FND 2016, Forskning i naturvetenskapernas didaktik (2016)
10. Chandler, P.: Dynamic visualisations and hypermedia: beyond the wow factor. Comput. Hum. Behav. **25**(2), 389–392 (2009)
11. Cope, B., Kalantzis, M.: Multiliteracies: Literacy Learning and the Design of Social Futures. Psychology Press (2000)
12. Michaelsen, L.K., Knight, A.B., Fink, L.D.: Team-Based Learning: A Transformative use of Small Groups. Greenwood publishing group (2002)
13. Montaez, A.: How science visualization can help save the world (2016). https://blogs.scientificamerican.com/sa-visual/how-science-visualization-can-help-save-the-world/
14. Oemig, P.A.: Promoting science literate identities through the use of trade books. In: R.D. Tim Spuck T. Rust (ed.) Best practices in STEM education: innovative approaches from Einstein Fellow alumni, pp. 399–419. Peter Lang, New York (2018)
15. Rau, M.A.: Conditions for the effectiveness of multiple visual representations in enhancing stem learning. Educ. Psychol. Rev. **29**(4), 717–761 (2017)
16. Rheingans, P.: Minor adventures in flipped classrooms, team-based learning, and other pedagogical buzzwords. In: Pedagogy of Data Visualization Workshop (2016)
17. Royal Swedish Academy of Engineering Sciences: NTA digital (2019). https://www.iva.se/projekt/nta-digital/
18. Schönborn, K.J., Anderson, T.R.: The importance of visual literacy in the education of biochemists. Biochem. Mol. Biol. Educ. **34**(2), 94–102 (2006)
19. Schönborn, K.J., Anderson, T.R.: Bridging the educational research-teaching practice gap. Biochem. Mol. Biol. Educ. **38**(5), 347–354 (2010)
20. Schönborn, K.J., Anderson, T.R.: Bridging the educational research-teaching practice gap: foundations for assessing and developing biochemistry students' visual literacy. Biochem. Mol. Biol. Educ. **38**(5), 347–354 (2010)
21. Schroeder, D., Keefe, D., Kowalewski, T., White, L., Carlis, J., Santos, E., Sweet, R., Lendvay, T., Reihsen, T.: Visualizing surgical training databases: exploratory visualization, data modeling, and formative feedback for improving skill acquisition. Computer Graphics and Applications (2012)
22. Scribner, S., Cole, M., Cole, M.: The Psychology of Literacy, vol. 198. Harvard University Press, Cambridge, MA (1981)
23. Shanahan, T., Shanahan, C.: Teaching disciplinary literacy to adolescents: rethinking content-area literacy. Harv. Educ. Rev. **78**(1), 40–59 (2008)

Chapter 21
Challenges and Open Issues in Visualization for Broad Audiences

Michael Böttinger, Helen-Nicole Kostis and Anders Ynnermann

Abstract As discussed in the last chapters, a lot of work has already been done by academia and practitioners with respect to developing and producing visualization tools and visual products specifically designed for broad audiences. While there have been significant efforts in the field, there are still open issues and new exciting areas to explore for visualization researchers and practitioners, such as the need to develop guidelines for engaging visualization and visual interfaces where humans are in the loop, and how to prepare the next generation of data-driven storytellers. Toward the end of this chapter, as we contemplate the next era of visualization, we reflect on past accomplishments and share a few thoughts about the future of our practice.

21.1 Simplicity in spite of Growing Data Size and Complexity

The broader interest in the field of data visualization is closely related to the increasing availability of computing resources for research starting in the late 1980s. The exponential growth of computing capacity caused a likewise exponential growth of the data produced with these systems—data that needed to be analyzed and transformed into knowledge. Both, the spatial and the temporal resolution of models are quickly getting larger, and more processes are taken into account, so that the number of variables increases, and the recent increase in the use of ensemble simulation techniques (see, e.g., Bittner et al. 2016 [1]) to capture uncertainty even adds a further dimension to the data. Similarly, technical developments lead to the quickly growing

M. Böttinger (✉)
DKRZ, Hamburg, Germany
e-mail: boettinger@dkrz.de

H.-N. Kostis
Universities Space Research Association & NASA Goddard Space
Flight Center, Greenbelt, MD, USA
e-mail: helen-nicole.kostis@nasa.gov

A. Ynnermann
Linköping University, Norrköping, Sweden

© Springer Nature Switzerland AG 2020
M. Chen et al. (eds.), *Foundations of Data Visualization*,
https://doi.org/10.1007/978-3-030-34444-3_21

size of data produced by digital imaging, e.g., by earth observing satellites or in medical applications.

Visualization has always been technically challenged by the ever increasing size of the data that needs to be handled. This trend is still ongoing, and the complexity and amount of information that needs to be encoded and eventually communicated to broad audiences increase accordingly. Therefore, we need to carefully design our visualizations as simple as possible without risking scientific credibility. In this context, several fields—such as ensemble visualization, and in particular visualization of data along with its related uncertainty—are still quite challenging, as the research community has been mostly focusing on how to improve data analysis tasks and to a lesser extent how to communicate such visual representations to broad audiences (cf. [20]). However, in application domains such as the weather and climate community, ensemble simulation techniques are regularly utilized for quite some time, and probabilistic information is accordingly visualized and communicated to the public on a regular basis. Some works in this application domain look into the suitability of different visualization approaches for communication of such results to broad audiences (e.g., [17]), but we think that more work is needed to structure the field, develop guidelines and respective workflows.

In the application domains, visualization systems are often used as tools for the analysis and communication of data, and default colormaps of the respective systems are all too often used for first quick results, but not further optimized for communication purposes later on. Within the visualization community, on the other hand, numerous excellent studies have been published in the field of color theory, colormap design, and perception. Recent works focus, for example, on the effect of colormaps on data analysis tasks, such as the study from Dasgupta et al. [8], the capability of colormaps to support feature detection [21]), or a mathematical underpinning for the evaluation of colormaps [2].

However, regarding the design of visualizations for specific audiences and having simplicity in mind, we may have to think even more deeply about the colormap we are using. For audience-oriented colormap design, we have to choose the *type of the colormap* (continuous vs. discontinuous), about the *number of hues or colors*, and about *psychological effects of specific colors*. In their recent climate change communication study, Schneider and Nocke [15] evaluated which emotions (such as concern, fear, and alarm) were induced by different colormaps for the same visualization. Especially in policy-relevant fields, we should strive to design our visualizations for science communication in a way that is neutral, i.e., not too alarming without being too conciliatory. The balance between faithfulness and embellishment that always needs to be found in visualization design is discussed in more detail in Sect. 11.2.2 of this book. An empirical study on using visual embellishments in visualization is described in Sect. 7.4.1.

21.2 Need for Authoring Tools

One of the crucial bottlenecks we have identified is the pronounced need for support of production workflows and authoring tools. The curation and production of visualizations for broad audiences presents challenges if for visualization, or data analysis, experts have to be actively involved in the production workflow. This realization calls for simplified tools for data handing, annotation, authoring, and storyboarding as well as the design of visual representations and interaction models. It is also recognized that massive distribution of data and software requires support organizations and tailored tools.

21.3 A Few Thoughts on the Future of Our Practice

Helen-Nicole Kostis

> Visualization is the method of seeing the unseen.
>
> *Tom DeFanti, Maxine Brown and*
> *Bruce McCormick [19]*

Even as we may contemplate the next era of visualization, there is significant value in comprehending the ground covered in the past three to four decades. To accomplish this, I will make brief mentions of some foundational texts and efforts that marked the field's beginnings. In 1987, Tom DeFanti, Maxine Brown, and Bruce McCormick penned the NSF funded *Panel Report on Visualization in Scientific Computing* [19]. This seminal report defined the field of visualization, described it as a domain, identified the opportunities, and predicted the looming challenges for the next 20–25 years. I have it as a to-do item to go back and read this report every 2–3 years. If you are now entering the field of data visualization, I urge you to read it as it will show you where our field started and, assuming you can imagine or remember the state of the art in computing and technology back then, will offer you an appreciation of what has been accomplished. Now, more than 30 years on, the report stands the test of time as the field is evolving. I often wonder what would the authors of the report see as the challenges over the next 20–25 years?

John Tukey's impactful publication [18] back in 1977 defined the approach of exploratory data analysis (EDA) and provided techniques and guidance on how to explore data. The work is more than a collection of techniques. "In essence is an attitude: the willingness to look for what can be seen, whether or not anticipated" [9, 18]. Fast forward to today, we find research studies and reports not only on exploratory and confirmatory visual methods, but also on how we perceive data graphics. Thanks to all this work over the years, we can now get the gist of how humans perceive data graphics in less than 30 min [10].

Thinking about progress and startling leaps, my mind goes to Dr. Larry Smarr, who in 1977 was trying to visualize the Collision of Two Black Holes in 3D space and was able to accomplish that by building a physical model from papier-mache [7]. He then led the initiative that formed the National Center for Supercomputing Applications (NCSA) that eventually served as a home base for Donna Cox's *Renaissance team* [5] who defined the field of cinematic scientific visualization and produced a series of works including the collision of two black holes [13]. Dr. Smarr over the years has spearheaded the establishment of state-of-the-art research facilities [3] and open science initiatives that propelled research and fostered collaboration between artists, technologists, and scientists that advanced the field of scientific visualization.

One of the differences between now and the 1970s and 80s is that the field is not new and the problems it is called to tackle are getting harder and more complex. As Donna Cox mentioned in [6] "The story of science of today that will be told to the generations of tomorrow will be paired with the advanced cyberinfrastructure and the discoveries made possible by advanced computing and digital technologies." Visualization will continue to be part of the story of science and is already embedded in other stories with societal, educational, and cultural dimensions.

Therefore, there will be research and development efforts to support exploratory, confirmatory, and explanatory visualizations or even ones that lay between or beyond these worlds for which we do not yet have a proper vocabulary. The challenges of developing visualizations for broad audiences may also lay outside the strict academic barriers that fields and academic institutions build on purpose and require us—the visualizers—to consciously engage and participate in two sets of praxes: **transparency** and **synthesis**. In reality, it is quite challenging to establish and even more so to teach them as they might be viewed more like social investments and cultural accomplishments. But, as we practice them, we will reach new ways of seeing the unseen.

Transparency allows the recipients of the visualization to question the data, to understand the choices made, to allow for alternate views to emerge, which enable repeatability, promote inclusion, democracy, and accountability. The praxis of transparency can generate critical insight and can inform and educate the recipients of the visualization. When we reference the source and make it publicly available, we support reproducibility and we give opportunity to other interpretations, visualization techniques, insights, and serendipitous discoveries. For example, data journalism and citizen visualization can play critical roles by teaching the public and students from an early age how to question the sources, the derived data-driven visuals, and the meaning conveyed. Transparency is also important as we need to illustrate limitations, assumptions, and uncertainty. These attributes not only provide valuable and educational context but may also engender new opportunities in our quest to overcome them.

Over the years, I have gained a tremendous appreciation of the practice of transparency and openness directives within NASA thanks in part to the process by which visualization productions are released to the public from the Scientific Visualization Studio [14]. Based on these practices and the importance of data biographies [12], the following framework is proposed:

Transparency and Openness Framework

Declare number of data sources
For each data source
Source economy
How the data was collected or derived
Who collected the data
Why was the data gathered
Source nature
Temporal/spatial
Scale (local, regional, global)
Dimensions
Source interpretation
Data issues/limitations/completeness
Data assumptions you had to make to develop the visualization
Data uncertainty
Visualization interpretation
Processing
Abstraction
Design
Key findings and limitations
Release and point to the source
Release Visualization Products
Create persistent reference to Released Product

Synthesis is about blending disparate information and knowledge in ways that yield novel insights or explanations [11, 16]. It is a distinct form of research and practice and can take place across disciplines and professional sectors. There is high significance in hyper-specialization [16] in a field for addressing increasingly sophisticated open problems. There is no doubt that there are hard problems to crack with hyper-specialization; synthesis, however, can allow the conceptualization of complex social, environmental, and scientific problems beyond one profession, discipline, and research methodology. And if we are able to conceptualize the problem, then we are one step closer to identifying what needs to be done to address the challenges.

It might be the case that the professional sectors that we practice within do not embrace these two praxes. There might be intellectual property issues antagonizing transparency, and if synthesis and transparency are not institutional values, then they most likely will not be embraced with rewards. As professionals, we are all looking for an appreciation of our work and our efforts. In certain cases, but not necessarily in all, success and metrics of success with respect to these praxes will have to be defined. It might be that success will take new forms and evolve our field in ways we cannot imagine currently.

And as I am exploring some of the open challenges on developing visualizations for broad audiences, I see an untapped power source that the field still needs: the artists.

Art: The Power of the Extrasensory

Artists have formal or informal training and talent to consciously see, hear, and craft messages and experiences through their art-making process. And because art is everywhere, through their practice and the art of storytelling, artists can bring complicated issues to the attention of a much wider audience [4] and carry data-driven visualized messages literally or metaphorically through digital distributions and social media outlets to find people where they are. Between 1969 and 1973, superhero of synthesis and pioneer of electronic art and visualization, Daniel J. Sandin developed the Image Processor (IP), a highly programmable analog computer for processing video images in real-time. At the same period, Sandin, an early advocate of open source culture—following the COPY-IT-RIGHT [22] distribution approach—worked alongside Phil Morton from the School of the Art Institute to document the development and operation of the IP, with the goal to enable artists and technologists to create their own low-cost replicas and copies, as instruments of artistic expression. Using the IP, Sandin collaboratively created visualized media and artworks that transcended existing types of works similar to *Spiral5 PTL* (Fig. 21.1). Artists always embraced technology, found inspiration in the sciences, and enabled the understanding and communication of sciences. Apart from the influence and impact artists had and continue to have on technology, sciences, and the field of visualization, there is something else that comes through art, the ***extrasensory***.

Artworks have the power to invoke our senses and craft memorable, emotional experiences that transport us to new dimensions and shape our world. Over the years, I had powerful encounters with artworks that defined my language and shaped my visual thinking. By way of illustration, the word *iceberg* immediately transports me to my encounter with Iñigo Maglano-Ovalle's installation *Iceberg (r11io1)*. In 2005, on a Sunday afternoon, I was walking through the galleries of the Art Institute of Chicago. I vividly remember entering a hall and all of sudden a 25-foot tall sculpture was floating above the window-lighted Morton Wing Staircase [7]. It was the *Iceberg (r11io1)*, a fabricated model of a real iceberg sighted off Labrador in 1998 that Maglano-Ovalle constructed between 2003 and 2004, based on topographical profiles provided by the Canadian Hydraulics Center. To this day, this encounter—a defeaning experience—which spurred my interest in exploring the field of data visualization, still embodies for me the iceberg concept, despite having watched in awe a real-life one floating in a lake in Patagonia. Similarly, thanks to my first experience with Dan Sandin's piece *Particle Dreams in Spherical Harmonics* (Fig. 21.2), I have irrevocably connected the term Virtual Reality with it. Those artworks, by transcending representations of data and the mediums they were produced in, heightened the senses and triggered a profound emotional connection that developed new meanings.

What is needed to bring the arts in visualization

The field of visualization has been learning from the fields of design and applied arts. Since it is a domain that is currently being taught either within the sciences or the arts, we may have to rethink how it is taught, especially as we continue the journey of developing visualizations for the general public. This will require to incorporate computational thinking into the arts-based visualization education and

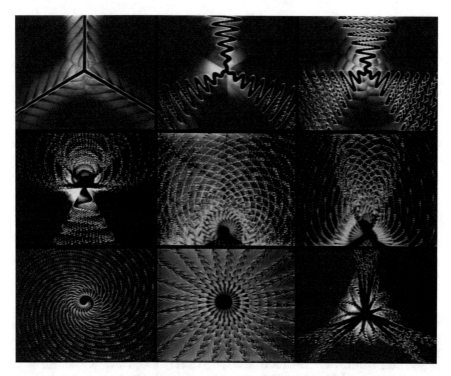

Fig. 21.1 *Spiral 5 PTL (Perhaps The Last)*, 1979. Live recording of performance before a small studio audience. Credits: Dan Sandin, Tom DeFanti, and Mimi Shevitz. *Spiral 5* is in the inaugural collection of Video Arts at the Museum of Modern Art (MOMA) in New York. It was the fifth of a series of performances of a piece called Spiral. It was performed live in front of audiences by people controlling digital computers and playing on the analog image processor. It is an abstract, mathematical form animation based on the linear spiral, in something you might call the visual music tradition. Image composition: Mary Rassmusen

visual thinking, i.e., audiovisual and visualization literacy, into the science-based programs – not as an afterthought, for a few elective course credits here and there, but purposefully, with the development of new integrated programs. Furthermore, if we would like our audiences to be visually and aurally literate, so that they can understand what they see and hear, and be critically minded, we should push toward inserting visual thinking and data literacy in K-12 education, similar to the effort for achieving writing literacy.

As visualization experts, it is part of our practice to simplify, clarify, and guide our audiences to the messages and truths revealed by research and data. And since the problems we are called to tackle become increasingly complex, this might require us to explore and develop new ways of *doing*, which will ask our audiences to respond with new ways of *seeing*.

Fig. 21.2 *Particle Dreams in Spherical Harmonics.* An interactive virtual reality installation in the StarCave of a colorful physical computer simulation of particles in space to involve the viewer–participant in the creation of an immersive, visual, and sonic experience. Content and application programming: Dan Sandin. Content and systems programming: Robert Kooima. Music and sound effects: Laurie Spiegel. Driver: Tom DeFanti. Image credit: Calit2 Communications team

Acknowledgements Helen-Nicole Kostis would like to thank Universities Space Research Association (USRA) and Goddard Earth Sciences Technology and Research (GESTAR), for supporting her participation in the Dagsthul seminar and time to contribute to this book through two Learning and Development (L&D) Awards. Kostis would like to express her gratitude to the late Dr. Bill Corso (former director of USRA/GESTAR) for his immense support over the years and for being a role model of a servant leader. A big heartfelt thank you to mentor and Superhero of Synthesis Dan Sandin for providing images of two groundbreaking works of art: *Spiral5 PTL* and *Particle Dreams in Spherical Hormonics*. Special thanks to all her colleagues at the Scientific Visualization Studio (SVS), and Dr. Nicholas White, Dr. Scott Miller, Dagmar Morgan, Jefferson Beck, Daria Tsoupikova, John Fujii, Evan Hirsch, Dr. Jessica Hodgins, and Dr. Dorothy Zukor. And last but not least, many thanks to Anastasios and Ellie Golnas and for their patience and love.

References

1. Bittner, M., Schmidt, H., Timmreck, C., Sienz, F.: Using a large ensemble of simulations to assess the northern hemisphere stratospheric dynamical response to tropical volcanic eruptions and its uncertainty. Geophys. Res. Lett. **43**(17), 9324–9332 (2016)
2. Bujack, R., Turton, T.L., Samsel, F., Ware, C., Rogers, D.H., Ahrens, J.: The good, the bad, and the ugly: a theoretical framework for the assessment of continuous colormaps. IEEE Trans. Vis. Comput. Graph. **24**(1), 923–933 (2018)

3. California institute for telecommunications and information technology. http://www.calit2.net/index.php
4. Cisneros, M.: Your are an artist (2018). https://www.meetup.com/Data-Visualization-DC/events/256745546
5. Cox, D.: Renaissance teams and scientific visualization: A convergence of art and science. In: Collaboration in Computer Graphics Education, SIGGRAPH'88 Educator's Workshop Proceedings, pp. 81–104 (1988)
6. Cox, D.: "The unofficial history of visualization" (2016). http://ncsa30.ncsa.illinois.edu/2016/09/the-unofficial-history-of-visualization-donna-cox/
7. Danto, G.: 'Iceberg (r11i01)'—by Iñigo Manglano–Ovalle (2005). https://www.bookofjoe.com/2005/03/cloud_prototype.html
8. Dasgupta, A., Poco, J., Rogowitz, B., Han, K., Bertini, E., Silva, C.T.: The effect of color scales on climate scientists' objective and subjective performance in spatial data analysis tasks. IEEE Trans. Vis. Comput. Graph. 26, 1–1 (2018). https://doi.org/10.1109/TVCG.2018.2876539
9. DiBiase, D.: Visualization in the earth sciences. Earth Min. Sci. 59(2), 13–18 (1990)
10. Elliott, K.: 39 studies about human perception in 30 minutes (2016). https://medium.com/@kennelliott/39-studies-about-human-perception-in-30-minutes-4728f9e31a73
11. Hampton, S.E., Parker, J.N.: Collaboration and productivity in scientific synthesis. BioScience 61(11), 900–910 (2011). https://doi.org/10.1525/bio.2011.61.11.9
12. Krause, H.: An introduction to the data biography (2019). https://weallcount.com/2019/01/21/an-introduction-to-the-data-biography
13. Lucas, T.: Black holes: The other side of infinity. Movie (2006)
14. Mitchell, H.G.: Scientific visualization studio. https://svs.gsfc.nasa.gov/
15. Schneider, B., Nocke, T.: The feeling of red and blue - a constructive critique of color mapping in visual climate change communication. In: Leal Filho, W., Manolas, E., Azul, A.M., Azeiteiro, U.M. McGhie, H. (eds.) Handbook of Climate Change Communication: Vol. 2. Climate Change Management, pp. 289–303. Springer, Cham (2018)
16. Sonnenwald, D.: Scientific collaboration: a synthesis of challenges and strategies. ARIST 41, (2007). https://doi.org/10.1002/aris.2007.1440410121
17. Stephens, E.M., Edwards, T.L., Demeritt, D.: Communicating probabilistic information from climate model ensembles - lessons from numerical weather prediction. Wiley Interdiscip. Rev.: Clim. Chang. 3(5), 409–426 (2012)
18. Tukey, J.W.: Exploratory Data Analysis. Addison-Wesley series in Behavioral Science. Addison-Wesley Publishing Company, Boston (1977)
19. McCormick, B.H., DeFanti, T.A., Brown, M.D. (Eds.), ACM SIGGRAPH Computer Graphics, 21, 6 (1987)
20. Wang, J., Hazarika, S., Li, C., Shen, H.: Visualization and visual analysis of ensemble data: a survey. IEEE Trans. Vis. Comput. Graph. pp. 1–1 (2018)
21. Ware, C., Turton, T.L., Bujack, R., Samsel, F., Shrivastava, P., Rogers, D.H.: Measuring and modeling the feature detection threshold functions of colormaps. IEEE Trans. Vis. Comput. Graph. 25(9), 2777–2790 (2019)
22. https://www.scribd.com/doc/4056835/Distribution-Religion. Accessed 2019

Printed in the United States
by Baker & Taylor Publisher Services